Vol. 32. **Determination of Organic Compounds: Metho**... rederick T. Weiss

Vol. 33. **Masking and Demasking of Chemical Reactions.** By D. D. Perrin

Vol. 34. **Neutron Activation Analysis.** By D. De Soete, R. Gijbels, and J. Hoste

Vol. 35. **Laser Raman Spectroscopy.** By Marvin C. Tobin

Vol. 36. **Emission Spectrochemical Analysis.** By Morris Slavin

Vol. 37. **Analytical Chemistry of Phosphorus Compounds.** Edited by M. Halmann

Vol. 38. **Luminescence Spectrometry in Analytical Chemistry.** By J. D. Winefordner, S. G. Schulman and T. C. O'Haver

Vol. 39. **Activation Analysis with Neutron Generators.** By Sam S. Nargolwalla and Edwin P. Przybylowicz

Vol. 40. **Determination of Gaseous Elements in Metals.** Edited by Lynn L. Lewis, Laben M. Melnick, and Ben D. Holt

Vol. 41. **Analysis of Silocones,** Edited by A. Lee Smith

Vol. 42. **Foundations of Ultracentrifugal Analysis.** By H. Fujita

Vol. 43. **Chemical Infrared Fourier Transform Spectroscopy.** By Peter R. Griffiths

Vol. 44. **Microscale Manipulations in Chemistry.** By T. S. Ma and V. Horak

Vol. 45. **Thermometric Titrations.** By J. Barthel

Vol. 46. **Trace Analysis: Spectroscopic Methods for Elements.** Edited by J. D. Winefordner

Vol. 47. **Contamination Control in Trace Element Analysis.** By Morris Zief and James W. Mitchell

Vol. 48. **Analytical Applications of NMR.** By D. E. Leyden and R. H. Cox

Vol. 49. **Measurement of Dissolved Oxygen.** By Michael L. Hitchman

Vol. 50. **Analytical Laser Spectroscopy.** Edited by Nicolo Omenetto

Vol. 51. **Trace Element Analysis of Geological Materials.** By Roger D. Reeves and Robert R. Brooks

Vol. 52. **Chemical Analysis by Microwave Rotational Spectroscopy.** By Ravi Varma and Lawrence W. Hrubesh

Vol. 53. **Information Theory As Applied to Chemical Analysis.** By Karel Eckschlager and Vladimir Štěpánek

Vol. 54. **Applied Infrared Spectroscopy: Fundamentals, Techniques, and Analytical Problem-solving.** By A. Lee Smith

Vol. 55. **Archaeological Chemistry.** By Zvi Goffer

Vol. 56. **Immobilized Enzymes in Analytical and Clinical Chemistry.** By P. W. Carr and L. D. Bowers

Vol. 57. **Photoacoustics and Photoacoustic Spectroscopy.** By Allan Rosencwaig

Vol. 58. **Analysis of Pesticide Residues.** Edited by H. Anson Moye

Vol. 59. **Affinity Chromatography.** By William H. Scouten

Vol. 60. **Quality Control in Analytical Chemistry.** By G. Kateman and F. W. Pijpers

Vol. 61. **Direct Characterization of Fineparticles.** By Brian H. Kaye

Vol. 62. **Flow Injection Analysis.** By J. Ruzicka and E. H. Hansen

Vol. 63. **Applied Electron Spectroscopy for Chemical Analysis.** Edited by Hassan Windawi and Floyd Ho

Vol. 64. **Analytical Aspects of Environmental Chemistry.** Edited by David F. S. Natusch and Philip K. Hopke

Vol. 65. **The Interpretation of Analytical Chemical Data by the Use of Cluster Analysis.** By D. Luc Massart and Leonard Kaufman

Vol. 66. **Solid Phase Biochemistry: Analytical and Synthetic Aspects.** Edited by William H. Scouten

(*continued on back*)

The Analysis of
Extraterrestrial Materials

CHEMICAL ANALYSIS

A SERIES OF MONOGRAPHS ON
ANALYTICAL CHEMISTRY AND ITS APPLICATIONS

VOLUME 81

A WILEY-INTERSCIENCE PUBLICATION

JOHN WILEY & SONS

New York / Chichester / Brisbane / Toronto / Singapore

The Analysis of Extraterrestrial Materials

ISIDORE ADLER

Silver Spring, Maryland

A WILEY-INTERSCIENCE PUBLICATION

JOHN WILEY & SONS

New York / Chichester / Brisbane / Toronto / Singapore

Library of Congress Cataloging in Publication Data:

Adler, Isidore, 1916–
The analysis of extraterrestrial materials.

(Chemical analysis, ISSN 0069-2883; v. 81)
"A Wiley-Interscience publication."
Bibliography: p.
Includes index.
1. Cosmochemistry. I. Title. II. Title: Extraterrestrial materials. III. Series.

QB450.A35 1986 522.02 85-17839
ISBN 0-471-87880-4

Printed in the United States of America

10 9 8 7 6 5 4 3 2 1

TO

My wife Annie
because of her patience and care,

my grandchildren Richard, Nicole, and Douglas
because I love them;

and Isaac and Janet Asimov
because they are so inspirational

PREFACE

A reading of history tells us about revolutions in knowledge that have occurred in the past. All of these were remarkable and some, in fact, awesome because of the extraordinary vision displayed. In many instances the results were a quantum leap forward in our understanding of the universe in which we live. Now in our own lifetime comes a development which in many aspects exceeds in magnitude anything that went before. It is the period that will be long referred to as the space age. In an historical context the elapsed time has been compared to the blink of an eye, but the accomplishments will affect mankind forever. This has been colorfully summarized in the NASA publication *A Meeting with the Universe,*

Only a single generation separates the flight of Sputnik 1 from the first spacecraft pictures of Saturn's moons, and the first geologist to explore the Antarctic is still around to talk with the first geologists to go to the Moon!

What has occurred in this astonishingly brief period? The catalogue of accomplishments would have easily exceeded anyone's belief as little as 40 years ago. We have now sent a probe to examine, close up, the surface of Mercury. On Venus we have penetrated by radar the ever present cloud cover, to learn about the surface in amazing detail; we have also learned about the dense atmosphere of carbon dioxide and the sulfuric acid clouds. In the process, we have discovered what is now referred to as the "runaway greenhouse effect" leading to the surprisingly high surface temperature of over 450°C, an observation that gives one pause about the uncontrolled dumping of carbon dioxide into our own atmosphere. One of the most remarkable feats has been the analysis of the soil of Venus by a Russian Venera spacecraft that survived the extreme conditions long enough to do this.

As remarkable feats go, however, the landing of men on the Moon will go down as one of the most extraordinary of all. Here, for the first time in recorded history, extraterrestrial materials were returned from identified locations elsewhere in the Solar System. By the end of the Apollo missions, over 800 pounds of assorted rocks, soils, and cores had been returned to Earth. The examination of these materials will constitute part of the subject of this book.

This is, of course, only part of the story. For example, we have now been able to examine Mars, the mysterious red planet of fable, in great detail. Both of the marvelous Viking spacecraft landed on widely separated parts of the Martian surface and then performed that most complex analysis of the Martian soil in an attempt to discover living organisms and organic matter. While we cannot say with certainty that Mars has neither life nor organic matter, we can lay to rest forever the perception by the astronomer Percival Lowell that Mars has canals, built by superior intelligence to carry water from the poles to irrigate arid areas.

Among the spectacular achievements of the last decade are the wonderful photographs sent back by the Pioneer and Voyager spacecraft. The pictures of the Jovian and Saturnian systems are incomparable. Both Jupiter and Saturn are unusual in that they radiate more energy than they receive. Their atmospheres show an amazingly complex chemistry that warrants a great deal of study. We have not only discovered new moons around Jupiter and Saturn, but we also learned that each moon has a character all its own. As an example, the Gallilean satellite Io is now known to be the most volcanically active body in the Solar System, pouring out what appears to be unbelievable quantities of sulfur and sulfur dioxide. It has been observed that Jupiter itself is rotating in a torus of sulfur produced by Io.

The flyby of Saturn has disclosed that the number of rings around the planet is in the thousands and it is thought that water is an important part of their composition.

Titan, the largest moon in the Solar System, in orbit around Saturn, has been found to have an atmosphere even denser than the Earth's. Furthermore, its composition of nitrogen and methane may perhaps represent the Earth's early atmosphere. It has been speculated that there are methane seas on Titan lapping at shores of frozen methane.

This has been a period when spectroscopy has played a very important role. The development of microwave, infrared, and X-ray astronomy has opened up new windows into the cosmos. Numerous organic molecules have been discovered in the interstellar medium, a series of observations that can have a considerable bearing on the understanding of the origin of life.

For the cosmochemist this has been a particularly exciting time. Such ventures as the lunar landings have offered the intriguing possibility of finding what Professor Wasserburg of Cal Tech has called the "Holy Grail"; samples of the primordial material from which our Solar System formed.

This has also been an exciting time for scientists involved in measurements. New standards have been set for inventiveness, ingenuity, and reliability. In this comparatively brief period of time, we have been able to acquire analytical data from planets as remote as a billion miles away. To do this, it was necessary to turn on instruments after years of flight through a hostile environment and to receive signals from vast distances transmitted at just a few watts of power. There is no question that the space age has had a great positive effect on the development of methods, instruments, and techniques. Some of the influence is direct, as a function of the developing technology. In other cases the reasons are subtle, reflecting the impetus provided, as Dr. Fletcher has stated, "by a steadfast mobilization of national will resources."

In this text I shall deal with the impact of the planetary exploration program and the study of extraterrestrial materials, describing new methods and their limitations. Part 1 of the book will consider extraterrestrial materials analyzed in terrestrial laboratories. Part 2 will involve the analysis of extraterrestrial materials under extraterrestrial conditions.

ISIDORE ADLER

Silver Spring, Maryland
March 1985

ACKNOWLEDGMENTS

There are many colleagues and associates to whom I owe thanks for help in preparing this book. A number of these colleagues and associates listened to me with great patience and made useful suggestions, all of which proved to be encouraging. Many were generous in supplying me with information such as reprints, references, and figures. The list is long and I tender my sincerest apologies if I fail to include all. Among these individuals are S. R. Taylor, B. H. Mason, R. Clark, J. A. Wood, T. B. McCord, G. J. Wasserberg, D. A. Papanastassiou, J. C. Laul, R. B. Singer, G. W. Lugmair, K. Keil, C. C. Schnetzler, L. Haskins, H. J. Rose. J. I. Trombka, L. I. Yin, B. C. Clark, J. A. Philpotts, and G. W. Wetherill. I want to particularly thank L. McFadden for her special efforts in scrutinizing the chapter on reflectance spectroscopy, S. Marans for reading and commenting on the chapter on meteorites, Glenn Davison for his help in preparing the figures, and the excellent staff of the Lunar and Planetary Institutes library facilities, and Fran Waranius and Stephen H. Tellier, who so kindly supplied me with references. Finally, I must claim sole responsibility for any deficiencies in the text.

I. A.

CONTENTS

PART 1

PART 2

The Analysis of Extraterrestrial Materials

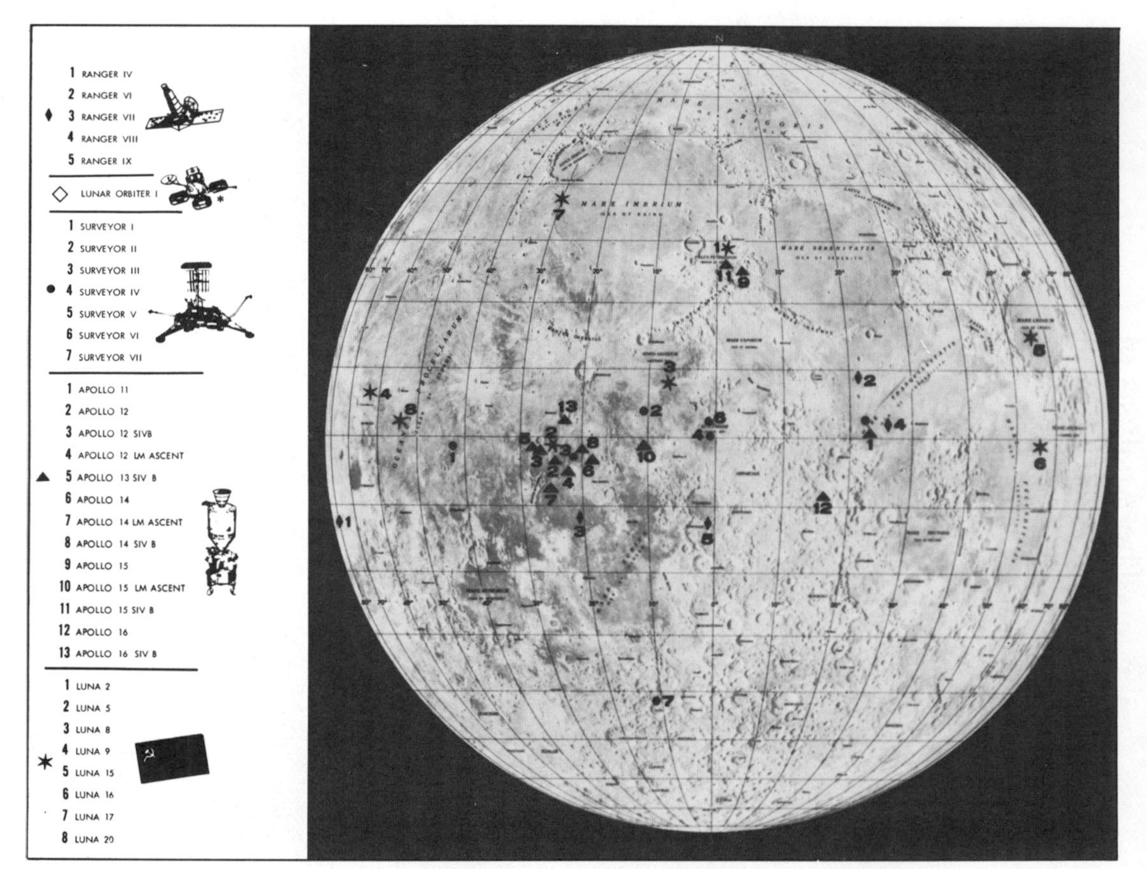

A graphic illustration showing the position of all space objects landed and impacted on the lunar surface. NASA 72-H-1375

An extraordinary view of the Martian surface. The Viking sampler arm shown in the view at the left was used to move the rock adjacent to it so that the soil beneath the rock could be sampled. The subsequent view on the right shows the vacant site after the maneuver. This was a truly remarkable accomplishment.

PART 1

CHAPTER

1

PLANETARY EXPLORATION; ITS IMPACT ON THE DEVELOPMENT OF ANALYTIC METHODS

INTRODUCTION

The planetologist or the cosmochemist who studies our Solar System faces a number of riddles that must be solved to understand it better. Beginning with the origin of the Solar System some 4.6 aeons (billion years) ago, what was the composition of the primordial solar nebula from which the Solar System evolved, and what was the sequence of events leading up to the formation of the Solar System we now inhabit? How does one explain the existence of the terrestrial planets like Mercury, Venus, Earth, and Mars, and the giant gaseous planets like Jupiter, Saturn, Uranus, and Neptune? Why is our own Moon depleted of volatile elements and enhanced in refractory elements? How did the Moon become depleted of the siderophile elements? Why are the moons around the giant planets so different? Why, for example, does Titan have an atmosphere and what sort of chemistry is going on in the atmosphere of its parent body, Saturn? What kind of atmospheric chemistry explains the colors of the Jovian atmosphere?

These are just a few of the many questions. It is obvious that one can go on and on. Many of these questions have newly risen as a result of the successful flights to these bodies.

Examining these questions points out the great significance of the analysis of extraterrestrial materials. As we shall see, the meteorites, the lunar rocks and soils, and the composition of planetary surfaces are excellent recorders of the ancient history of the planets and the Solar System. The determination of the chemical composition of these materials can provide important comparisons to cosmic, solar, and terrestrial abundances. Radiometric and chemical analysis of particular nuclides and elements produce accurate dates for important geological events.

Detailed examination of the microstructures of the rocks and analysis of the coexisting minerals can yield definitive information about initial compositions and cooling histories. The lunar rocks, for example, were able to provide detailed information about radiation exposure histories leading to knowledge about the Sun's behavior. In brief, a reading of the rocks is similar to the discovery of ancient historical written records. Suitable analytical techniques can be said to supply the Rosetta stone for deciphering these records.

If the planetary exploration program is unique regarding analytical studies, this singularity came from the fact that in some instances there were special demands for sensitivity and precision, and in other instances because the climate in which scientists operated was unique. For example, there was a need to deal with special materials, often never seen before, in which the search for anomalies was the key. It appears obvious that if this is true, it is necessary to establish their reality—that they are not artifacts of the method employed. Such requirements placed constraints on investigators to evaluate the methods applied, in terms of reliability, sensitivity, and usefulness in the studies undertaken.

A second factor that had considerable influence in the development of analytical methods was the nature of the planetary science community. As a class, these investigators came to the program through a rigorous selection process. As a group, they were unusually competent and very vocal. Many of them were engaged in similar studies. In a sense they were practicing science "in a fishbowl." Their work and their results came under rapid scrutiny and debate, which once again produced a great impetus for the refinement of their methods.

There were interesting examples of feedback and interaction. The study of lunar samples borrowed heavily on methods that had been used traditionally in meteorite studies. In time, advances in techniques that evolved from the lunar sample program were then applied to improved examination of meteorites.

The methods for radiometric dating improved in a spectacular way because of the need for establishing an evolutionary history of the Moon and meteorites. This led to advances in mass spectrometry and in standards for operating under literally contamination-free conditions. Both have had a great impact on studies of trace elements.

There were also substantial advances in methods for the remote analysis of Solar System bodies both from orbit and in situ, an aspect of the exploration program that created an interesting situation. We saw many instances of a too easy acceptance by the theorists of numbers quoted by the experimenters, particularly if the numbers appeared to fit some model that had been proposed.

CHAPTER

2

THE FIRST ET'S, THE METEORITES

The study of extraterrestrial material did not begin with the space program; rather, it began with the study of meteorites. The principal question from the very beginning was, where did they come from? Speculation about their origin can be found in the ancient Chinese, Greek, and Latin literature. It was not always recognized or accepted that these objects had fallen from the sky, and until the beginning of the nineteenth century the idea was greeted with considerable skepticism. However, the insight of the ancient Greek philosopher Anaxagoras is fascinating; he wrote that they were heavenly bodies that had become detached from their place in the heavens and then had fallen to the Earth. Equally perceptive was E. F. F. Chladni, a German lawyer and physicist. Chladni wrote a book published in 1794 with the title *Observations on a Mass of Iron Found in Siberia by Professor Pallas, and on Other Masses of the Like Kind, with some Conjectures Respecting their Connection with Certain Natural Phenomena*. The work drew a connection between fireballs and meteorites and emphasized that the only reasonable explanation for these objects would be extraterrestrial origin. As so often happens, these ideas were not immediately accepted, but the skepticism began to evaporate after a detailed report about the L'Aigle meteorite fall in April 1803. The observer, J. B. Biot, authenticated the event so thoroughly that little doubt was left about Chladni's hypothesis. As B. P. Glass (1) points out in his book, Chladni's work and the confirmation by Biot proved to be a landmark in the history of meteoritics.

Following the acceptance of extraterrestrial origin, the initial meteoritic research was concerned with the nature and chemical composition of the falls. Some gross chemical analyses were published by Berzelius as far back as 1806. It is noteworthy that some of his analyses, done by the classical wet methods of his day, are remarkably still valid today.

With the development of the polarizing microscope in 1860 came one of the most powerful tools for examining the microstructure of the meteorites. Mason (2) has written that the contribution by G. Tschermak in Vienna of his book, *Die Mikrokopische Beschaffenheit der Meteoriten*

Erlautert Durch Photographische Abbildungen, in 1855, is still unequaled by any other book on the subject.''

There were also a number of very important analyses of minor and trace elements made by Goldschmidt and his co-workers (3) and by the Noddacks (4).

Recent years have seen a tremendous increase in meteorite studies. These activities have included new methods of instrumentation, new methods of analyses, and a new understanding of the significance of some of the exotic elements.

SIGNIFICANCE OF METEORITE STUDIES

J. T. Wasson (5) in his book on meteorites has discussed the special relationship of the meteorites to any attempt to learn more about the origin and evolution of our Solar System. He points out that meteorites carry information about a number of Solar System processes. To begin with, they appear to carry clues to events that occurred in the solar nebula prior to the formation of the planets. Second, some meteorites may have come from a planetary interior and thus contain information about internal processes similar perhaps to those that occurred in the interior of the Earth or other planets. Meteorites may also show evidence of the effect of collisional interactions, that is, shock and fragmentation processes, and finally, information about interactions with solar and galatic cosmic rays.

Wasson goes on to say that, until the Apollo 11 mission to the Moon in 1969, meteorites were the only extraterrestrial samples available for study. When they were examined, the lunar samples from the six Apollo missions and the Luna 16 and 20 probes proved to be products of igneous processes. In addition, many of the samples had obviously experienced additional alteration by shock and reheating as a result of bombardment of the lunar surface by the infall of meteorites and other objects. As a consequence, although the lunar samples provide important information about the early history of the Moon, it is still the meteorites that continue to be unique sources of data about processes occurring in the solar nebula.

CLASSIFICATION OF METEORITES

The accumulation, with time, of a large number of meteorites has led to various classification schemes, the object being to group them in a way that would make comparative investigations more systematic. One of the more commonly used classifications is that of Mason, which in turn is based on a scheme of Prior (6), and further revised by Mason (2) in 1967. The search for effective classifications continues and new ones are being proposed. Mason himself has commented on classification schemes as being difficult to achieve because there is a lack of agreement on criteria and how to apply them. In general, the criteria that have been used include factors such as chemical composition, internal structure, and color. To quote Mason, "The possible permutations and combinations are many and meteorite classification becomes simple or complex depending on the inclination of the classifier."

A clear and easily read treatment of what meteorites are has been prepared by J. A. Wood (7). Meteorites are divided into three broad classes: stones, irons, and stony irons. It is now apparent that about 90% of those seen to fall are the stony meteorites, although the bulk of the meteorites seen in museums are metallic. This simply reflects the fact that the metals are the specimens most likely to survive weathering on the Earth's surface and are also the ones most easily identified as meteoritic. Stony meteorites look like terrestrial rocks. In contrast to the stones, the proportion of irons is about 6% and the stony irons about 1.5%.

The chondrites, so named because of the presence of chondrules in the rocks, are the most common. The chondrules have been described by Mason (2) as spheroidal bodies commonly about 1 mm in diameter and consisting of the minerals olivine or orthopyroxene. For the most part, the chondrules have microstructures and are made up of crystalline grains of varying composition. The olivine chondrules may consist of single crystals or numerous crystals and appear granular. There are also more exotic varieties such as barred chondrules, consisting of alternate layers of olivine and a dark interstitial material. In some instances the chondrules are bordered by nickel–iron and troilite (Fig. 2.1).

Since this class of meteorites is by far the most common, the existence of the chondrules is particularly significant. Any consistent theory of the origin of the meteorites must explain their origin. There are numerous views, but no general agreement. One viewpoint has the chondrules deriving from the solar nebula. The other view is that chondrules are secondary objects, formed in a planetary body. The one area of accord is that these are very early objects originating in the primordial planetary matter.

Among the chondrites we find a subgroup called the carbonaceous chondrites. These are among the most valuable and interesting classes of meteorites. They are dark in appearance and often fragile. Some of the carbonaceous chondrites contain water and hydrocarbon compounds including amino acids and fatty acids (see the section on organic matter). The optical properties of the amino acids provide strong evidence for abiogenic origin. With the recent discovery by microwave astronomy of

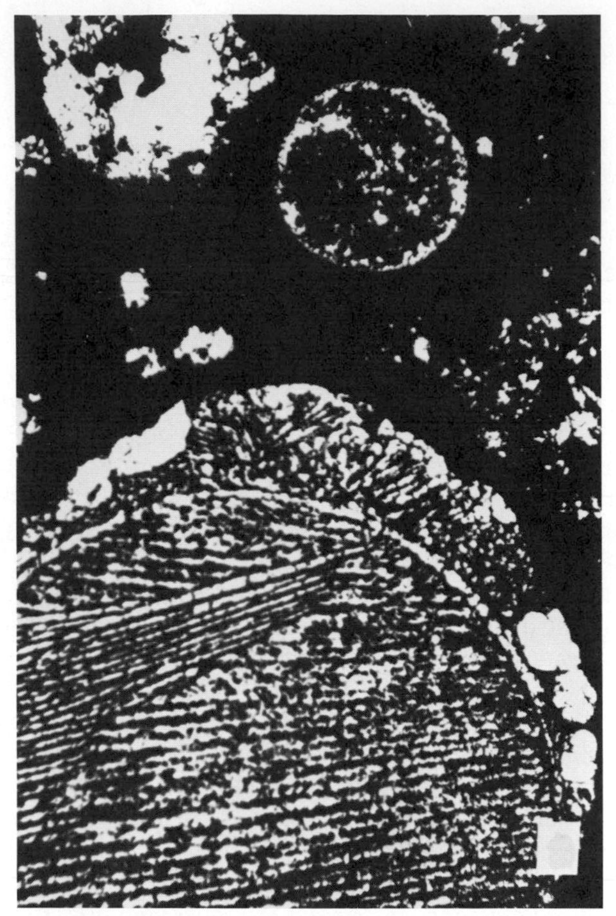

Figure 2.1. Examples of chondrules found in meteorites. In this instance the meteorite is the Pueblito de Allende meteorite. The large chondrule is an example of a barred-pyroxene–glass chondrule with a thin rim. The smaller chondrule is a fine-grained olivine, pyroxene, glass chondrule.

interstellar organic molecules, it is possible that these carbonaceous chondites may contain traces of interstellar dust.

Another class of stones is the achondrites, so called because they lack chondrules. They have been described as igneous rocks or planetary "soils." Some achondrites are basaltic both in composition and texture. They may, perhaps, represent lava flows on the surface of their parent bodies. Most of them are brecciated rocks (crushed and recemented angular fragments in a finer grained matrix).

Among the achondrites are subclasses called eucrites, howardites, and a rare group, the ureilites. Until recently the eucrites were considered to be monomict breccias—derived from a single rock source. By comparison, the howardites were defined as polymict breccias—derived from more than one source rock. A second distinguishing feature was the lower magnesium content of the pyroxene in the eucrites. These distinctions have become confused with the discovery of polymict eucrites in the recently discovered Antarctic meteorites—in fact, the polymict eucrites are now considered a new class of the stony meteorites.

The rare ureilites (of which the number of specimens is 11, with 6 from Antarctica) appear to bear a resemblance to some ultramafic rocks from Earth's lower crust or upper mantle. Particularly interesting are the minute crystals of diamond and lonsdalite (hexagonal diamond). The presence of the diamond is considered to be evidence of shock, resulting from collision in space.

The iron meteorites, while smaller in number than the stones, are actually larger in terms of mass. They consist almost entirely of nickel–iron metal and the siderophile elements (elements concentrating in the metallic phase). The lithophile elements—those elements concentrating in the silicate phases—are almost entirely missing. Wood (7), in his book *Meteorites and the Origin of the Planets,* poses the question "How could such objects have formed?" Because their composition is so highly specialized and totally dissimilar to the abundance patterns of condensible elements in the Sun, they cannot be representative of the early condensed primordial material, but rather the products of some later fractionation process. A mechanism described by Wood has the following scenario.

The first solid material formed in the Solar System would have a generalized composition, with some of the nickel–iron present as metallic alloys and sulfide minerals (one sees these in chondritic meteorites). If such material experienced subsequent melting, it would not form a homogeneous fluid. The silicates and oxides would have mingled to form a liquid or magma, while the metals and sulfides would have formed another, immiscible with the first. If there had been a sizable object having an appreciable gravitational field, then the denser metal–sulfide liquid would have sunk to the center of the object and coalesced into a molten core. The lighter silicate magma would have formed a mantle. Following this, cooling and crystallization and then a shattering collision could have left orbiting core fragments with the properties of iron meteorites.

The iron meteorites are divided into three classes, based on the nickel content and the microstructures. The two major minerals making up the irons are kamacite and taenite. There are also minor minerals such as troilite (iron sulfide); schreibersite (iron, nickel, phosphide), and cohenite (iron carbide). The three classes are:

1. The hexahedrites, which contain kamacite as the principal mineral; the nickel ranges from about 4% to 6%. A characteristic of these meteorites is that in polished sections they show fine linear striations called Neumann bands, which are thought to be the mechanically produced structures caused by shock, conceivably by collision.
2. The octahedrites contain larger amounts of nickel, about 6% to 14%, and both kamacite and taenite. These meteorites have a very characteristic pattern in cross-section, the Widmanstatten structure, an intergrowth of the two minerals (Fig. 2.2). As a further distinction, these meteorites are also classified on the basis of the widths of the kamacite bands.
3. The ataxites involve those irons where the percentage of nickel is quite high, 12% to 14%, and the Widmanstatten structure is replaced by a fine intergrowth of kamacite and taenite.

The final group of meteorites are the stony irons, a minor group. The stony irons are divided into two major groups according to the nature of the silicate minerals. Detailed studies have convinced investigators that the stony irons are mixtures.

CHEMICAL COMPOSITION OF METEORITES

Although meteorites are no longer the only extraterrestrial matter available, they are still a key source for yielding data on the elemental composition of our Solar System. The story they have to tell has been neatly summarized by Wood (8) in his accounting of the Allende meteorite (found near Pueblito de Allende in 1969).

To date, the Allende meteorite is the most intensively studied of all the meteorites. It is unique for the following reasons:

1. Its fall was documented and the collection of specimens almost immediate.
2. The Allende became available just as studies were about to begin on the returned Apollo samples, thus providing a very important training for the analysis of the lunar samples.

In appearance, the meteorite was a dense, hard rock, made up of dark gray material, free of cavities or visible porosity. Its specific gravity of 3.67 made it heavier than terrestrial rock. On a smaller scale, a broken specimen showed inclusions and chondrules ranging in size from about 1 to a few millimeters across, embedded in a dark matrix.

Figure 2.2. Widmanstatten pattern in an iron meteorite. Shown is an etched surface of the Carbo, Mexico meteorite. The bands are kamacite lamellae. The large inclusion is troilite. Courtesy of R. S. Clarke, Smithsonian Institution, Washington D.C.

A comparison of its chemical composition with that of elemental obtained spectroscopically, abundances in the solar atmosphere, is shown in Figure 2.3. The scales are logarithmic, and the abundance of each element is normalized to 1 million silicon atoms. The line drawn is at 45°. Elements that fall to the left and above the line are more enhanced in the Sun relative to the meteorite, whereas the elements to the right and below the line are enriched in the meteorite relative to the solar abundances. Note that most of the 69 elements studied lie on this 45 degree line. To quote Wood, "It is as if a mass of solar material had ripped out of the Sun and allowed to cool and condense."

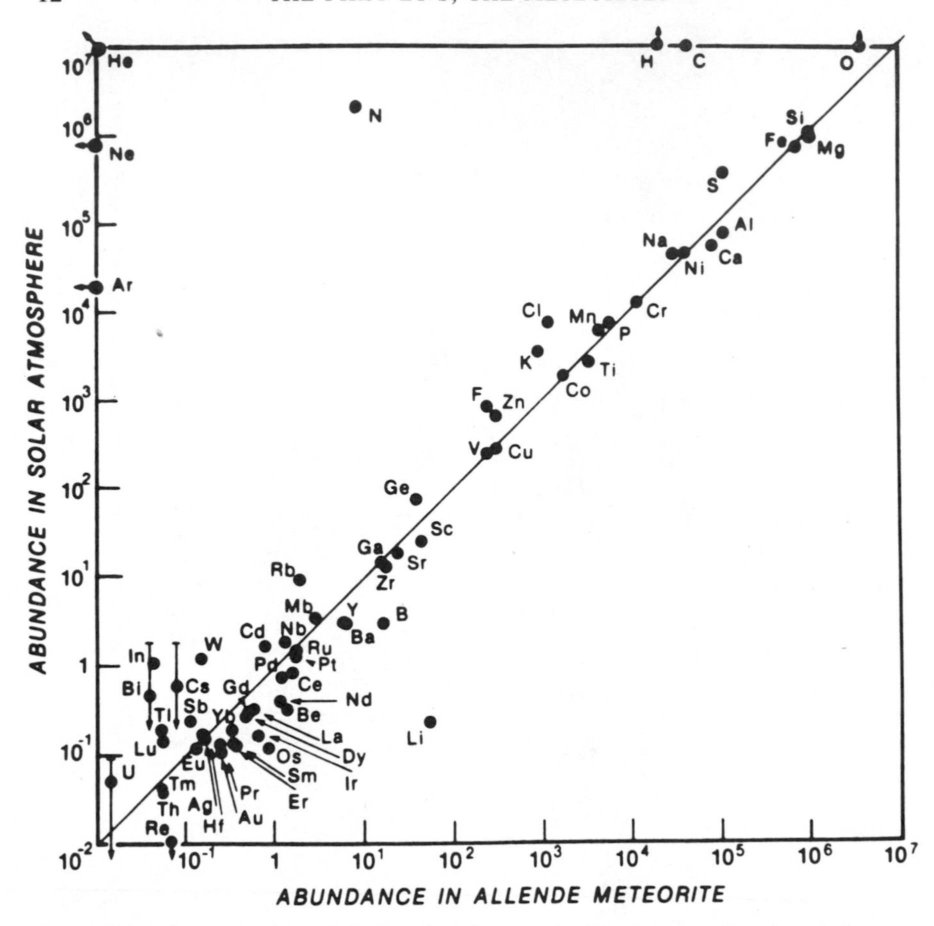

Figure 2.3. Concentrations of 69 chemical elements in Allende, plotted against their concentrations in the solar atmosphere (using logarithmic axes). In both cases, abundances are relative to 1 million silicon atoms.

There are, however, some departures from this relationship, but there is a cosmic rationale. For example, hydrogen, carbon, nitrogen, oxygen, and the noble gases (seen to the left of the 45° line) are either so volatile, or form compounds that are so volatile, that they were not likely to condense in the early, hot, inner Solar System. They would, as a consequence, be depleted in the condensed solar material. Lithium, seen to the right of the line, appears to be depleted in the Sun relative to the condensed material, because of destruction by thermonuclear processes in the hot stellar interior. Again to quote Wood, "After 4.6 billion years of attrition in the Sun, this element has been depleted in the solar atmosphere."

Having established the elemental abundances and their relationship to solar abundances, can one determine the age of the meteorites? What do we mean by the age of a meteorite or a rock? The age is considered to be the time elapsed since the rock crystallized. It represents the time since the decay products (the daughter or daughters) began to accumulate. There are many radiometric methods to measure this time, and these will be discussed under age determination. In the case of the Allende, for example, the amount of argon-40 generated by the decay of the potassium-40 yielded an age of 4.57 aeons, very close to the accepted age of the Solar System.

Observations such as the age and the unfractionated chemical composition of the Allende are taken as convincing evidence that it is a specimen of primitive planetary material. This is generally true of chondritic meteorites, of which the Allende is an example.

Wood goes on to say that most of the excitement of the Allende comes from the inclusions, particularly the calcium–aluminum rich ones. Close examination shows them to be separate mineral systems that have formed independently of the matrix material in which they are embedded. Based on the observations it is felt that these inclusions, initially distributed in space after formation, subsequently accreted together as mineral dust into planetary objects of some type.

The refractory inclusions are interesting minerals that are not very common on the Earth. Furthermore, they had not, until recently, been recognized as abundant meteoritic phases. The minerals all have in common high concentrations of calcium, aluminum, and titanium relative to the mean abundances in the bulk Allende meteorite. It is to be noted that these are the most refractory of the major elements in meteoritic or planetary matter.

To understand the significance of this, one needs to go back to the early models for the formation of planetary matter. For example, in 1962 A. G. W. Cameron (9) proposed a model (later modified) for the formation of the Solar System, where the gravitational collapse of a cloud of intersteller gas and dust produced both the proto-Sun and a rotating accretion disk or nebula that was also of solar composition.

As a consequence of the gravitational compression of the nebula, it heated to the point where temperatures in the region of the terrestrial planets became sufficiently high to vaporize completely the interstellar dust incorporated into the plasma. Following this, the hot nebula began to cool by radiation and, as one might expect, the first grains to form were those richest in the most refractory elements: calcium, aluminum, and titanium.

There is an excellent correspondence between observations and theory. L. Grossman (10) employed assumptions based on the elemental abundances in the Sun and gas pressures in the range of 0.0001 to 0.001 atm (as called for in the Cameron model). He was able to predict a condensation sequence based on chemical thermodynamics. These predictions covered both the expected mineral species and the order in which they would appear (Fig. 2.4). The arrows in the figure show the direction in which the minerals, in continuous contact with the vapor, would transform as systematic cooling proceeded. On the other hand, if cooling occurred rapidly, then a complete transformation would not occur. Rather high-temperature minerals like a spinel ($MgAl_2O_4$), melilite [$Ca(Al,Mg)(Si,Al)_2O_7$], or perovskite ($CaTiO_3$), would be frozen out. This is consistent with the occurrence of spinel, melilite, and perovskite in the calcium–aluminum inclusions in the Allende.

The Allende contains another class of minerals which one finds lower down in the condensation sequence. These are much more abundant. The minerals are olivine [$(Mg,Fe)_2SiO_4$] and pyroxene [$(Mg,Fe)SiO_3$]. The major elements are magnesium, iron, and silicon, and their concentrations are more characteristic of the solar atmosphere. Generally, the minerals appear in the form of chondrules whose shape and texture show evidence of melting. It is widely accepted that they are frozen droplets of planetary material, melted and splashed onto the accreting parent body of the meteorite as a result a collision.

One of the most exciting developments in the study of the meteorites comes from the work of Clayton and co-workers (11) on the determination of the oxygen isotopes in meteorites. A review of this work is given by Wood (8).

The three principal oxygen isotopes are oxygen-16 (99.756%), oxygen-17 (0.039%), and oxygen-18 (0.205%). Each of these isotopes is stable and presumably the percentages quoted are primordial, going back to the very beginning of the Solar System. Clayton and co-workers have developed the Figure 2.5, which shows the proportions of the oxygen isotopes in various planetary and meteorites samples. Included are terrestrial, lunar, and meteoritic samples.

The phenomenon that makes oxygen so interesting is the way in which the various isotopes partition themselves in mineral phases. It is known that the various isotopes do not enter mineral sites with the same ease. The vibrational energy of an oxygen atom depends on its mass, and it appears that a given mineral site prefers the isotope of oxygen producing the lowest vibrational energy at that particular position. This leads to a

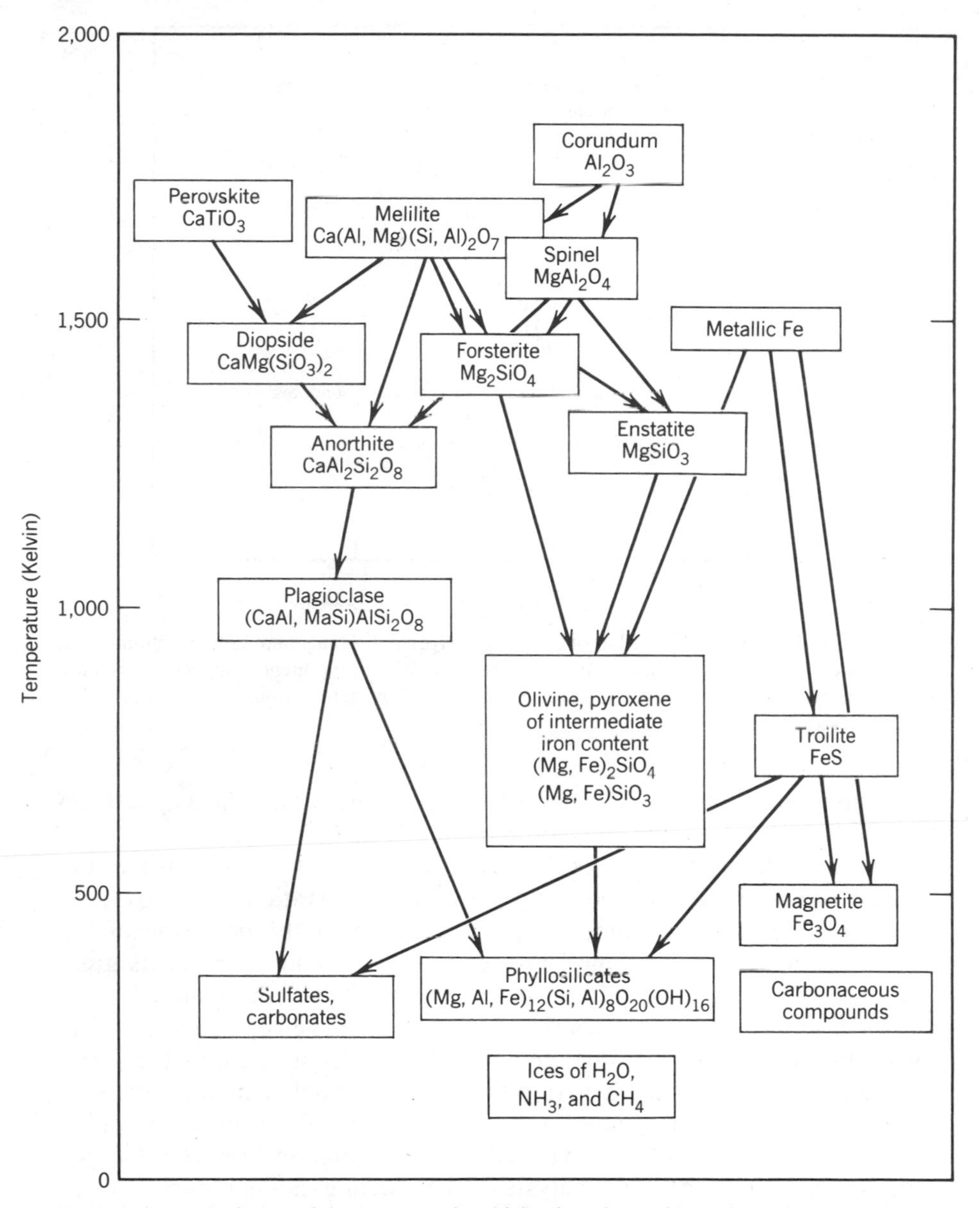

Figure 2.4. A depiction of the sequence in which minerals would condense from a cooling gas of solar composition. The arrows indicate that a continuing reaction with the cooling residual would transform the minerals from the upper boxes into those minerals in the box below. Rapid cooling would prevent complete transformation, accounting for the presence of the spinel, melilite, perovskite, and the like in the Allende CAI's (8).

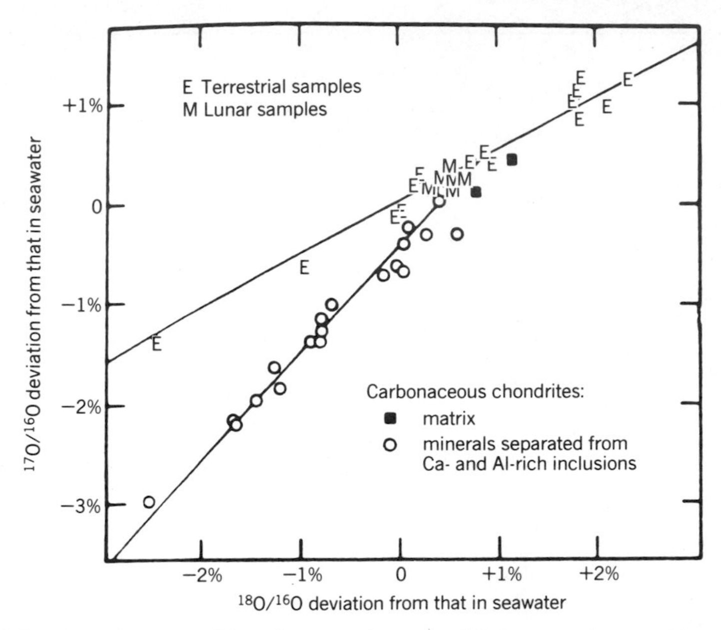

Figure 2.5. Isotopic composition of oxygen in terrestrial, lunar, and meteoritic materials. The lower curve, pertaining to carbonaceous chondrite components, appears to require mixing of two oxygen components which predate the solar system. Source: R. N. Clayton et al., *Science* **182** (1973) (11).

partitioning of oxygen isotopes between minerals and the vapor from which they condense or between the mineral and its melt.

Because the partitioning is a function of the isotope mass, one would expect a greater effect for the oxygen-18/oxygen-16 ratio than the oxygen-17/oxygen-16 ratio. In fact, the prediction is that the deviation of the isotope ratios from "normal" in the minerals would be twice as great because there is a mass difference of 2 units versus 1 mass unit.

Such an effect is seen clearly in the upper curve of Figure 2.5. The variation in the oxygen-18/oxygen-16 ratio is twice as great as the variation in the oxygen-17/oxygen-16 ratio. This is not only true for terrestrial samples but for lunar samples as well, a very significant observation.

The lower curve demonstrates a surprising relationship. The data for the curve comes from the analysis of the calcium–aluminum-rich inclusions chondrules and minerals from the Allende and similar meteorites. In the lower curve the slope is 1 : 1, an observation not explainable by equilibrium partitioning. For the present, the most reasonable explanation is that mixing has occurred in various proportions of two different oxygens, each with their own isotopic composition.

Wood offers the following as a mechanism to explain the lower curve. The early Solar System contained two discrete components, each with oxygen of a characteristic isotopic composition. The various other isotopes, including oxygen, were created in a variety of astrophysical settings such as the "Big Bang," fusion reactions in stellar interiors, supernova explosions, and the like. In the light of this, one can conceive of the various components of intersteller gas and dust contributing to the early Solar System, each consisting of atoms from different astrophysical sources and therefore containing elements of different isotopic compositions. This is one reasonable explanation for two different kinds of oxygen in the primordial Solar System.

But now, however, there is a surprise. Isotopic variability for the other chemical elements is not observed. With only a few exceptions, elements other than oxygen are quite invariable in their isotopic compositions in either meteorites or meteoritic components. It appears that the oxygen components of the early nebula did not come from batches of intersteller gas and dust that were never completely mixed. Rather, the two oxygens came to the nebula in each of two components that must have been well mixed.

Note that oxygen is the only element in both the dust and gas, whereas the other elements were concentrated in one component or the other. About one-seventh of the Solar System's oxygen would occur as oxides in the dust and the remainder in the gas phase as water or carbon monoxide gas. Consider that the oxygen in the gas was close to terrestrial oxygen in isotopic makeup, and that the dust was enriched in oxygen-16. The condensation process at high temperature, which produced the calcium–aluminum inclusions, would lead to mixtures with varying proportions of the two oxygens. The other elements concentrated in the gas or dust would have remained isotopically homogeneous.

The observations just discussed cast a new light on the high-energy events in the early nebula. It is now apparent that the nebula did not get hot enough to totally vaporize and then recondense the dust. Had this occurred, the oxygen isotopic composition would have been completely homogenized. This is not the case, as the figure shows. It is intriguing that Cameron has recently modified his theory to state that the compression of the nebula did not heat it sufficiently to vaporize the dust particles after all. He now postulates a relatively cold nebula and local high-energy events distilling off the more volatile components, leaving a residue enriched in calcium and aluminum. In the process this causes a partial exchange of oxygen between the calcium–aluminum inclusions and the surrounding gas.

There is another highly significant development in meteorite studies that will be mentioned briefly here and discussed in greater detail later. Like the oxygen isotope work, it not only bears on the early Solar System but carries information on pre-Solar System events and pre-Solar System raw materials. This new development comes from G. J. Wasserburg and co-workers, who in 1976 (12) published a report on the evidence of the presence of aluminum-26 at some time in the distant past in the Allende meteorite, in particular in the calcium–aluminum inclusions.

Aluminum-26 has a relatively short half-life (720,000 y) as it decays to magnesium-26. For this nuclide to have appeared in the Allende means that it was created shortly before or at the time of formation of the Solar System. It is Wasserburg's contention that a supernova, which among other things would have created the aluminum-26, must have occurred sufficiently close to the forming Solar System to contribute freshly synthesized nuclides. It is the argument of Cameron and Truran that this is not at all coincidental. The supernova not only contributed the radioactive aluminum but also the shock wave that initiated the formation of a number of stars of which our own Sun is one.

ANALYSIS OF METEORITES

The analysis of meteorites is not a simple matter. This becomes particularly evident in the studies of more exotic elements and the determinations of those elements present in what appears to be diminishingly small concentrations. It is in these areas that the study of extraterrestrial materials has contributed so much to advances in analytical chemistry.

There are a number of practical difficulties in meteorites analysis like those encountered in rock analysis, except that in many instances the analysis of meteorites is even more difficult. The chondrites, for example, contain not only gray rocklike silicate material but other phases like sulfide and metallic minerals as well. There are serious questions involving authenticating the meteorite, the effect of the external environment, and general representative sampling.

Systematic studies of elemental abundances in meteorites began with the published work of V. M. Goldschmidt in 1923 (3). In the years following, Goldschmidt and co-workers made many determinations of specific elements in meteorites. His results and those of others such as Noddack and von Hevesey were summarized in a paper *Geochemisch Verteilungsgesetze der Elemente,* published in 1937. Goldschmidt used these data to prepare the first comprehensive table of elemental abundances. Notably, he introduced the practice of referring atomic abundance to silicon in order to relate terrestrial and meteoric elemental abundances to solar abundances. This is still standard practice today.

There have been numerous compilations since. To mention but a few, there are those of Ringwood (13), Urey (14), Cameron (15), Mason (16), and Goles (17). Goldschmidt's tables are by now of great historical interest. Some of his values for the major elements are very close to the presently accepted values, while his minor and trace element values are either within an order of magnitude or closer.

Many new and improved analytical procedures have been developed since the early work. A brief list includes emission and X-ray spectrography, colorimetry, and some of the more recent methods such as neutron activation and isotope dilution. One very important development is the electron-probe microanalyzer, which has truly added another dimension by providing the capability of in situ analysis of the microscopic phases making up the meteorite sections.

Wood (7) states that even today the analysis of chondrites by wet chemical methods is considered more art than science. Currently, there is only a small number of analysts whose work is acceptable without question.

In practice, the analysis of the stony material is customarily presented as oxides of the elements rather than in terms of elemental concentrations, similar to the practice followed in rock analysis. This is a convention, because in actuality the elements are part of complex minerals involving several different elements. Table 2.1 shows a typical chondritic analysis determined by H. B. Wiiks (18) of the Geological Survey of Finland.

Presenting the data in this way shows that iron occurs in three different forms. The FeO is part of the silicate minerals; the Fe is in the nickel–iron grains, and the remainder in the FeS.

With the accumulation over time of more and more "good" analyses, with compositions expressed on a metals-only basis, the chondrites appeared to be nearly identical, differing only in the degree of oxidation of iron. In some instances, the ratio of metallic to FeO was high and the reverse true in other cases.

Based on such analyses, the early conclusion was that all chondrites had a common origin. It was assumed that initially all the iron had been oxidized and had then undergone various degrees of reduction. Wood states that today we know that not all chondrites are identical in composition and could not have been derived from a common parent. This follows from the fact that not only the degree of oxidation but also the absolute abundance of iron is variable.

Table 2.1 Analysis of the Richardton, North Dakota Meteorite[a]

Species	wt %	Species	wt %
SiO_2	34.3	Fe	18.3
MgO	22.2	Ni	1.6
FeO	9.9	Co	0.1
Al_2O_3	2.6		
CaO	1.4	Metallic grains	20.0
Na_2O	1.0		
Cr_2O_3	0.6	FeS	6.0
P_2O_5	0.5		
H_2O	0.5	Sulfide grains	6.0
MnO	0.4		
C	0.2		
K_2O	0.1		
TiO_2	0.1		
Total stony material	73.5		

[a] From Wood (7).

The situation is summed up in Figure 2.6. Plotted on the *y* axis is the ratio (Fe as metal)/(total Fe present). This is a measure of the degree of oxidation of the iron. Plotted along the *x* axis is the molar ratio of total Fe/Si. This reflects the amount of Fe present (on a metal-only basis). We observe the following: The points representing individual chondrites fall into several clearly defined groupings. Some of the groups have the same average values of Fe/Si as the others, notably the C and H groups. These could have conceivably been derived from one parent material by different degrees of reduction. On the other hand, the L and LL groups seem to have come from another parent material. There are other small differences in metals-only composition such as in Mg/Si and Na/Si. Thus it appears that chondrite group comes from different batches of planetary material (7).

In addition to the major elements cited above, the rarer elements also play an extremely important role in connecting the meteorites to the early material of the Solar System. Some of the early trace element work was done by the Noddacks (4). While some of their values are no longer considered valid, their general conclusions still are. They showed that the meteorites have a more generalized composition than the Earth's crust, that is, close to the Sun's composition.

The Noddacks found them to contain more elements, such as the lithophiles, chalcophiles, and siderophiles. Wood tells us that the natural pro-

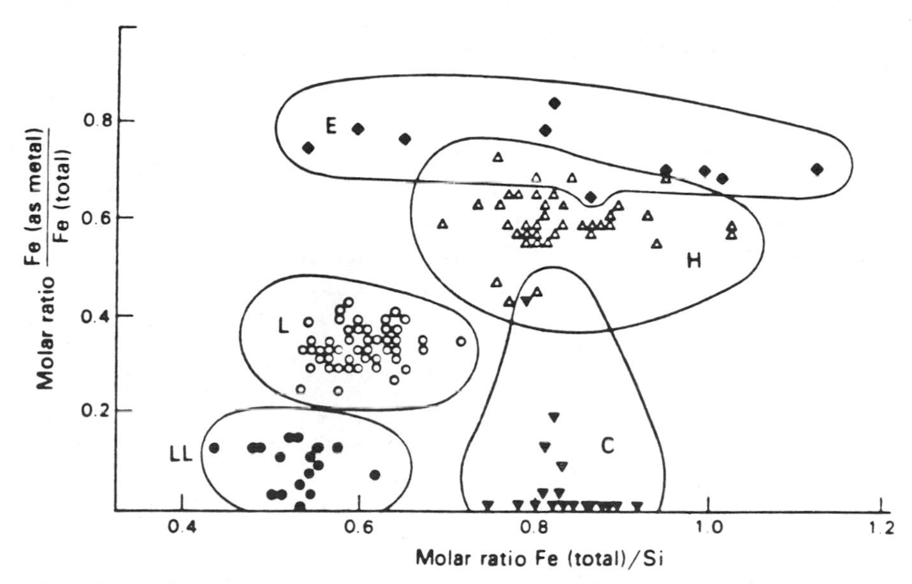

Figure 2.6. Differences in the overall abundance (Fe/Si) and state of oxidation (Fe metal/ total Fe) in chondrites. On the basis of these and other chemical differences, chondrites can be separated into five groups (see circled groups). Source: J. A. Wood (8), *Meteorites and the Origin of Planets,* 1968; permission, McGraw Hill.

cesses operating in and on the earth, such as weathering, sedimentation, and metamorphism, strongly tend to differentiate the three groups from one another. Thus we find the crustal rocks generally are composed of the lithophile elements (Se, Sr, Rb, Ba, Thand U), while the chalcophile elements (Cu, Zn, Sn, Pb, Ag, Hg, Cd, In) concentrate in a few areas as ores. The siderophile elements (Be, Ga, Ru, Pd, Os, Ir, Rh) generally are missing from the crust and presumably are found in Earth's core. By contrast, these three groups of elements are found mixed together in the meteorites. The conclusion drawn therefore is that these chondrites show little evidence of differentiation, and so perhaps have remained unchanged since the early days of planetary formation.

The undifferentiated character of the chondrites has been shown in Figure 2.3, when we discussed the Allende meteorite. A comparison between a differentiated and undifferentiated body is further enhanced by comparing the Earth's crust with the solar atmosphere. It has been observed that the siderophiles and the chalcophiles are enhanced in the Sun relative to the Earth.

Clearly, from the above discussion the chemical analysis of meteorites is of great importance. There are now a wealth of data and, as indicated above, there have been a number of systematic efforts to sort out "superior analyses."

A very useful book was published by Mason in 1971 (19). *Handbook of Elemental Abundances in Meteorites* is the output of the activities of the working group on extraterrestrial chemistry of the International Association of Geochemistry and Cosmochemistry. The work on the elemental abundances was organized by Dr. Brian Mason (21) as part of an interdisciplinary effort. Not only does the volume include extensive data on the abundances of the elements in meteorites but also notations on the relative quality of the data. The object of the book was to "provide a standard reference on the subject for years to come."

The Mason compilation is very comprehensive, including discussions by some of the most prominent workers in the field, of some 63 elements, the rare gases and the rare earths. Also included is an index of meteorites listing over 400 meteorites that are referenced in the text. In the main, the Mason volume is an excellent source book for workers in the field of meteoritics.

While a complete review of the Mason book is beyond the scope of this monograph, I shall give an example to demonstrate how the volume is arranged. For illustrative purposes I shall select one of the more common elements.

Among the common elements is Si, which is discussed by C. B. Moore (20). As we have seen above, Si is a major element in stony and stony-iron meteorites. The major minerals present in these meteorites are silicates. In classical wet chemical methods silicon is normally determined by gravimetric techniques. Recently, however, these methods have been replaced by X-ray fluorescence and neutron activation, which provide direct analysis for silicon. Because of the high relative and constant abundance and the high-quality analyses, silicon is used as the basis of comparison of atomic abundances. The abundances of individual elements are compared to 10^6 atoms of silicon. In general, the results obtained by conventional chemistry, X-ray fluorescence, and 14 MeV neutron activation agree quite well. The source for these data are Mason (21), Mason and Wiiks (16), von Michaelis and co-workers (22), Vogt and Ehmann (23), and Ehmann and Durbin (24).

A list of findings based on systematic studies follows:

1. Many elements show relatively small variations with respect to silicon abundances, but careful selection and evaluation of data indicate significant fractionation between chondrite groups.
2. The question of Si/Mg, Si/Ca, and Si/Ti fractionation has been examined by Ahrens (25–27), Ahrens and Cameron (28), and DuFresne

and Anders (29) in some detail. They have reported that the Si/Mg ratios in carbonaceous (1.11–1.16), ordinary (1.23–1.30), and enstatite (1.48–1.58) chondrites are significantly different.

3. The Si/Ca, Si/Al, and Si/Ti ratios show even greater variation.

4. The agreement between Si values from wet analyses and instrumental techniques is also very good for the achondrites.

5. The amount of silicon in iron meteorites is difficult to determine. Early analysis, which reported as much as 1.2% of SiO, were in error. The large sample size taken (up to 100 g) could contain hidden silicates. In addition, silica was disolved from the soft soda-lime chemical glassware used at the time.

6. Most modern instrumental methods for the detection of trace concentrations of silicon in iron meteorites have not proved useful. Neutron activation methods have a phosphorus interference, and colorimetric methods are interfered with by germanium.

7. The electron probe has proven to be a useful device for determining low concentrations of silicon in metallic meteorites. The advantage is that it avoids including silicate inclusions in the sample being analyzed. Beaulieu and Moore (30, 31) have used a colorimetric method to detect 4–100 ppm in iron meteorites. Some 145 iron meteorites were examined by them.

8. Finally, the isotopic composition of silicon in the Melrose chondrite was studied by Reynolds (32) and Reynolds and Verhoogen (33). They found that this meteoritic silicon had a δSi^{30} value of -1.0% as compared with an olivine standard. Epstein and Taylor (34) reported that silicon in a hyperstene achondrite and enstatite achondrite was generally lighter than silicon in most terrestrial and lunar rocks.

ELECTRON-PROBE STUDIES

No instrument has contributed more to the study of meteorites than the electron microprobe. K. Keil, in a review paper "Applications of the electron microprobe in geology" (35), states that

> In recent years, no other research tool has revolutionized geology as much as the electron microprobe, and its impact on earth sciences can only be compared with the introduction of the polarizing microscope to geology more than 125 years ago.

Keil cites the following figures: Until 1964, 14 years after the invention of the probe, only about 60 papers on the subject had been published, and more than half dealt with meteorite research and related topics. By 1971, when his review article was written, there were some 600 papers dealing with the application of the microprobe to mineralogy, petrology, geochemistry, meteoritics, and other associated disciplines. Although I have not made a specific determination, it is safe to say that at this time the number has climbed to several thousand. Furthermore, a great deal of impetus has come from the lunar sample program and the increased activity in the study of meteorites.

In a general way, a major part of geochemical research concerns itself with an understanding of the distribution of chemical elements in multiple assemblages. To properly understand a naturally occurring system, it is necessary to establish the composition of coexisting phases as accurately as possible.

Separating a rock mass or a meteorite into its component phases for analysis is often tedious and difficult. It is frequently very difficult to establish just how successful the separation procedure has been. Furthermore, the very process of separating the various phases distorts or even destroys potentially useful information about the textures, the structures, gradients, and so on. Bulk analysis of separated phases can only give average compositions but little information about detailed variations within a phase.

Reviews of the use of the electron microprobe in meteorite studies have been written by Fredriksson and Reid (36) and Keil (35). They point out that the electron probe is ideally suited for minimizing these problems. Phases are examined in situ without the need for separations, thus the important spatial considerations are maintained intact.

The history of electron probe studies in meteoritics is a very impressive one, having led to the discovery of a number of new minerals and an elucidation of the relationships between coexisting minerals and mineral formation. A few selected examples will be given.

The electron microprobe analyzer is by now not a new instrument. It has undergone considerable development and, like so many analytical instruments, usually has a dedicated computer that makes the handling of large amounts of data and the problem of data reduction a routine matter. The basic instrumentation and its application to the analysis of materials has been described in numerous publications. A thorough treatment of the subject can be found in a recent text by K. J. F. Heinrich, *Electron Beam X-Ray Microanalysis* (37).

The analysis of geological samples poses a number of problems that have been summarized by Keil (35) in his review. Most geological samples are poor electrical and thermal conductors. It is necessary to deal with problems of electrical charging, and in some instances loss of elements due to volatilization or even migration. It has been found that of the various ultrathin coatings to be evaporated on the sample surface, carbon is the most suitable. It increases the conductivity of the surface without appreciably increasing the absorption of the incident electrons or the emitted X rays. Where there is danger of element loss by volatility, the problem is dealt with by decreased exposure time to the electron beam, reduced beam power, or even defocusing the beam to reduce the surface loading.

Because rocks are generally complex mineral assemblages and often contain phases of different degrees of hardness, the preparation of a polished totally flat surface, free of relief, is quite difficult. There are also problems with smearing and selective plucking of some phases. Thus special pains are required in sample preparation and, above all, an awareness of the possible errors introduced.

One of the unique aspects of studying the rock-forming minerals, and this applies to terrestrial as well as extraterrestrial materials like lunar rocks and the stony meteorites, is that they are mainly made up of low-atomic-number elements. For example, 98.5% of the earth's crust is made up of oxygen (46.6%), silicon (27.2%), aluminum (8.1%), iron (5.0%), magnesium (2.1%), calcium (3.6%), sodium (2.8%), and potassium (2.6%). Because of the low atomic numbers of these elements, their characteristic K lines have long wavelengths and are somewhat more difficult to measure than the harder X-ray lines of the heavier elements dealt with in many metallurgical problems or in the metallic meteorite studies. Thus, geological samples require instrumental capabilities that permit the detection and measurement of soft X rays. A basic requirement is for a device that minimizes the absorption effects of these soft X rays, and uses detectors specially designed for soft X-ray measurements.

Among the more significant advances in the area of microprobe applications is the development of software programs for the performance of quantitative analysis. Such programs are not only suitable for dispersive analysis, but for the nondispersive mode as well. There are numerous programs that have been evaluated for general applicability and accuracy. For example, Beamann and Isai (38) have reported that there are something like 40 programs available for converting raw microprobe X-ray intensities into chemical composition. Of these, there are at least two that are widely used in meteorite studies. The technique developed by Bence and Albee (39) is highly empirical, depending on the determination of correction factors based on the use of well-characterized simple minerals, oxides, and synthetic components such as mineral end members. On the other hand, Beaman and Isai have concluded that the computer program MAGIC is the most generally applicable. MAGIC, which translates into Microprobe Analysis General Intensity Corrections, was first described by Colby (40) in 1967. It has by now achieved wide-scale use and has been modified by various users for many different computers. Such programs are continually being updated, and the contribution of such techniques in data handling and reduction has been quite revolutionary.

The impact of the electron microprobe on meteoritics can be clearly demonstrated by the following table, which is abstracted from the review paper of Keil (29). There are a surprising number of new minerals that have been discovered, many of which are new species never observed before. From 1962 to the time of the Keil publication (1973), the number of meteoritic minerals had increased from 40 to more than 80. Of these, 19 were new phases unknown on the earth.

The ability to characterize the composition of these phases was due to the power of the electron microprobe. Many of these phases have been described as exotic, containing unusual components such as native elements, phosphides, nitrides, carbides, silicides and rare sulfides, phosphates, and silicates that originated due to the low oxygen partial pressures at the time of formation. A partial list of these new minerals is shown in Table 2.2. The list has grown since the Keil review. It is apparent that the electron microprobe analyzer has made the hunt for new minerals a fruitful and exciting area. It is apparent also that a better understanding of the formation of such a variety of minerals should lead to a better understanding of the chemistry of the early Solar System.

The use of the electron microprobe in the hunt for new minerals in extraterrestrial materials is only a small part of the story. As I have stated, the composition of coexisting minerals in meteorites and lunar rocks can provide important clues to the conditions of origin: temperatures, pressures, oxygen partial pressures, cooling rates, and the like. A substantial number of studies have been performed on coexisting phases in meteorites. There are numerous literature references (see references at the end of this chapter). The following titles of recently published papers should supply some notion of the variety of electron microprobe studies being performed:

"Elemental abundances in chondrules from unequilibrated chondrites: evidence for chondrule origin by melting of pre-existing materials" (41);

"A 3.6-b.y.-old impact-melt rock fragment in the Plainview chondrite: implications for the age of the H group chondrite parent body regolith formation" (42);

"The matrices of unequilibrated ordinary chondrites: implications for the origin and history of chondrites"(43);

Table 2.2 New Minerals (Meteoritic and Lunar) Discovered and Characterized with the Use of the Electron Probe Microanalyzer

Mineral Class	Mineral	Chemical Composition	Occurrence
Elements, alloys, nitrides, phosphides and silicides	Barringerite	$(Fe_{0.58}Ni_{0.42}Co_{0.003})P$	Ollague pallasite
	Carlsberite	CrN	Iron meteorites
	Haxonite	$Fe_{23}C_6$	Iron meteorites
	Lonsdalite	C	Canon Diablo, Goalpara iron meteorite
	Perryite	$(Ni,Fe)5(Si,P)_2$	Enstatite chondrite meteorites
	Sinoite	Si_2N_2O	Enstatite chondrite meteorites
Sulfides, arsenides, antimonides tellurides, bismuthides	Brezinaite	Cr_3S_4	Tucson iron meteorite
	Djerfisherite	$K_3(Cu,Na)(Fe,Ni)_{12}S_4$	Enstatite chondrite meteorite
Oxides and hyroxides	Armalcolite	$(Mg,Fe)Ti_2O_5$	Mare Tranquillitatis, Moon
Phosphates and arsenates	Brianite	$Na_2CaMg(PO_4)_2$	Meteorites
	Panethite	$Na_2(MgFe)_2(PO_4)_2$	Meteorites
	Stanfieldite	$Ca_4Mg_3Fe_2(PO_4)_2$	Mesosiderites and pallasites
Silicates	Krinovite	$NaMg_2CrSi_3O_{10}$	Iron meteorites
	Majorite	$(Mg_{2.98}Na_{0.10})(Fe_{1.02}Al_{0.23}Cr_{0.03}Si_{0.78})Si_{3.00}O_{12}$	Coorara meteorite
	Merrihueite	$(K,Na)_2(Fe,Mg)_5Si_{12}O_{30}$	Mezo-Madras meteorite
	Pyroxferroite	$(Fe_{0.84}Ca_{0.13}Mg_{0.02}Mn_{0.02})SiO_3$	Mare Tranquillitatis, Moon
	Ringwoodite	$(Mg,Fe)_2SiO_4$	Tenham stone meteorite
	Roedderite	$(Na,K)_2Mg_2(Mg,Fe)_3Si_{12}O_{30}$	Indarch stone meteorite
	Tranquillityite	$Fe_8^{2+}(Zr+Y)_2Ti_3Si_3O_{24}$	Mare Tranquillitatis and Mare Procellarum, Moon
	Ureyite	$NaCrSi_2O_6$	Iron meteorites

> "Graphite-Magnetite aggregates in ordinary chondritic meteorites" (44);
>
> "Rare-earth and other trace elements in rim and interior portions of a peculiar Allende chondrule" (45);
>
> "On the distribution of Zn in Allende chondrules, inclusions, and matrix" (46).

Obviously, these titles are only a fraction of the total, but they do illustrate the importance of the probe studies.

Zonal structure: The small size of the electron beam makes the instrument ideally suited for the study of zonal structures in the meteorite minerals. The method recommended by Keil is to maintain a static beam while the specimen is moved in preselected steps relative to the beam while recording the X-ray signals.

A considerable amount of effort has gone into the study of metallic meteorites. For example, the zoning and the diffusion gradients between the low nickel alloy kamacite and the high nickel alloy taenite has been investigated. Some of the earliest quantitative analyses were done on the iron–nickel alloys in the iron meteorites. The diffusion gradient studies between the kamacite and the taenite were used to infer the cooling rates of iron meteorites. Such measurements led to estimates of cooling rates ranging from about 0.2 to 100°C/m.y. and the speculation that the maximum depth of origin was about 200–250 km in the parent body. One particularly interesting example cited by Keil is the analysis of two early Chinese bronze weapons with blades made of meteoritic iron, showing the characteristic growth of kamacite–taenite (47).

Geochemistry of elements in individual mineral grains: geochemical studies have usually involved the determination of elements in bulk rock samples. The attempts to determine geochemical relationships by determining the elemental composition on separated minerals are very much less common. In large part, this arises from the difficulty of separating pure mineral phases. As in the other instances described, the electron-probe microanalyzer makes it possible to do such analyses on polished thin sections without the requirement of separation. As Keil points out, the limitation comes from a limited sensitivity to trace elements. There are examples in meteorite studies where such investigations have been done. Goldstein (48) determined the distribution of germanium in the metallic phases of iron meteorites at concentrations ranging from about 9 to 2000 ppm. Silicon and chromium distributions were determined in irons by Wai and Wasson (49). Titanium, a litophilic element in terrestrial rocks, was found by Keil (50) to be chalcophilic in highly reduced meteorites such as enstatite chondrites and achondrites (concentrated in the troilite).

The large-scale use of the electron-probe microanalyzer in meteorite studies provided extensive experience for its application in lunar sample studies, as we shall see in the section on lunar samples.

THE ION MICROPROBE

It had been recognized for some time that an in situ capability for performing precise isotope analysis would be a great boon to the geochemist and the geochronologist. With the advent of the ion microprobe it seemed that at last the in-place analysis of isotopes would become a reality, but it remained an elusive goal until recently. Now under the thrust supplied by the space program a development that offers some noteworthy successes and considerable hope for the future has occurred.

A recent report by Huneke, Armstrong, and Wasserburg (51) titled *Fun With Panurge: High Mass Resolution Ion Microprobe Measurements of Mg in Allende Inclusions,* provides an excellent summary of the situation. It is pointed out that reliable and efficient isotopic measurements on minerals by ion microprobe mass spectrometry require the following instrumental improvements:

1. sufficiently high mass resolving powers to minimize isobaric interferences by molecular and multiply charged ions;
2. adequate control of the electrostatic and magnetic fields for efficient and routine data acquisition at these high mass resolutions; and
3. an ion detector and pulse-counting system with operating characteristics that enable accurate counting up to megaherz rates.

Because of the need for precision and efficiency, operation occurs in a peak jumping mode to maximize the useful yield of sputtered ions and to enable data acquisition on a time scale that is short compared to variations in analytical conditions. This subject has been treated in another paper by the same authors (52) in 1982. It is the viewpoint of the authors that the above properties had not been demonstrated in toto for any ion microprobe to this point in time.

PANURGE referred to in the paper is a modified CAMECA IMS3F ion microprobe, a second generation ion microprobe/ion microscope designed for high resolution mass spectrometry. The instrument as acquired is unique in that the performance specifications were arrived at by consultation between the Cal Tech investigators and the officers of CAMECA Inc. Details about the instrument can be found in publications by Lepareur (53) and Slodzian (54). Details about the performance of the enhanced instrument are included in the papers by Huneke, Armstrong, and Wasserburg (50) and Armstrong, Huneke, and Wasserburg (51). In brief, these investigators have developed so precise an instrument and careful techniques that the only limiting factor to the precision is the counting statistics.

The feasibility of using the ion microprobe for isotopic analysis of Mg at the microscopic level, using an instrument of low mass resolution, has been successfully demonstrated. In fact, the ion microprobe was employed to verify the discovery of the excess of ^{26}Mg in Ca–Al-rich inclusions in the Allende meteorite. Importantly, the ability to analyze selected 10-μm areas was used to demonstrate the existence of regions within crystals with $^{26}Mg/^{24}Mg$ enhanced by about 40%, and that the ^{26}Mg content in a single crystal was constant at $^{26}Mg/^{27}Al = 5 \times 10^{-5}$ (55, 56).

A large number of other studies have been done on the Mg isotopes using the ion microprobe. The references for these are included in Huneke, and associates (51).

In view of the extensive prior work on the Mg isotopes, Armstrong and co-workers felt that a good test of the capabilities of the PANURGE/IMS3F could be done by measurement of Mg and Si isotopic composition, as well as Mg/Al in single crystals from the Ca–Al-rich inclusions in the Allende meteorite. Two inclusions designated as WA and C1 were chosen because it was possible to compare the ion probe measurements against the precise results obtained by conventional mass spectrometry performed by Lee and co-workers (57), Esat and associates (50), Wasserburg and co-workers (59), and Clayton (60). The measurements were reported as a demonstration of high resolution ion microprobe capability on specific examples.

As for results, the Allende WA and C1 inclusions provided outstanding illustrations of the two major Mg isotopic effects that had been discovered in the Allende inclusions. The following observations have been reported: Mg in minerals where the Al/Mg is high is generally enriched in ^{26}Mg, a daughter product of the ^{26}Al decay ($T_{1/2} = 7 \times 10^5$y).

The Mg/Al ratio in the Ca–Al inclusion varies from 0 to 7×10^{-5}, and in WA equals 5.1×10^{-5}. The correlation of ^{26}Mg with Al in different minerals has been interpreted to provide evidence for the in situ decay of ^{26}Al, and thus for the presence of ^{26}Al in the solar nebula. The implication that follows is that ^{26}Al was produced in a nucleosynthetic event less than a few million years before the formation the Ca–Al inclusion found in the Allende.

The second observation is that in a few inclusions, including the C1 chondrule, Mg is very heavily fractionated and exhibits a minor nonlinear anomaly of unknown nuclear origin (61). The presence of strongly fractionated Mg, and the nuclear anomalies of unknown origin correlates with unusual iotopic compositions of nearly every other element studied. A particular example is the heavily mass-fractionated Si with a possible nuclear anomaly in ^{29}Si (60). The origin of elements that are fractionated with unknown nuclear sources is not established; however, their presence demonstrates an isotopic heterogeneity in the solar nebula.

These facts represent some important findings. The distribution of ^{26}Mg and the fractionation of Mg and Si among different minerals of an inclusion and within a single crystal must relate to the origin of the isotopic components and the formation history of the Ca–Al inclusions. The contribution of the ion microprobe technique to these studies is considerable. It provides a means to study the spatial distribution of ^{26}Mg and the fractionated compositions on a micron scale. This has made it possible to establish areal distributions on a microscopic scale using a 3-μm focused primary ion beam to sputter materials from selected regions and, by the appropriate use of instrumental apertures, to further restrict measurements to small regions within a larger sputtered area.

The nature of the technique is such that simultaneous measurements of Mg and Si composition can be made on the same microvolumes. The investigators in this paper (51) report that their results are certain beyond experimental error. By a careful evaluation of the magnitude and variability of instrumentally produced mass fractionation, they have determined that the ion microprobe can indeed be used for the studies described above.

ISOTOPIC ANOMALIES

One of the more interesting stories to come out of the space exploration period is that of the search for isotopic anomalies in the meteorites and lunar samples. The way in which the story has evolved is highlighted in a statement made by J. T. Wasson (5) in his book on meteorites published in 1974.

Probably the single most important conclusion resulting from the study of the isotopic composition of relatively nonvolatile meteorite elements is that meteorites are samples of normal solar system material. This conclusion is based on the fact that, as of now, all nonvolatile elements have the same isotopic compositions in meteorites as in terrestrial and lunar samples. In fact, the recent history of isotopic investigations of meteorites is a story of anomalies discovered and anomalies lost.

The point of this statement is that many of the early claims of anomalous isotope concentrations were subsequently discredited by more painstaking work.

It is interesting to report, however, that since the Wasson statement was made there have been a number of highly significant discoveries of anomalous isotope abundances. This comes as the result of remarkable advances in mass spectrometry, refined chemical separation methods, refined mineral preparation procedures, and in mineral identification and analysis (62).

An account of the search for isotope anomalies has been summarized in recent review articles by Lewis and Anders (63), Lee (64), Wasserburg, and Papanastassiou and Lee (65). By all estimates, our Solar System is considerably younger than the universe, some 4.5 aeons compared to 10 to 15 aeons. Logically then, it must contain older matter with a prior history, but until recently no evidence for relict matter had been found, either in meteorites, the Earth, or the Moon. It was pointed out by Lee (64) that the degree of isotopic homogeneity of the various elements found in terrestrial, lunar, and meteoritic samples was indeed quite remarkable in view of the fact that the production of different isotopes of the same element often requires markedly different physical conditions and different sources. Thus the degree of homogeneity found can only indicate effective mixing processes in the early solar nebula, before the planetary bodies formed.

Recognition of presolar matter is one of the major problems. Isotopic composition seemed to be a very promising approach. We know that stars are essentially sites where the chemical elements undergo transmutations, followed by the ejection into intersteller space, and that ultimately they find their way into the next generation of stars. It is also understood that the ejected matter can vary from star to star as a function of the star's mass, temperature, and evolution. Lewis and Anders make the important point that "isotopic composition is a particularly durable hallmark, since it can be changed by few processes (short of nuclear reactions) and then only to a limited predictable extent."

Among the various areas of research in meteorites, the search for isotopic anomalies is being pursued with great vigor. This is evident when one considers the number of recent literature references (66). One of the most significant findings previously discussed comes from the work of Clayton and co-workers (60) on the carbonaceous chondrites. They have reported that these meteorites have oxygen isotopic compositions substantially different from those found in terrestrial and lunar rocks. These anomalies are not explainable by nebular processes, and the suggestion has been made that they represent the injection of presolar material from outside the nebula early in its history. Thus, if nothing else, this offers a promise of providing insights into processes outside the nebula.

Up to now, isotopic anomalies have been found in a number of other elements, such as Mg, Ca, Sr, Sm, Ba, and Nd. The noble gases have been particularly interesting and their relationships particularly complex. In their case, as Dodd points out (66), an extremely complex pattern is emerging; it will be a while before a comprehensive theory emerges.

While it is beyond the scope of this volume to treat this subject in detail (it is quite extensive), some examples will be cited for illustrative purposes.

The question of the oxygen isotopic anomalies has already been discussed above (see section on the Chemical Composition of Meteorites). Let us examine the relationship between the oxygen and some of the other elemental isotopes. If the oxygen-16-rich component did in fact enter the primordial solar nebula in discrete extrasolar grains, one might reasonably expect then to find a correlation between the oxygen anomalies and anomalies in the silicon and magnesium ratios. This has logically led to a study of silicon and magnesium isotopic ratios. There are, however, as yet no conclusive findings. The few silicon data available suggest that silicon is generally isotopically normal. While considerable variation has been found in silicon-29/silicon-28 and silicon-30/silicon-28 ratios in the Allende meteorite, they have been attributed to mass fractionation rather than a peculiar nucleosynthesis.

As for the magnesium, the situation is complicated by the fact that magnesium-26, which could have arrived in extrasolar grains, could also result from the decay of aluminum-26, a relatively short-lived radioisotope (half-life = 0.72 m.y.). It has been found that magnesium-26 excesses in the Allende aggregates correlate rather poorly with the oxygen-16 excesses. The conclusion has been drawn that much of the magnesium-26 came from the in situ decay of the aluminum-26. While these findings are a disappointment to those looking for extrasolar grains, it does provide a powerful source of heat for the thermal metamorphism encountered in some meteorites (66). In recent work reported by Clayton and Mayeda (67) and Wasserburg and others (61), it was found that not all of the magnesium-26 was radiogenic. Two particular Allende aggregates showed that magnesium-26 and oxygen-16 anomalies did correlate in detail. Further, these aggregates also showed significant anomalies in calcium.

The situation continues to be complicated. If one focuses on the carrier or carriers of the isotopic anomalies, the data from the oxygen and the magnesium appear to point in different directions. Whereas the oxygen-16 in the aggregates suggest that the extrasolar component entered the Solar System in discrete, solid particles, the isotopic composition of the magnesium, though distinctive, shows no correlation with mineralogy. An explanation that has been offered by such investigators as Clayton and Mayeda and Wasserburg and co-workers is that the low oxygen-16 content of the minerals melilite and anorthite in the Allende aggregates reflect an oxygen exchange with a resevoir of normal composition after the aggregates had formed in an oxygen-16 rich resevoir, an as yet unproved hypothesis.

The noble gases have still another story to tell. They are unique in that they are affected by all the methods of isotopic fractionation that have been discussed. The isotopic anomalies that they display are associated with the matrices of carbonaceous chondrites rather than the high-temperature aggregates. The interpretation of rare gas isotopic data involves the evaluation of the isotopic contributions from numerous processes. To begin with, the noble gases are highly volatile and unreactive. Because of this, it is highly likely that they did not condense, not even in the most primitive of the meteorites. Their abundance in the meteorites is but a minute fraction of the solar abundance. What little gas there is is found tightly bound in the meteorites, and the release occurs only at high temperatures as the host minerals either melt or decompose.

The various isotopes are analyzed by a process of controlled-step heating, a method employed by Black and Pepin in some of the initial studies (68). In this manner, it was possible to separate the rare gases into several components. The radiogenic components were found to consist of the products of in situ decay of radioisotopes. Some of these components can only be found in very old materials. One noteworthy example is xenon-129, formed from the short-lived isotope iodine-129 (half-life 17 m.y.). Another group of noble gas isotopes, the cosmogenic ones, are those generated by cosmic ray interactions with the elements in the meteorites.

In addition to the components formed in the meteorite subsequent to its formation are those trapped in the meteorite during formation. These have been labeled the trapped or primordial noble gases and are divided, on the basis of isotopic composition, into the planetary and solar components. The planetary component is isotopically similar to the terrestrial atmosphere and is present in all chondrites in abundances that vary with the petrologic type. The solar component is isotopically similar to solar gas and is characteristic of brecciated chondrites and achondrites. There is strong evidence, such as a correlation with charged particle tracks, that suggest an implantation by either solar flares or the solar wind. However, this is not the entire matter. When the above components are subtracted, some materials show residual gases that are difficult to explain. There is, for example, a peculiar xenon component discovered by Reynolds and Turner (69) in the Renazzo chondrite. There is also an anomalous neon component "neon E," abundant in the silicate fraction of the Orgueil meteorite (70, 71).

The question of which mineral phases carry these rare gases is a vital one and is the subject of considerable investigation. Some success has been achieved (72–75).

The evidence is that certain of the various noble gas components in carbonaceous chondrites can be found in different phases from several sources. It has been proposed by Anders (76) that the neon E and an anomalous Kr–Xe component are extrasolar, coming from highly evolved stars. An unusual Xe component (heavy) may be indigenous or exotic and have come as a result of the fission of a now extinct superheavy element within the Solar System or as an alternative, nucleosynthesis in a supernova. It is obvious that this is a very exciting and fertile field for investigation. At this time there are more questions than answers.

The discovery and study of anomalous isotopic distributions has had some very significant spinoffs, an example of which is the development of new rock dating methods. One particularly outstanding method based on Sm/Nd has contributed greatly to geochronology. In the brief period since its development it has proved to be a useful addition to the arsenal of methods for providing dates for some of the earliest events in the Solar System. The method not only has value in the study of extraterrestrial materials, but for terrestrial materials as well.

The basis for Sm–Nd ages is as follows: Several investigators had shown that significant variations exist in Sm/Nd ratios between mineral phases in various meteorites as well as in lunar rocks. For example, Tanaka and Masuda (77) found substantial differentiations of the rare earth elements in some inclusions in the Allende meteorite. In an abstract published in *Meterotics* in 1974, G. W. Lugmair (78) stated that since both Nd and Sm are refractory elements with similar properties, the natural radionuclide ^{147}Sm (half-life = 1.06×10^{11} y) could yield important radiometric age information, particularly where the Rb–Sr isochron is ill defined because of subsequent metamorphic events or in samples with low Rb/Sr ratios. The successful application of the Sm/Nd ratios depended on careful analysis because the variations in Sm/Nd were quite small and the corresponding variation in the radiogenic ^{143}Nd was of the order of parts per thousand. Thus, using the information that ^{147}Sm decays to ^{143}Nd by the emission of an alpha particle at the rate given by the half-life constant, Lugmair was able to develop the method and perform isochron dating. (For additional details see Chapter 5.)

The first application reported by Lugmair was the determination of an internal isochron for the basaltic achondrite called Juvinas. The ratios Sm/Nd were determined by isotopic dilution in liquid aliquot samples. An isochron age of 4.56 aeons was obtained, with a possible error of 0.08. This agreed closely with the Rb/Sr isochron age of this meteorite obtained by Allegre and co-workers (79). The results were quite significant, they demonstrated that on a sample that showed no evidence for any major metamorphic events after it had formed, the Sm–Nd and Rb–Sr methods agree very well. In the dating of rocks the finding of concordance by independent methods is particularly exciting.

The Sm–Nd method is now widely accepted as a powerful dating method. There have been numerous applications to meteorites and lunar samples. Analytically, this potential could only be exploited by the development of greatly improved methods of sample handling and analysis.

ORGANIC COMPOUNDS AND ORGANIZED MATTER

The presence of organic matter in carbonaceous chondrites has been of considerable interest for a period extending over the last two decades. Considerable excitement was engendered when not only organic matter but organized elements were observed. This subject was introduced very briefly by Dodd (66) and discussed in greater detail by Wasson (5).

Because these organic compounds were found to be of considerable complexity and because there was also evidence of organization, a question arose about the relationship between these compounds and biogenic origin. To some investigators, the question was whether or not this provided evidence that life existed at the very beginning of our Solar System. There were numerous studies and considerable controversy. A summary given by Dodd is that "no irrefutable evidence of biogenic origin has been advanced." In fact, Anders and co-workers demonstrated that the suite of organic compounds in the carbonaceous chondrites could be duplicated by abiogenic processes (80).

The interest in the organic compounds is a continuing one, with the major effect being directed towards understanding the mechanisms for formation. Early reports about the organized elements stated that the morphology suggested formation by organisms; to some investigators this was evidence of life in the primitive solar nebula. On investigation, some of these objects were shown to be terrestrial contaminants such as pollen. Others had a chemical origin; at least one was shown to be a hoax. In no instance was it possible to demonstrate extraterrestrial biological origin.

With regard to the complex organic compounds, alternate mechanisms have been proposed to accommodate a nebular origin. A number of investigators have proposed a Fischer–Tropsch synthesis involving a mixture of carbon monoxide, water, and ammonia heated to moderate temperatures in the presence of a catalyst. The products would be a mixture of stable and metastable organic compounds with the relative amounts depending on the reaction conditions.

A second and competing concept is a Miller–Urey synthesis in which a mixture of methane, ammonia, and water is subjected to brief bursts of energy at relatively low temperatures. This latter mechanism involves hot atom chemistry with the products controlled mainly by kinetics rather than equilibrium.

In the main, the determination of the synthetic processes that have produced meteorite organic matter is a difficult undertaking and as yet requires considerable further research.

Among the more interesting observations is the presence of amino acids in the carbonaceous chondrites. Particularly exciting is their presence in the recently discovered Antarctic meteorites, specimens that have survived relatively free of contamination. Ponnamperuma (81) has reported the presence of a number of nonprotein amino acids and their enantiomers, in almost equal amounts. In this instance, the conclusion has been drawn that there is strong evidence that they are prebiological in nature. The presence of bases that occur in nucleic acids and small amounts of dipeptides may have some bearing on the general question of whether processes of chemical evolution have occurred elsewhere in the Solar System.

ANTARCTIC METEORITES

No account of the meteorites can be complete without reference to Antarctic meteorites. This development in the history of the meteorites has had an impact nearly comparable to the return of lunar samples. An excellent review has been written by Ursula B. Marvin of the Harvard–Smithsonian Observatory for Astrophysics (82). The article points out that in the past decade the Antarctic ice cap has yielded more than 5000 fragments of meteorites. These do not represent individual falls, however, because many are fragments of the same meteorite.

While the vast majority of these finds are stony, a few have proved to be entirely new species. Among these specimens, about 36 were fragments of metallic nickel–iron meteorites; of these one was found to contain diamonds, making it only the second of its kind in the world. One of the most intriguing aspects of the Antarctic meteorites was the discovery of specimens that may have come from the Moon and Mars, as opposed to the asteroid belt.

On the basis of the evidence presented for the lunar meteorite, the case for lunar origin is unusually strong—so strong that when this finding was presented to a large group of scientists at the Lunar and Planetary Conference held in 1983, the acceptance of the idea was essentially unanimous. Since this volume deals with analysis, let us examine the evidence.

The particular specimen is known as the Allan Hills 81005 meteorite, named for the site in which it was found in 1982. The specimen as found was small (about 31 g) and partially covered with a frothy greenish-tan crust. A section of the sample revealed a remarkable similarity to the lunar highland breccias (welded soil). Prominently displayed were white clasts of anorthosite—a common highland rock, calcium and aluminum rich—making up the bulk of the lunar highland crust. Significantly, these minerals, which do occur on the Earth and Moon, are rarely seen in meteorites. The pyroxene minerals (calcium, magnesium iron silicates) in the clasts had lunar ratios of FeO/MnO, significantly different from those in achondrite meteorites. Another piece of strong evidence was the amount and distribution of the rare earth elements, the implanted noble gases as well as the isotopic ratios of oxygen, which were all typically lunar rather than meteoritic.

The find of the lunar meteorite is not without mystery. There is at the moment no clear picture or viewpoint about the dynamic process that could remove a sample from the lunar surface without producing a highly shocked rock, something that is missing in the appearance of the sample.

A second and perhaps even greater puzzle involves the "martian meteorites," two of which have been found in Antarctica. These two are part of the class of meteorites known as the SNC meteorites, having been found in Shergotty, India in 1865; Nakhla, Egypt in 1911; and Chassigny, France in 1815. The find in Antarctic has now strengthened the proposal that these bodies originated on Mars. As in the case of the lunar meteorites, let us examine the evidence for a Martian origin, which in a dynamic sense is even more difficult to understand. The subject has been clearly reviewed by S. P. Maran in an article titled *Do rocks fall from Mars?* (83).

The nine martian meteorites are igneous in nature and obviously formed from a molten rock or magma. When they were dated radiometrically, a surprising crystallization age of 1.3 b.y. was obtained, in sharp contrast to the usual meteorite age of 4.6 b.y.

An inference that was drawn from the crystallization age is that these samples cooled from molten rock about 1.3 b.y. ago. Furthermore, cosmic ray exposure ages tell us that these samples were exposed to cosmic ray bombardment as small fragments prior to landing on the earth.

The question that arises is what planetary body could have been sufficiently hot internally to have produced magmas as recently as the SNC meteorites call for? The Earth is, of course, still volcanically active, but in view of the fact that the SNC meteorites had chemical and isotopic compositions that were meteoritic, it was not likely that the Earth was the source. In addition, these SNC meteorites showed a history of extended flight through space. As for the Moon, all evidence is that it had stopped being volcanic about 3 b.y. ago. Mars, on the other hand, has immense volcanoes on its surface, disclosed by the recent Pioneer and Voyager flights. It is quite possible that Mars had active volcanism as recently as 1.3 b.y. Thus attention is focused on Mars as a possible source.

Adding to the evidence is the presence of an unusual glassy component called maskelynite, which could have formed from the feldspar minerals in the igneous rock in response to shock produced by impact. Finally, it has been reported that the noble gases trapped within the meteorites are similar in composition to the Martian atmosphere reported by the Viking researchers.

As Marvin points out,

> despite all the theoretical and intuitive objections to a Martian origin—the uncertainties of the evidence and difficulties of proof—excitement is mounting at the possibility that, without waiting for a mission to return samples from Mars, we may already have Martian samples at hand.

REFERENCES

1. B. P. Glass, *Introduction to Planetary Geology,* Cambridge University Press, New York, 1982.
2. B. Mason, *Meteorites,* Wiley, New York, 1962.
3. V. M. Goldschmidt, *Norske Videnskaps-Akad. Skrifter, Math-Naturv. Klass,* **4** (1937).
4. I. Noddack and W. Noddack, *Naturwiss.,* **18,** 757 (1930).
5. J. T. Wasson, *Meteorites,* Springer, New York, 1974.
6. G. T. Prior, *Mineral Mag.,* **19,** 51 (1920).

7. J. A. Wood, *Meteorites and the Origin of Planets,* McGraw-Hill, New York, 1968.
8. J. A. Wood, *The New Solar System,* Cambridge University Press, New York, 1982.
9. A. G. W. Cameron, *Icarus,* **1,** 13 (1962).
10. L. Grossman, *Geochim. Cosmochim. Acta,* **36,** 597 (1972).
11. R. N. Clayton, *Science,* **182,** 485 (1973).
12. G. J. Wasserburg, T. Lee, and D. A. Papanastassiou, *Meteoritics,* **12,** 377 (1977).
13. A. E. Ringwood, *Rev. Geophys.,* **4,** 113 (1966).
14. H. C. Urey and H. Craig, *Geochim. Cosmochim. Acta,* **4,** 36 (1953).
15. A. G. W. Cameron in L. H. Ahrens and A. G. W. Cameron, Eds., *Origin and Distribution of the Elements,* Pergamon, New York, 1968.
16. B. Mason and H. B. Wiik, *Geochim. Cosmochim. Acta,* **28,** 533 (1964).
17. G. G. Goles, *The Handbook of Geochemistry,* Springer, New York, 1969.
18. H. B. Wiik, *Geochim. Cosmochim. Acta,* **9,** 279 (1956).
19. H. B. Mason (Ed.), *Handbook of Elemental Abundances in Meteorites,* Gordon and Breach, New York, 1971.
20. C. B. Moore, Silicon, in H. B. Mason, Ed., *Handbook of Elemental Abundances in Meteorites,* Gordon and Breach, New York, 1971, p. 125.
21. B. Mason, *Amer. Mus. Novitates,* No. 2223, 1 (1965).
22. H. Von Michaelis, J. P. Willis, A. J. Erlank, and L. H. Ahrens, *Earth Planet Sci. Letters,* **5,** 387 (1969).
23. J. R. Vogt and W. D. Ehmann, *Geochim. Cosmochim. Acta,* **25,** 373 (1961).
24. W. D. Ehmann and D. R. Durbin, *Geochim. Cosmochim. Acta,* **32,** 461 (1968).
25. L. H. Ahrens, *Geochim. Cosmochim. Acta,* **28,** 411 (1964).
26. L. H. Ahrens, *Geochim. Cosmochim. Acta,* **29,** 801 (1965).
27. L. H. Ahrens, *Geochim. Cosmochim. Acta,* **31,** 861 (1967).
28. L. H. Ahrens and A. G. W. Cameron (Eds.) *Origin and Distribution of the Elements,* Pergamon, New York, 1968.
29. E. R. DuFresne and E. Anders, *Geochim. Cosmochim. Acta,* **26,** 1085 (1963).
30. P. Beaulieu and C. Moore, *31st Annual Meeting, The Meteoritical Society,* Cambridge, Mass. (1968).
31. P. Beaulieu and C. Moore, *32nd Annual Meeting, The Meteoritical Society,* Houston, Texas (1969).
32. J. H. Reynolds, *Proc. Conf. Nuclear Processes in Geol. Settings,* **64,** 224 (1953).
33. J. H. Reynolds and J. Verhoogen, *Geochim. Cosmochim. Acta,* **3,** (1953).
34. S. Epstein and H. P. Taylor, Jr., *Science,* **167,** 533 (1970).

35. K. Keil, in C. A. Anderson, Ed., *Microprobe Analysis,* Wiley, New York, 1973.
36. K. Fredriksson and A. M. Reid, in P. H. Abelson, Ed., *Researches in Geochemistry,* Wiley, New York, 1967.
37. K. F. J. Heinrich, *Electron Beam X-Ray Microanalysis,* Van Nostrand Reinhold, New York, 1981.
38. D. R. Beamann and J. A. Isai, *Anal. Chem.,* **42,** 1540 (1970).
39. A. E. Bence and A. L. Albee, *J. Geol.,* **76,** 382 (1968).
40. J. W. Colby, *Advances in X-Ray Analysis,* Vol. 11, Plenum, New York, 1968, p. 287.
41. J. L. Gooding, K. Keil, T. Fukuoka, and R. A. Schmitt, *Earth Planetary Sci. Lett.,* **50,** 171, 1980.
42. K. Keil, V. Fodor, P. M. Starzyk, R. A. Schmitt, D. D. Bogard, and L. Husain, *Earth Planetary Sci. Lett.,* **51,** 235, 1980.
43. G. R. Huss, K. Keil, and G. J. Taylor, *Geochim. Cosmochim. Acta,* **45,** 33 (1981).
44. E. R. D. Scott, G. J. Taylor, A. E. Rubin, A. Okada, and K. Keil, *Nature,* **291,** No. 5816 (1981).
45. W. V. Boynton and D. H. Hill, *Lunar and Planetary Science X111, Abstrats,* 1982, p. 63.
46. E. H. Cirlin and R. M. Housley, *Lunar and Planetary Science X11,* 1982, p. 104.
47. R. J. Gettens, R. S. Clark, and W. T. Chase, *Occasional Pap.* 4, Freer Gallery of Art, Washington, DC (1971).
48. J. I. Goldstein, *J. Geophys. Res.,* **72** (1967).
49. J. T. Wasson, and C. M. Wai, *Geochim. Cosmochim. Acta,* **34,** 169 (1969).
50. K. Keil, *Earth Planetary Sci. Lett.,* **7,** 243 (1969).
51. J. C. Huneke, J. T. Armstrong, and G. J. Wasserburg, *Geochim. Cosmochim. Acta,* 1635 (1983).
52. J. T. Armstrong, J. C. Huneke, and G. J. Wasserburg, in K. J. F. Heinrich, Ed., *Microbeam Analysis,* San Francisco Press, 1982, p. 202.
53. M. Lepareur, *Rev. Tech. Thomson-CSF,* **12** (1980).
54. G. Slodzian, *Workshop on Secondary Ion Mass Spectrometry (SIMS) and Ion Microanalysis,* NBS Spec. Publ. 427, U.S. Govt. Printing Office, Washington, DC 1975.
55. J. G. Bradley, J. C. Huneke, and G. J. Wasserburg, *J. Geophys. Res.,* **83** (1978).
56. I. D. Hutcheon, I. M. Steele, J. V. Smith, and R. N. Clayton, *Proc. Lunar Sci. Conf., 9th,* 1978, p. 1345.
57. T. Lee, D. A. Papanastassiou, and G. J. Wasserburg, *App. J. Lett.,* **3,** 41 (1976).

58. T. M. Esat, D. A. Papanastassiou, and G. J. Wasserburg, *Geophys. Res. Lett.,* **5,** 9 (1978).
59. J. Wasserburg, D. A. Papanastassiou, and T. Lee, *Communications presentees au XXII Colloque International d'Astrophysique,* Universite de Liège, Belgium (1979).
60. R. N. Clayton, T. K. Mayeda, and S. Epstein, *Proc. Lunar Sci. Conf., 9th,* 1978, p. 1267.
61. G. J. Wasserburg, T. Lee, and D. A. Papanastassiou, *Geophys. Res. Lett.,* **4,** 299 (1977).
62. G. J. Wasserburg, F. Terra, D. A. Papanastassiou, and J. C. Huneke, *Earth Planetary Sci. Lett.,* **35,** 294 (1977).
63. R. S. Lewis and E. Anders, *Sci. Am.,* **249** (2) (1983).
64. T. Lee, *Rev. Geophys. and Space Phys.,* **17**(7), 1591 (1979).
65. G. J. Wasserburg, D. A. Papanastassiou, and T. Lee, *Early Solar System, Processes and the Present Solar System,* LXX111 Corso, Soc. Italiana de Fisca, Bologna, Italy (1980).
66. R. T. Dodd, *Meteorites,* Cambridge University Press, Cambridge, 1981.
67. R. N. Clayton and T. K. Mayeda, *Geophys. Res. Lett.,* **4** (1977).
68. R. O. Pepin, L. E. Nyquist, D. Phinney, and D. Black, *Proc. of the Apollo 11 Lunar Science Conf.,* **2,** 1435 (1970).
69. J. H. Reynolds and G. Turner, *Geophys. Res.,* **69,** 3263 (1964).
70. D. Black, *Geochim. Cosmochim. Acta,* **36** (1972).
71. P. Eberhardt, *Earth Planet Res. Lett.,* **24,** 347 (1974).
72. R. S. Lewis, B. Srinivasan, and E. Anders, *Proc. Nat. Acad. Sci.,* **72,** 268 (1975).
73. J. Gros and E. Anders, *Earth Planet. Sci. Lett.,* **33,** 401 (1977).
74. U. Frick and R. K. Moniot, *Meteoritics,* **11,** 281 (1976).
75. U. Frick and U. Chang, *Meteoritics,* **13,** 465 (1978).
76. E. Anders, *Phil. Trans. Roy. Soc. Lond.,* **285,** 23 (1977).
77. T. Tanaka and A. Masuda, *Icarus,* **19,** 523 (1973).
78. G. W. Lugmair, *Meteoritics,* **9,** 369 (1974).
79. C. J. Allegre, J. L. Birk, and S. Fourcade, *Meteoritics,* **8,** 323 (1973).
80. E. Anders, R. Hayatsu, and M. H. Studier, *Science,* **182,** 781 (1973).
81. A. Shioyama, C. Ponnamperuma, and K. Yani, *Proc. Fourth Symposium on Antarctic Meteorites,* **15** (1979).
82. U. B. Marvin, *New Scientist,* **97,** 710 (1983).
83. S. P. Maran, *Sky Reporter* (Nov. 1983).

CHAPTER

3

THE LUNAR RECEIVING LABORATORY

One of the more interesting and unusual undertakings in the lunar sample program was the construction of the Lunar Receiving Laboratory (LRL), known as Building 37 at the Manned Spacecraft Center (now the Johnson Space Center in Houston). It became one of the most unique installations in the world, built to provide a central facility for performing the receiving functions for the men and materials returning from the Moon. The planners were faced with the major problem of establishing an analytical protocol for completely unknown samples whose composition and physical form could only be guessed at.

The completed building was divided into three main functional areas: the sample operations area, the crew reception area (and its support laboratory), and the administrative area. We shall concern ourselves with that part of the laboratory devoted to the sample studies. Importantly, the LRL was also used for preflight preparations. The various functions include the following:

1. Preflight preparation of the geologic hand tools and the sample return containers.
2. The opening and unpacking of the sample return containers.
3. Preliminary physical chemical tests.
4. Sample cataloguing.
5. Biological quarantine clearance tests.
6. Biological isolation of the crew and the material.
7. Time-dependent studies.
8. Repackaging and distribution of the samples to the investigators.
9. Storage of undistributed samples.
10. Sample data storage and retrieval.

Functions 5 and 6 were considered the most vital and seriously influenced the design of the LRL, as well as the philosophy that guided its operation. Based on a study of the National Academy of Sciences,

NASA, working in cooperation with a number of government regulatory agencies, undertook to protect the terrestrial biosphere against the introduction of foreign pathogens, however remote the possibility. There was, in fact, an International Treaty on Principles Governing Activities in the Exploration and Use of Outer Space, Including the Moon and Other Celestial Bodies, which dictated the nature of the steps to be taken. Thus the scientists and technicians were required to perform the preliminary studies of the samples behind biological barriers.

Figure 3.1 is a diagrammatic sketch of the vacuum laboratory, one of the pivotal installations in the LRL. In addition to its use in the preparation of the tools and containers for outbound flight, it was also employed for the reception of the extraterrestrial samples and to funnel the samples into other parts of the LRL and to the scientific community. The compo-

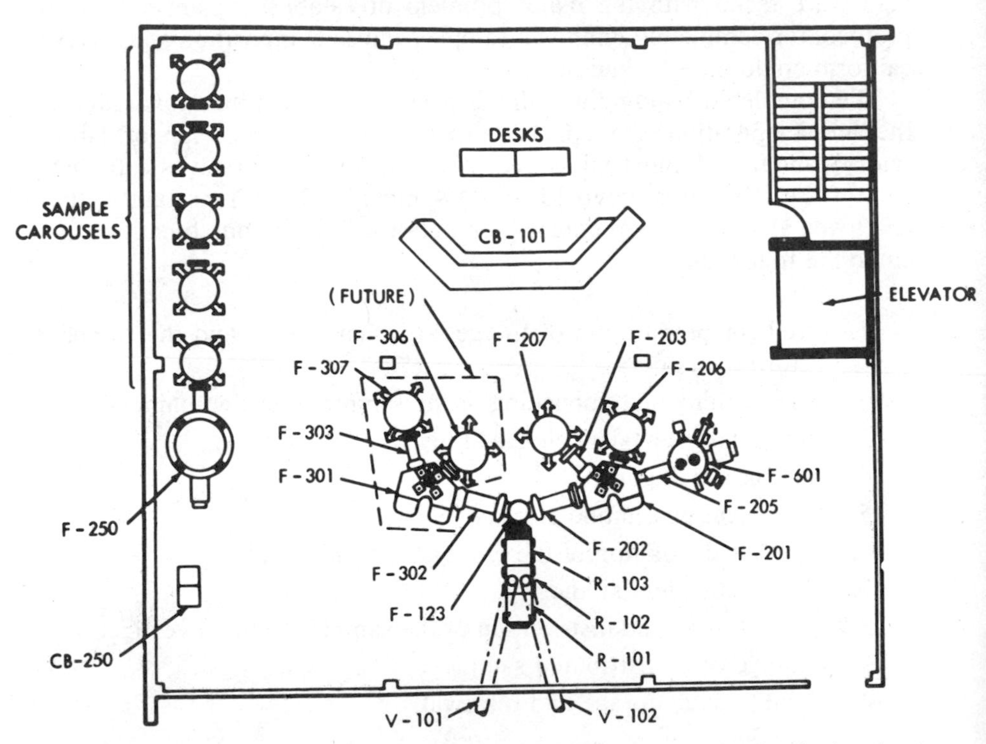

Figure 3.1. Vacuum laboratory: V-101 and V-102 are the transfer tubes for physico-chemical test samples. F-201 is the 1×10^{-6} high-vacuum glove chamber; F-601 is the 1×10^{-11} ultrahigh-vacuum chamber; F-206 is the tool carousel; F-207 is a sample carousel; F-250 is a conditioning chamber.

nent vacuum systems of this laboratory were quite likely the only ones of their kind in the world, having been designed entirely for the extraordinary function of dealing with the lunar samples.

It was a large installation occupying two floors. The first floor contained the rough pumps for the vacuum chambers, the electrical distribution system, the power supplies for the instruments, and the utilities distribution system. The second floor contained the vacuum chambers, the storage carousels, the consoles, and the conditioning system.

As an example of how this laboratory functioned, consider F-250, the conditioning chamber that was used for the preflight operations such as the outgassing of the sample return containers, as well as high temperature, vacuum sterilization (to prevent back contamination). Decontamination was performed on the various items or tools prior to storage in the tool or sample carousels. All transport into and out of the conditioning chamber was through vacuum locks.

The system used for atmospheric decontamination is shown in Figure 3.1 by R-101, R-102, R-103, V-101, and V-102. The cabinet R-101, which served as an airlock for the vacuum system, was equipped with high-intensity ultraviolet lamps. R-102 had a peracetic acid wash system for decontaminating the outer surfaces of the various containers as they moved through the primary vacuum complex. The peracetic acid was removed by a sterile water wash while in R-103, and the containers were dried by a flow of warm, sterile nitrogen.

Figure 3.2 shows the details of the chamber F-201 in which the sample containers were opened and the processing of the lunar samples begun. The chamber contained a residual gas analyzer, a Cahn electrobalance, load cells, six automatic cameras, a Leitz binocular microscope, a video monitor, and assorted instruments to facilitate sample handling. All operations were performed by an operator working through arm gloves into the vacuum chamber, consisting of inner and outer Teflon gloves with the inner space pumped.

We note in Figure 3.1 that there were two carousels attached to F-201. The details of these carousels are seen in Figure 3.3. One of these was used to store tools and the outer containers for the samples moving to the radiation laboratory. The other was employed in storing repackaged lunar samples. The nature of the configurations was such that it was possible to interchange carousels without bringing either the carousel or the lock up to atmospheric pressure.

Figure 3.4 shows the ultrahigh-vacuum chamber that was used to subdivide the samples brought back from the Moon under lunar ambient pressure. The chamber was designed to operate at pressures of 10^{-11} to 10^{-12} Torr. By the use of a mechanical manipulator and a rotary table, the

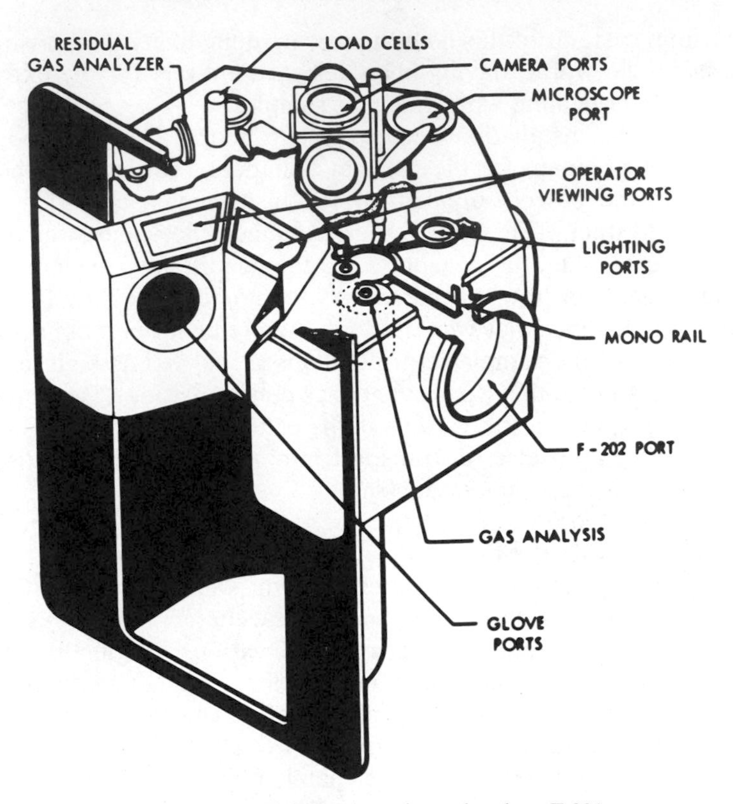

Figure 3.2. High-vacuum glove chamber, F-201.

operator was able to subdivide and store the samples in individual containers that were appendages on the main chamber. The arrangement was such that the appendage containers had their own pumping system; thus the lunar samples could be shipped outside the LRL at pressures of about 10^{-11} Torr.

Figure 3.5 demonstrates the arrangement of the Gas Analysis Laboratory, built for the express purpose of effluent gas analysis on the returned lunar samples, both at ambient and elevated temperatures. Analyses were performed by a variety of mass spectrometers. The high-temperature evolution of gas involved programmable resistance heating, which took the samples from ambient to 600°C. Beyond that point, induction heating brought the samples to 2000°C. Temperatures were controlled to about 1% of the indicated values.

The LRL included a densely instrumented physical chemical test area consisting of a spectrographic laboratory, a chemical laboratory, and a

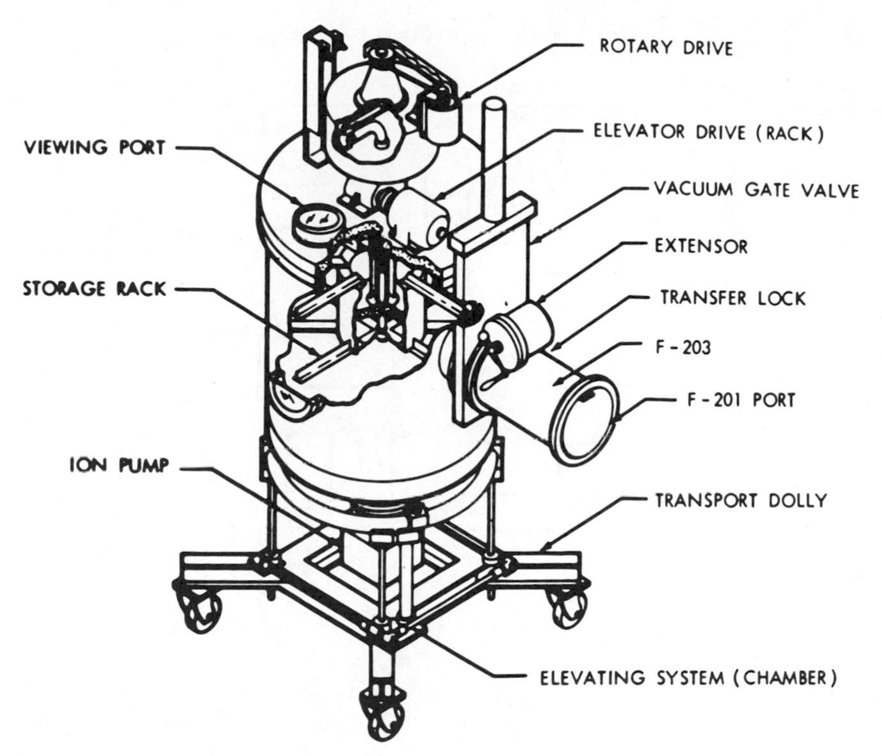

Figure 3.3. Carousel details; F-206.

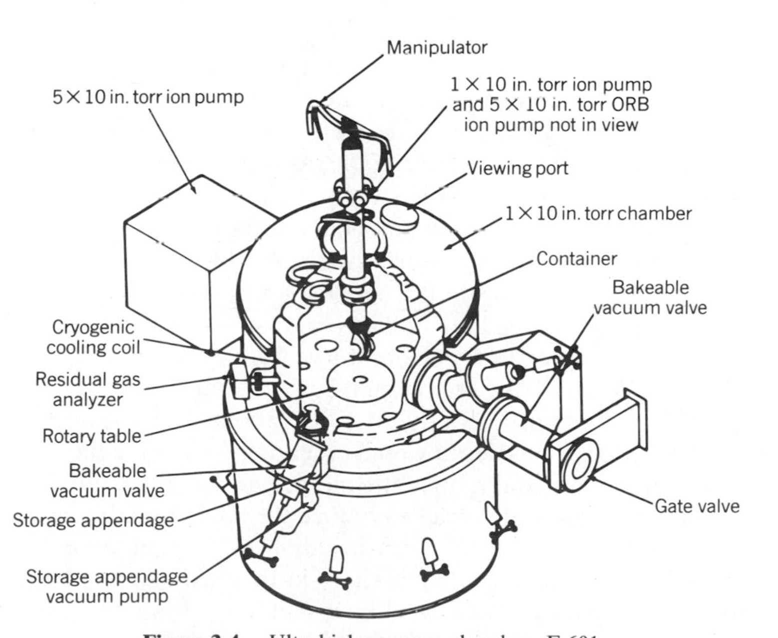

Figure 3.4. Ultrahigh-vacuum chamber, F-601.

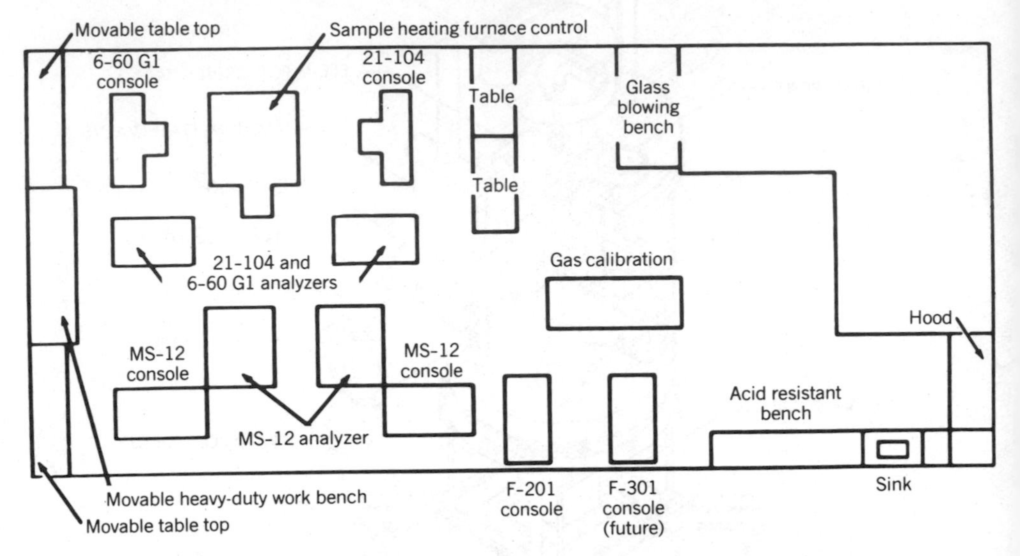

Figure 3.5. Arrangement of the gas analysis laboratory, Room 3-302.

dark room. Inside the installation were a large number of double-sided, gas-tight, positive-pressure, nitrogen atmosphere biological barrier cabinets, with glove ports for two operator positions. The samples were moved into this cabinet system through a transfer tube from the vacuum laboratory.

The spectrographic facility included a specially built X-ray fluorescence spectrometer designed to operate in one of the gas-tight cabinets and a 3.4-m Ebert stigmatic spectrographic. The excitation source for the emission spectrograph was housed in a gas-tight cabinet, while the spectrograph itself was outside the gas tight enclosure.

Among the various unique components of the LRL was the Radiation Counting Laboratory (RCL) shown in Figure 3.6. The counting facility itself was underground, while the support areas were aboveground. The first floor aboveground included offices for staff and visiting scientists, as well as a standards preparation laboratory. Above the counting laboratory was a system for removing the radon from the gamma ray counting room.

The underground portion of the RCL was about 40 ft below the surface. It was estimated that this depth would essentially eliminate the cosmic ray nucleonic component and reduce the mu meson flux by about 80% inside the measurement room. In order to reduce the backgrounds, the counting room was surrounded on all sides by about 3 ft of material, carefully selected for low inherent radioactivity—in this instance the mineral dunite (olivine), supported by a steel liner plate.

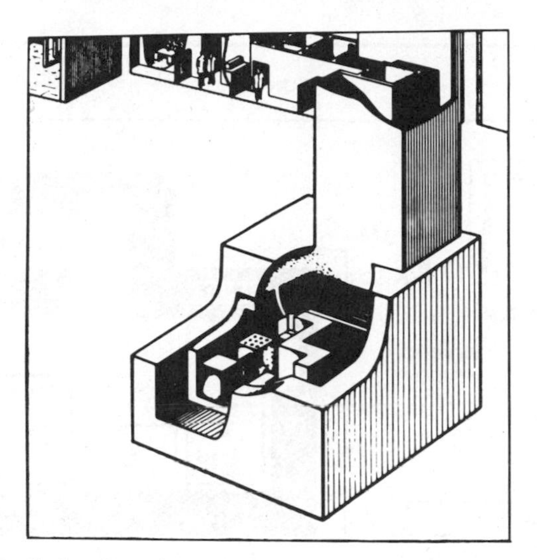

Figure 3.6. The Radiation Counting Laboratory of the LRL is shown here in cutaway. It lies about 40 ft below the surface.

The radon removal system was made up of chilled charcoal beds and absolute filters. The system was redundant, so that continuous operation was possible during filter regeneration or replacement. Entrance to the counting was through an airlock, with the room maintained at a positive pressure to prevent radon leakage into the counting room.

An important aspect of the facility was the shielding, shown in Figure 3.7. There were two internal shields, a main counter shield and an anticoincidence shield. The main counter shield had lead walls over 20 cm thick and an internal volume of 6 m^3. The radiation detectors inside their anticoincidence shield are shown in Figure 3.8. There were two sodium iodide detectors oriented at 180° enclosed in a 25-cm-thick inner anticoincidence mantle of polyvinyl toluene. Surrounding these was an inner mantle of a lithiated lead shield and an outer coincidence mantle. In principle, the inner mantle was used to improve the peak to total ratio by rejecting events in the main detectors where only a partial conversion of the gamma rays occurred. The outer mantle was sensitive only to meson interactions. The passive lead shield contained a small concentration of lithium in order to shorten the lifetime of meson produced neutrons.

At this point, the author would like to point out that the RCL is of historical interest. It was at the time the most advanced low-level counting facility, utilizing all the latest technology. Detection of gamma radiation today makes use of solid state detectors such as Ge–Li and intrinsic

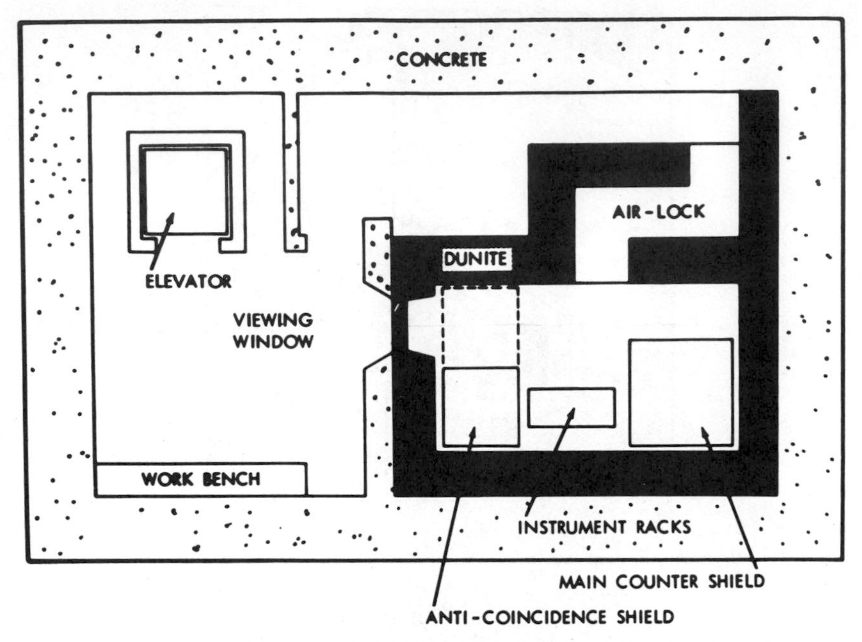

Figure 3.7. Radiation counting laboratory; underground chamber floor plan.

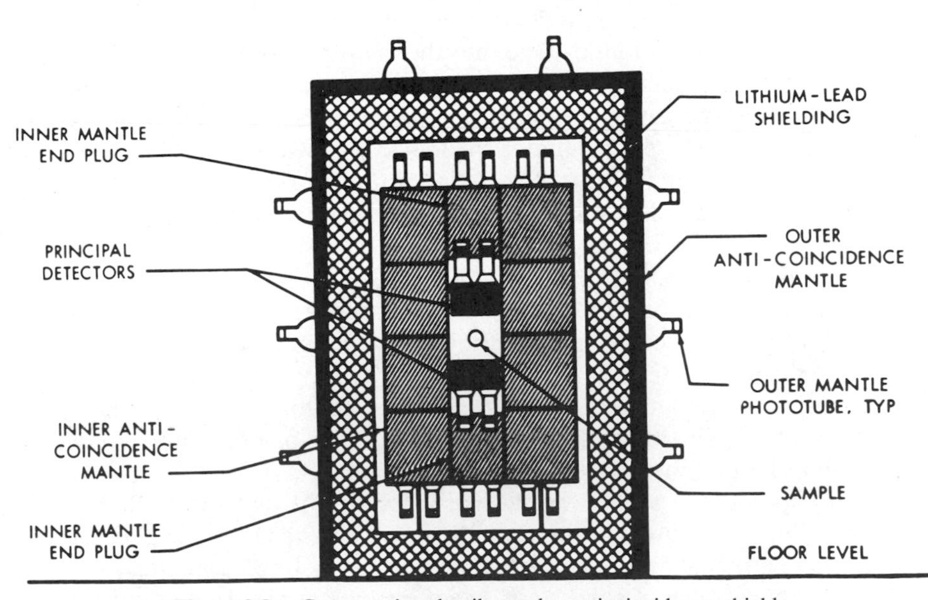

Figure 3.8. Construction details on the anticoincidence shield.

Ge, both with energy resolution far superior to that of the sodium iodide detectors of the RCL.

The sample flow through the LRL makes an interesting story. It was stated above that the LRL's functions began before the actual Apollo flights. The Apollo Lunar Sample Return Container (ALSRC) was prepared at the LRL for the outbound flight. The preparation was a serious affair, involving extensive cleaning, degassing, sterilization, sealing, and leak checking. Upon the completion of the listed activities, the containers were delivered in protective coverings to the launch site.

On return from the Moon during the early missions, the containers were placed into biological isolation containers aboard the recovery vessels. Once delivered to the LRL, the containers were immediately taken to the vacuum laboratory, introduced through an airlock, and the processing begun. Many of the investigators at the LRL doing the initial investigations were also the principal investigators performing the more detailed studies. Their initial observations, supplying very useful background data, were reported to the rest of the scientific community.

The initial flow of the samples is shown in Figure 3.8. These procedures were based on the concern for biological barriers and quarantine. Many of the requirements were waived on later missions in response to the observations that no viable organisms existed. The key concern was for speed. Note that in step 3 of the flow special samples were removed for measurement. One of these went to the RCL, where the objective was to search for short-lived radionuclides produced by cosmis rays. In view of the great historic importance of the samples, they were thoroughly photographed from six different views. At this stage also, some chips were removed for biological and physicochemical studies.

The procedure followed for the gamma-ray analysis is shown in Figure 3.9. Gamma-ray analysis of irregularly shaped samples required standards that are as close as possible in size, shape, density, and radioactivity. Furthermore, it is essential to orient the samples and standards in an identical manner. These requirements were met in the manner shown in the figure. The need for biological isolation made the task more difficult. The standards were fabricated by mixing rock and metal powder with radioisotopes, simulating the density and radioactivity of each lunar sample.

The LRL was arranged to perform two types of gas analysis: the analysis of gas evolved at ambient temperatures from the lunar samples and their containers as they were opened in the vacuum glove box and the measurement of occluded and interstitial gases generated at high temperatures. Figure 3.10 is a schematic representation of the glove chamber gas handling and collection system for ambient-temperature gas.

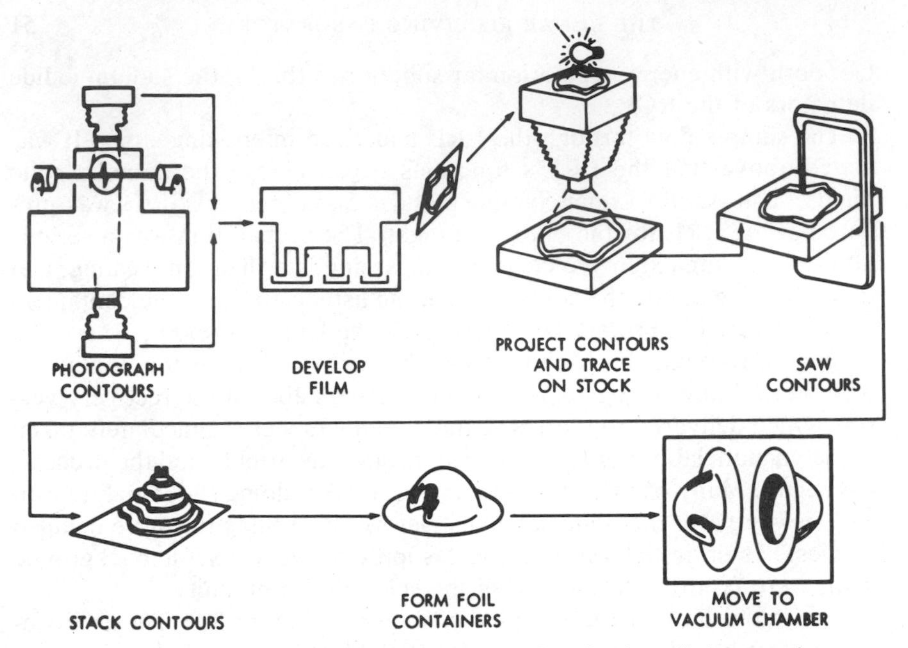

Figure 3.9. Procedure for fabricating the inner container for the radiation counting procedure.

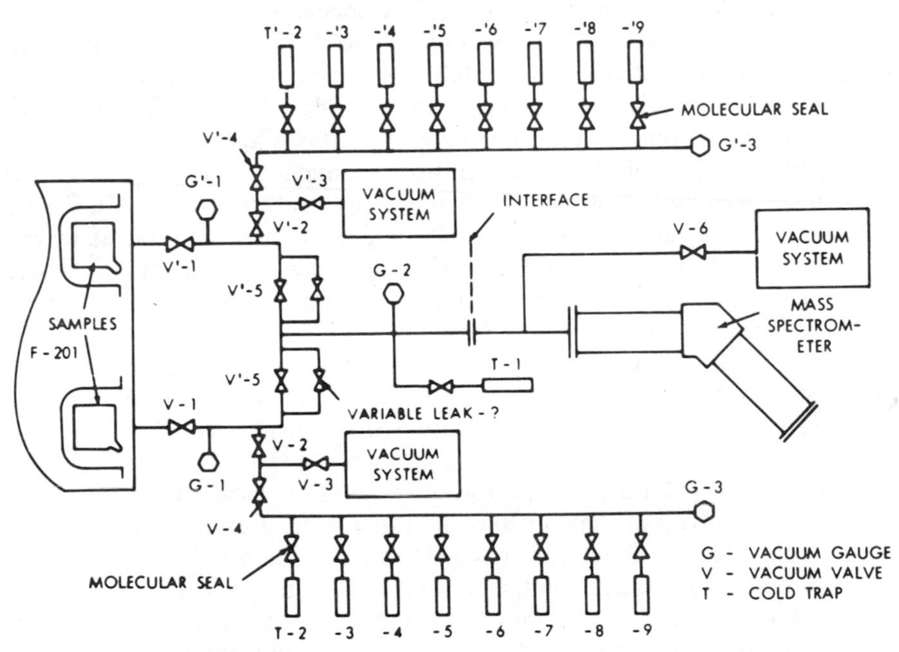

Figure 3.10. Schematic diagram of the glove chamber gas handling and collection system.

This system was used to analyze the gases in the Apollo Lunar Sample Receiving Outer Container and the individually sealed field sample bags containing the general purpose lunar samples. The various operations were scheduled to be performed manually at the glove chamber while in voice contact with the mass spectrometer console operator stationed in the Gas Analysis Laboratory. A typical procedure is outlined in Figure 3.11. The sample path selected depended on the nominal pressure observed in the bag. T2 through T9 in Figure 3.10 are liquid nitrogen adsorbate traps, which were attached to a removable manifold, so that where necessary a new manifold could be attached for additional sample collection. The samples of gas trapped in this arrangement were transported to the gas analysis laboratory for detailed analysis on the large mass spectrometer.

The occluded and interstitial gases were evolved by heating the sample. Analyses were performed for rare gases, inorganic gases, organic volatiles, and organic pyrolysis products as a function of temperature.

The search for organic gas components, very important initially, became less so when it became evident that little or no organic matter was to be found. The plans were to use two different spectrometric techniques. In the first, the gas went directly to a medium resolution mass spectrometer (resolution between 3000 and 4000). The connecting lines were kept at temperatures of 300°C or higher. To minimize surface effects, the lines

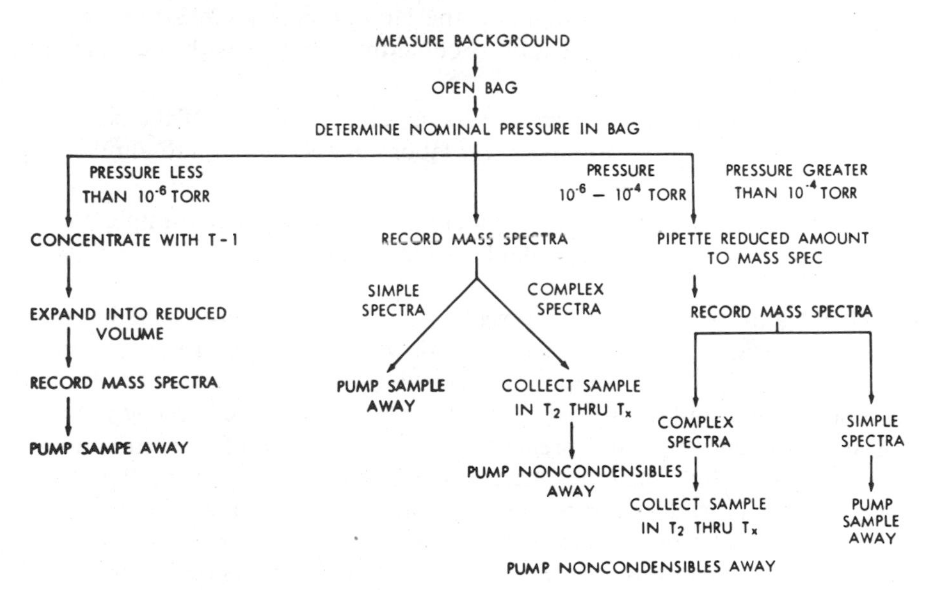

Figure 3.11. Glove chamber gas analysis procedure.

were coated internally with ceramic or other inert material. The second method used a combination gas chromatograph and mass spectrometer, including a molecular separator. The purpose for the system was to obtain simplified mass spectra and more reliable qualitative and quantitative data.

It is quite obvious that considerable care went into the design of the LRL. There was a great attention to detail, and a large effort was made to build a state of the art laboratory. The overriding philosophy, however, was to perform only preliminary or screening operations.

CONTAMINATION CONTROL

The matter of contamination control during and after the collection of the lunar samples was of particular concern. It was felt that, unless great care was taken, contamination would make some measurements difficult if not impossible to do; thus it was essential to list all possible contaminants. In practice, a contamination control program with the following objectives was established:

1. To keep to a minimum the release of all foreign matter on the lunar surface.
2. To control and identify all of the materials used in the hand tools, the sample containers, the laboratory systems, and elements of the system likely to come into contact with the samples.
3. A thorough precleaning, sterilizing, and outgassing of the sampling tools, return containers, laboratory systems, and similar items.
4. Maintenance of an environmental history on each sample in order to identify unavoidable contaminants.

One major worry was the possible contamination produced by the spacecraft engine exhaust during touchdown. The nature of the exhaust, as well as its spatial distribution, became the subject of a special study. In a similar vein, the leakage from the space suit represented an even more serious problem, since it represented a moving source of contamination.

Because the lunar samples were to be handled with tools and the samples were to be returned in special containers, the materials from which these were made were selected with particular care. It was essential to minimize, or where possible to eliminate, all the elements of geochemical interest for example, those isotopes used in radiometric dating such as lead, uranium, thorium, and potassium. Where elastomers were required, fluorocarbons such as Teflon or Viton-A were selected because they are distinctly recognizable by mass spectrometry.

The design of suitable containers for the return of the lunar samples involved considerable effort. The different types of vacuum containers are shown in Figure 3.12. There was, for example, the large "rock box" for returning samples under a modest vacuum of about 10^{-6} Torr. There was also a high vacuum container designed to maintain the ambient lunar pressure (thought to be about 10^{-12} Torr). The large box had a double seal, the outer one an elastomer type and the inner one made with crushed indium. It was stipulated that, given enough time, the individual rock samples were to be enclosed in Teflon bags prior to being placed in the large containers. Should this not be possible because of time or difficulties in manipulation, the samples of rocks would be placed in the container directly.

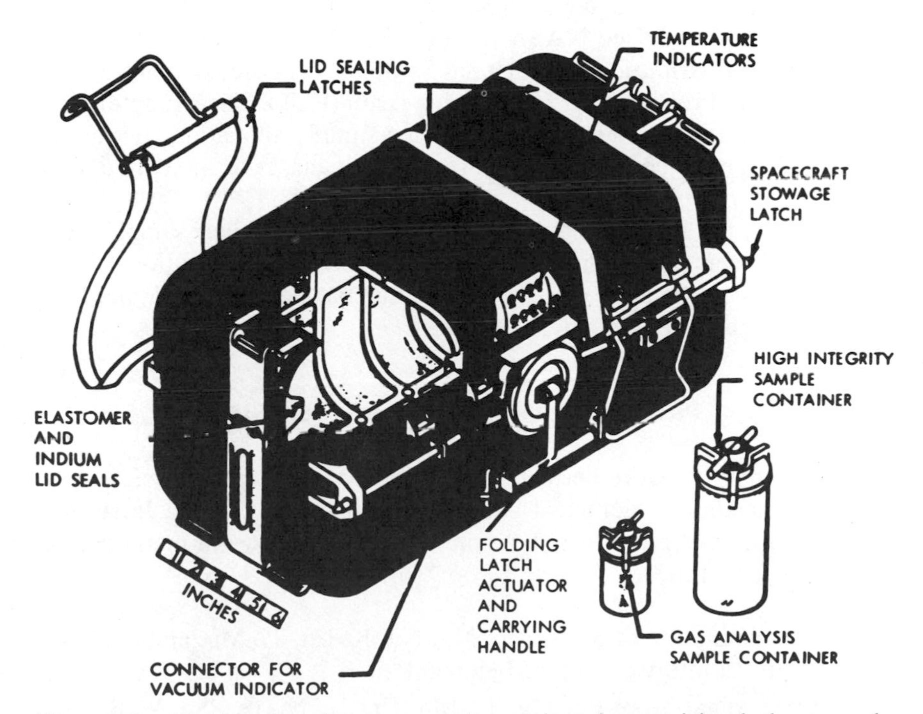

Figure 3.12. Various types of vacuum sample containers for containing the lunar samples after collection. The small containers have been designed to maintain the lunar ambient pressure. After sample storage in the small containers, they are then stored in the larger box for transport.

EXPERIMENTER TEAMS

Similar care was exercised in the choice of investigators. By the time of the return of the lunar samples, some 180 principal investigators had been selected by the NASA office of Space Science and Applications to perform a great variety of measurements. Approval was contingent on a number of factors such as the merit of the proposal submitted, the established reputation and competence of the scientist, and the nature of the sample requirements. The ultimate makeup of the investigating teams included scientists from America, Canada, Japan, Finland, Switzerland, and Australia. The institutions represented included a number of major universities, research institutions, and a small number of companies. The variety and number of research proposals would take several pages to list. These have been described in detail by Adler and Trombka (1).

The design of the LRL as described here evolved in large part in response to the advice of a series of committees. One well-recognized group was called the Lunar Sample Analysis Planning Team (LSAPT). LSAPT not only advised NASA on the LRL but also helped to develop plans for the distribution of the lunar samples. Another committee was designated the Preliminary Examination Team (PET). Their function was to obtain the basic data used to determine sample allocation. It was, on the whole, a well-coordinated effort, which can serve as a model for future programs.

In view of the historic significance of the return of lunar samples, it is worthwhile reporting the results of the preliminary investigations. However, because of the large amount of data that has become available, I will confine myself to the chemical results.

PRELIMINARY CHEMICAL ANALYSIS

Chemical analyses were performed principally by optical emission spectrography inside the biological barrier. The instrument was a Jarrel-Ash Ebert with a dispersion of 5.2 Å/mm. The three separate methods employed were as follows:

1. Determination of Si, Al, Fe, Mg, Na, K, Ca, Ti, Mn, and Cr, using Sr as an internal standard element.
2. Determination of Fe, Mg, Ti, Mn, Cr, Zr, Ni, Co, Sc, V, Ba, and Sr, using Pd as the internal standard.
3. Determination of Li, Rb, Cs, Pb, Cu, and other volatile elements using Na as the internal standard element.

An estimate of the precision was given as about ± 10% of the amount present. The accuracy of the result was controlled by using international rock standards such as G-1, W-1, Sy-1, BCR-1, DTS-1, PCC-1, G-2, AGV-1, and GSP-1 for calibration. Additional calibration points came from the analysis of terrestrial basaltic rocks from Hawaii and the Galapagos, chondritic meteorites from Forest City, Iowa and Leedey, Mezö-Madras, and achondrites from Sioux County, Nebraska, and Johnstown, Colorado. The spectrographic line identifications were checked against the spectra of individual elements, the M.I.T. wavelength tables, and the U.S. Bureau of Standards tables. The spectrographic plates were examined to establish the presence or absence of all elements that had spectra in the wavelength region covered.

Three samples were sterilized and then brought out of the biological barrier for analysis by atomic absorption for Fe, Mg, Ca, Ti, Na, K, and colorimetrically for Si.

The preliminary chemical analysis contributed the following interesting facts:

1. There was a marked similarity in composition of the various samples. The more significant variations were shown by some of the trace and minor elements such as Ni, Zr, Rb, and K. The major elemental constituents were Si, Al, Ti, Fe, Ca, and Mg. The minor elements were Na, Cr, Mn, K, and Zr.
2. The lunar rocks showed an unusually high concentration of refractory elements such as Ti, Zr, and Y. A comparison with the chondritic meteorites showed the lunar material to have higher concentrations of Ca and Al, but lower concentrations of Fe and Mg.
3. Zr, Sr, Ba, Y and Yb were enriched by about two orders of magnitude in the lunar samples relative to the chondrites.
4. K and Rb were present in the lunar rocks in amounts similar to those in chondrites, whereas Ni and Co were depleted. Ni was not detected in some of the rocks (less than 1 ppm), although the Fe was high.
5. The concentration of Zr was unusually high in the lunar rocks.
6. The volatile elements such as Pb, Bi, and Tl, if present, were below the limits of detection of the spectrographic technique. This was also true for the elements of the Pt group and Au and Ag.
7. The ratios of Rb to Sr were low, similar to those found in terrestrial oceanic basalts. Ba, Cr, and Sc were relatively abundant.

In summary, the lunar rocks on detailed analysis did show differences from terrestrial or meteoritic samples.

PRELIMINARY RARE GAS ANALYSIS

The gas analysis was done in the special facilities created for the purpose. The instrument employed was a 6-in. magnetic deflection mass spectrometer with a sensitivity of about 2×10^{-10} CC/mV for He, Ne, and Ar and about 3×10^{-13} CC/mV for Kr and Xe. The samples were prepared both under nitrogen and air in the following manner: chips of rock were taken, weighed, wrapped in aluminum foil, and sterilized by heating to temperatures between 125 and 150°C for periods of from 5 to 24 h. It was observed that the heating released less than 1% of the gases present.

The samples were melted by radiofrequency heating in a molybdenum crucible. The released gases were purified by the use of a hot titanium getter. The heavier noble gases were condensed, and the gases were introduced into the mass spectrometer for measurement under static vacuum conditions in three fractions: (1) He and Ne, (2) Ar and Kr, and (3) Xe.

After each sample was analyzed, a second heating was done to make certain that all the gas had been released and to serve as a blank correction. The entire procedure was standardized by introducing into the mass spectrometer a calibrated amount of different rare gases.

The rare gases showed three patterns related to the three rock types: the breccias, the igneous rocks, and the fines. The breccias and the fines contained very high concentrations of rare gases, particularly in the surface and subsurface materials. From the determined isotope ratios and the amounts of rare gases, the conclusion was drawn that the solar wind was the principal source. The igneous rocks showed substantially smaller quantities of rare gases, due either to the loss during the formation of the rocks or because they came from within the surface where there had been little or no exposure to the solar wind. The temperature-release experiments on the breccias and the fines demonstrated that the noble gases were tightly bound rather than surface adsorbed.

Based on the measurements of the Preliminary Observation Team the following conclusions were published.

1. There was an accumulation of substantial amounts of rare gases of solar composition.
2. There was an enrichment of ^{20}Ne relative to ^{22}Ne, predictable on the basis of nuclear processes occurring in the Sun.
3. The ratio of ^{4}He to ^{3}He in the soil and breccia was approximately 2600, in accordance with theoretical estimates.
4. The isotopic patterns for Xe in the fines and breccias was similar to that of the trapped Xe in carbonaceous chondrites, except for the addition

of a small amount of the lighter Xe isotopes (presumably due to spallation reactions) and a deficiency of ^{134}Xe and ^{136}Xe. The ratio of the ^{129}Xe to ^{132}Xe in the lunar material was essentially the same as that in the carbonaceous chondrites.

5. Data for the Kr in the fine lunar material was less precise, but again there was a resemblance to that found in the carbonaceous chondrites.

PRELIMINARY AGE DETERMINATIONS

Because several of the crystalline rocks contained radiogenic ^{40}Ar, noble gases produced by spallation reactions, and potassium, it was possible to perform a K–Ar dating measurement and a determination of cosmic ray exposure. Seven of the rocks yielded consistent ages of $3.0 \pm 0.7 \times 10^9$ y. The radiation exposure ages varied from 10×10^6 y to approximately 160×10^6 y.

GAMMA-RAY MEASUREMENTS

Gamma-ray measurements were performed on eight lunar samples in the Radiation Counting Laboratory. Included was a sample of fines, a rock from the contingency sample, and five rocks from the documented sample box.

In view of the newness and complexity of the operation, counting was not begun in the RCL for about 4 days after the samples reached the LRL. As a consequence, the radioactive species with short half-lives (less than 4 days) became undetectable. In addition, the high activity from the uranium, thorium, and their daughter products caused considerable interference in the determination of the weak gamma-ray components. The results are summarized below:

1. Twelve radioactive species were identified, some tentatively. The shortest half-lives among species observed were for ^{52}Mn (5.7 days) and ^{48}V (16.1 days).

2. The K concentrations was variable and close to that found in the chondrites (0.085 wt. %). The U and Th were near the values for the terrestrial basalts, with the ratio of U to Th about 4.1. One remarkable difference was that the ratio of K to U was unusually low, much lower, in fact, than similar ratios for terrestrial rocks and meteorites.

3. The cosmogenic ^{26}Al was generally high. These measurements, coupled with the rare gas analysis, indicated a cosmic ray exposure of

several million years. A more complete summary can be found in the text of Adler and Trombka (1) and in a preliminary report published in *Science* (2).

PRELIMINARY ORGANIC CHEMISTRY

The question of organic molecules in the lunar samples was a particularly significant one. Accordingly, a simple survey method, as well as a more elaborate one, was used to estimate the abundances in the lunar materials. For the screening, a pyrolysis–flame-ionization detector that was capable of supplying an estimate of the total organic matter in the samples, regardless of the type and origin, was used. The second method involved a computer-coupled, high-sensitivity mass spectrometer capable of giving detailed mass spectometric data on the volatile or pyrolyzable organic matter. Based on the mass spectral data, combined with the vaporization characteristics, an assessment of the relative contributions of terrestrial contaminants, as opposed to indigenous lunar organic, was made. The published values obtained from the preliminary investigation gave the organic content as less than 10 ppm.

CONCLUSIONS

The major findings based on the preliminary examination are listed below. These are shown here:

1. The mineralogy and texture of the rocks divides them into two genetic groups: fine- and medium-grained crystalline rocks of igneous origin and breccias.
2. The crystalline rocks differ from terrestrial rocks and meteorites.
3. The appearance of the rocks suggests a strong erosional process, but different from terrestrial processes.
4. The chemistry suggests that the crystalline rocks were formed under highly reducing conditions (low partial pressure of oxygen, water, and sulfur).
5. The absence of secondary hydrated minerals suggests the absence of surface water at the landing site during any part of the rocks' exposure.
6. There is evidence of shock or impact.
7. The rocks show glass-lined surface pits, due to impact by small particles.

8. The fines and the breccias contain large amounts of noble gases, and the isotopic and elemental data indicate the solar wind as a source.

9. K/Ar dates give crystallization ages of between 3 and 4 aeons.

10. Indigenous organic matter is very low (less than about 1 ppm).

11. The chemical composition of the rocks and fines is similar.

12. There is an enrichment of such refractory elements as Ti and Zr and a depletion of the alkali and some volatile elements.

13. Elements normally enriched in iron meteorites such as Ni, Cd, and the Pt group were either absent or in very low abundance.

14. The ratio of K to U is unusually low compared to terrestrial rocks.

15. High ^{26}Al abundances indicate a long exposure to cosmic-rays.

16. There is no evidence of biological matter.

The more complete picture that has emerged from the detailed studies will be described in the following chapters.

REFERENCES

1. I. Adler and J. I. Trombka, *Geochemical Exploration of the Moon and Planets,* Springer, New York, 1970.
2. Preliminary Examination Team, *Science,* **165,** 3899 (1969).

SUPPLEMENTAL REFERENCE

Lunar Receiving Laboratory, MSC Building 37, Apollo Missions, *Preliminary Report,* Manned Spacecraft Center, Houston, 1966.

CHAPTER

4

LUNAR SAMPLES: DETAILED ANALYSIS

INTRODUCTION

A major point to keep in mind in the discussion of the detailed studies of the lunar samples is the feedback between the meteorite program and the lunar sample program. Initially, many of the analytical techniques were either borrowed or taken directly from the analysis of meteorites. In turn, the lunar sample studies have produced evolutionary developments in analytical techniques which are presently being applied to the analysis of meteorites. With regard to the lunar samples, it is fair to say that every variety of technique was applied in an effort to extract the maximum information from these invaluable samples. Furthermore, these studies continue as still more advanced methods become available.

At present, such an enormous amount of data has emerged that it is sometimes difficult to develop a perspective. In an effort to do so, I shall go back to one of the early and important preliminary studies, the NASA 1965 Summer Conference on Lunar Exploration and Science (NASA SP-88). The geochemistry working group proposed a number of objectives for the manned landings and sample collection and detailed their scientific significance. This group recognized that the single most important objective of the Apollo missions was the return of lunar samples. Although the scientific yields from the observation of the astronauts on early Apollo missions were of great importance, they would be less important than the yields from the investigations of scientists on the returned samples. The greatest limitations would be set by the amount of sample available for study after the completion of the mission.

The objectives of the program of chemical, mineralogical, and isotopic studies were to acquire information on the nuclear processes in the early Solar System, the origin of planetary bodies, and the geological evolution of the Moon. The array of measurements on the lunar material called for would include the determination of the abundances of all the elements, their isotopes, and the nature of their coexistence in the various phases making up the lunar material.

A detailed list of some of the fundamental questions about the Solar System and the Moon posed by the working group are given below:

1. The conditions under which the Moon and other planets formed in the early Solar System could be inferred from the geochemistry of certain groups of elements. In particular, the abundance of relatively volatile elements such as Hg, Tl, Zn, Cd, Bi, and Pb might provide an important clue to the temperatures during the Moon's accretion. The fractionation of Rb and Sr in the early Solar System could also be related to its high temperature fractionation processes.
2. The abundance of such elements as Li, Be, and B are sensitive to nuclear processes in the early Solar System. It was similarly important to determine whether or not the H/D and the $^{40}K/^{41}K$ ratios are the same in all parts of the Solar System.
3. The abundance of ^{129}Xe in the lunar materials might indicate whether or not the Moon formed before or after the carbonaceous chondrites.
4. The distribution of Rb, Sr, U, Th, and Pb combined with isotopic data on Sr and Pb would provide information on the nature and sequence of events on the Moon's surface.
5. The concentration of U, Th, and K would provide information for modeling the Moon's thermal regime.
6. Rare-earth distribution patterns might help to clarify the differentiation histories of the lunar materials as compared to terrestrial and meteorite samples.
7. The analysis of the isotopic compositions of the rare gases such as Ne, Ar, Kr, and Xe and the radioactive and stable nuclides such as ^{10}Be, ^{36}Cl, ^{59}Ni, and ^{50}V should provide useful data toward the understanding of solar wind history and nuclear bombardment.

This is a partial list against which the analytical results are to be compared. It does provide a base line for evaluating the accomplishments of the exploration program.

MAJOR, MINOR, AND TRACE ELEMENTS

The analysis of the lunar samples involved, for the most part, a large number of conventional techniques such as classical wet methods, optical emission spectroscopy, X-ray fluorescence spectrometry, atomic absorption spectrometry, neutron activation methods, mass spectrometry, spark source mass spectroscopy, and the electron probe. There was from the very outset a strong emphasis on the use of either nondestructive methods or methods requiring minimal samples. The objective was to preserve as much of these precious samples as possible to make them available either for other types of studies or for confirmatory analysis. For the most part, these methods have been described in numerous publications.

One of the more commonly used techniques was X-ray fluorescence spectrometry. Compston and co-workers (1) used the methods developed by Norrish and Chappell (2), where the matrix effects were corrected for either by direct measurement or the calculations of the absorption coefficients. They used powdered samples directly, contrary to their usual practice of pelletizing so that the samples could be recovered intact. To deal with the problem of preparing the sample as a fine powder, pains were taken to minimize contamination. Thus the samples were carefully ground in agate, so that the only possible adventitious element added would be silicon, one of the ubiquitous and major components.

Another approach was taken by the team at the U.S. Geological Survey (3). In order to preserve samples, they worked with a sample of 60 mg and used the semimicro X-ray fluorescence equipment. They prepared their sample by fusing 60 mg of lunar material with 940 mg of lithium tetraborate. After the bead cooled, it was brought to 1200 mg with powdered cellulose and then ground to less than 350 mesh.

Smales and associates (4) used the XRF method of Norrish and Hutton (5), which involves the use of a fusion mix consisting of lithium tetraborate, lithium carbonate, and lathanum oxide, including 0.8% lithium nitrate. The XRF was only one of a large battery of analytical techniques employed by the Smales team. In addition to the X-ray emission method, they also used spark source mass spectroscopy (SSM) and emission spectrography (E). To obtain more accurate results for a smaller number of elements they used activation analysis (AA); instrumental neutron activation analysis (INAA); radiochemical neutron analysis (RNAA); radiochemical, high-energy gamma, activation analysis (RGAA); and mass spectrometric isotope dilution analysis (MSID).

The use of independent methods made it possible to make comparisons among the methods. Their paper, for example, shows excellent agreement between XRF and INAA. The activation analysis proved to be particularly attractive in the lunar sample program because of the inherent accuracies and the nondestructive nature of the methods. A comprehensive review of the use of neutron activation analysis for geological materials can be found in a paper by J. C. Laul (6) published in 1979.

In the general realm of small sample analysis and the ability to analyze samples nondestructively, the electron probe microanalyzer is ideal. It was used to provide not only major and minor elemental data but also made possible the accumulation of modal data. In fact, many of the samples supplied by the curatorial facility in Houston were in the form of mounted and polished specimens, which were moved from laboratory to laboratory for studies. Among the methods mentioned above, emission spectroscopy, activation analysis, and mass spectrometry were also used for trace element analysis.

A complete account of the various methods cannot be given here in detail. Over 1000 pages of text were devoted to the analysis of the Apollo 11 samples alone. Reports have been published annually as proceedings of the successive lunar conferences, as well as in various journals.

Wherever ambiguities arose in the interpretation of the data, these did not come from weaknesses in the analytical results. They came, rather, from large problems in sampling and from the problems produced by the obvious mixing of materials from different sources.

CHEMISTRY OF THE LUNAR REGOLITH

The following picture of the lunar regolith is based on the description published by Papike, Simon, and Laul (7). It is defined as a thick layer of unconsolidated debris that forms the interface between the Moon and its space environment. It is formed from lithic sources by both destructive and constructive processes. The destructive process involves the continuing comminution of the rocks by ongoing bombardment of the lunar surface by infall from space. The constructive process is the formation of agglutinates consisting of comminuted rock, mineral, and glass fragments bonded together with glass, again as a consequence of impact.

Papike and co-workers (7) point out that these processes can lead to a steady-state soil that can remain in dynamic equilibrium until the system becomes disturbed either by burial or mixing with new soil. The mixing process is considered to be fairly local in scale; the vertical dimension is of the order of meters and the horizontal dimension of the order of several kilometers. Lateral transport is generally accepted as being relatively inefficient, and thus most of the lunar soil is locally derived. The content of exotic components, which have been studied at some length (defined as coming from much greater distances), is under 1%.

An interesting observation is that the highland soils are distinct from the mare soils, both chemically and petrologically. Furthermore, soils samples collected near mare-highland contacts are mixtures of both lithologies.

At this time the core samples, collected at a number of sites, are still not well understood. It has been difficult to establish how many depositional and erosional events occurred and over what time span.

The average chemical compositions of the fine component regolith (less than 1 mm) are shown in Tables 4.1 and 4.2. These results were obtained by instrumental neutron activation. The errors cited based on counting statistics are ± 0.5–3% for TiO_2, Al_2O_3, FeO, MnO, Na_2O, and Cr_2O_3, and ± 5% for MgO, CaO, and K_2O.

There are noticeable differences between the highlands as represented by the Apollo 16 samples 64501 and 67461 and the mare samples—for example, the Apollo 11 soils 10084. The mare sample is considerably richer in Ti than all the other samples except the Apollo 17 samples 70009. A second distinction is that the Apollo 16 sample from deep in the highlands is relatively enriched in Al and Ca and has a low Mg and Fe content.

When the chemistry and the petrological evidence is combined, it becomes obvious that the fine soils reflect the nature of the underlying bedrock. As examples, the Apollo 11 soils with high Ti have a number of characteristics of the Apollo 11 basalts. Soil sample 12001 is similar to the Apollo 12 olivine and ilmenite basalts. Samples 64501 and 67461 are similar to the anorthositic rocks characteristic of the highlands. There is an

Table 4.1. Major Element Composition of Soils from the Apollo Landing Sites[a]

	Apollo Site								
	11	12		14	15		16		17
Sample No.	*10084*	*12001*	*12033*	*14163*	*15221*	*15271*	*64501*	*67461*	*70009*
SiO_2	41.3	46.0	46.9	47.3	46.0	46.0	45.3	45.0	40.4
TiO_2	7.5	2.8	2.3	1.6	1.1	1.5	0.37	0.29	8.3
Al_2O_3	13.7	12.5	14.2	17.8	18.0	16.4	27.7	29.2	12.1
FeO	15.8	17.2	15.4	10.5	11.3	12.8	4.2	4.2	17.1
MgO	8.0	10.4	9.2	9.6	10.7	10.8	4.9	3.9	10.7
CaO	12.5	10.9	11.1	11.4	12.3	11.7	17.2	17.6	10.8
Na_2O	0.41	0.48	0.67	0.70	0.43	0.49	0.44	0.43	0.39
K_2O	0.14	0.26	0.41	0.55	0.16	0.22	0.10	0.06	0.09
MnO	0.21	0.22	0.20	0.14	0.15	0.16	0.06	0.06	0.22
Cr_2O_3	0.29	0.41	0.39	0.20	0.33	0.35	0.09	0.08	0.41
Σ	99.8	101.0	100.8	99.8	100.5	100.4	100.3	100.8	100.5

[a] Values given in wt.%.

Sources: J. C. Laul and J. J. Papike, *Proc. 11th Lunar Conf.* (1980) (6*a*); J. C. Laul et al. *Proc. 9th Lunar Conf.* (1978) (6*b*).

Table 4.2. Trace Elements in Soils from the Apollo Landing Sites[a]

	Apollo Site								
	11	12		14	15		16		17
Sample No.	*10084*	*12001*	*12033*	*14163*	*15221*	*15271*	*64501*	*67461*	*70009*
Rb	3.2	23	14	14.6		5.7	2.0		
Ba	170	430	600	800	240	300	130	60	120
Pb	1.4		4.0	10		2.8			
Sr	160	140	160	170	120	130	170	170	210
La	15.8	35.6	50	67	20.5	25.8	10.8	4.7	7.9
Ce	43	85	133	170	54	70	28	12	28
Nd	37	57	85	100	36	45	19	7.2	23
Sm	11.4	17.3	22.8	29.1	9.7	12	4.8	2.0	8.1
Eu	1.60	1.85	2.45	2.45	1.30	1.50	1.05	1.00	1.76
Tb	2.9	3.7	4.9	5.9	2.0	2.6	1.0	0.45	1.9
Dy	17	22	30	36	12	—	6.0	2.8	11.4
Ho	4.1	5.0	7.2	8.6	2.9	3.9	1.4	—	2.9
Tm	1.6	1.8	2.6	3.2	1.1	1.4	0.55	0.25	—
Yb	10.0	13.0	17	21	6.9	8.5	3.4	1.6	7.1
Lu	1.39	1.85	2.45	3.00	0.97	1.20	0.49	0.22	1.1
Eu/Eu	0.37	0.30	0.30	0.22	0.38	0.35	0.60	1.38	0.59
Y	99	—	160	190	86	—	—	—	—
Th	2.1	5.40	8.50	13.3	3.0	4.6	1.85	0.83	0.95
U	0.54	1.7	2.4	3.5	—	1.2	0.4	—	0.23
Zr	320	—	760	850	—	390	—	—	—
Hf	9.0	11.8	16.6	23	6.7	8.6	3.3	1.6	6.6
Nb	118	—	44	46	—	25	—	—	—
V	70	110	100	45	80	80	20	20	100
Sc	60	40	36	22	21	24	8.0	7.8	57
Ni	200	310	210	330	273	220	380	—	—
Co	28	43	34	33	41	41	20	9	32
Cu	10	7.2	8	8	—	9	—	—	—
Zn	23	—	14	34	—	21	—	—	44
Li	10	18	24	27	—	—	—	—	—
Ga	5.1	4.2	3.1	8.3	—	4.4	—	—	6.3
Au, ppb	2.4	2.6	—	5.4	—	4	14	—	3
Ir, ppb	6.9	11	—	14	—	9	12	—	—

[a] Values given in ppm except where noted.

Sources: J. C. Laul and J. J. Papike, *Proc. 11th Lunar Conf.* (1980) (6*a*); J. C. Laul et al, *Proc. 9th Lunar Conf.* (1978) (6*b*).

example of an exotic component in the Apollo 12 sample 12033, where the pyroxene chemistry is similar to the Apollo 14 KREEP basalts. Parenthetically, we note that these two sites are geographically quite close.

Attempts to explain the regolith as the result of melted or partially melted material from the lunar interior have been unsuccessful. It has been necessary to invoke a mixing process, but this is not simply done. In order to produce a consistent model, it is necessary to know the nature of the components (the end members). Other factors to be considered are the extent of lateral mixing and the addition of meteoritic components, both, however, thought to be small.

From a practical point of view, one of the questions is how representative was the astronauts' sampling of the surface in view of the sampling difficulties. Taylor (8) points out the remarkability of such a good match between the regolith and the whole rock components at the mare sites.

In the following sections we will examine the chemistry in somewhat greater detail, and we will see how chemical analysis provides an insight into understanding the lunar surface.

CHEMICAL COMPOSITION VERSUS GRAIN SIZE

Papike and co-workers, (7) after an extensive survey, have reported that a major chemical discontinuity appears in the very fine soil fraction (less than 10 μm), which makes up some 10–15% of the bulk soil. This fraction appears to be enriched in a highland component, that is, it is more feldspathic than the coarse fraction of the soils. The relationship is shown in Figure 4.1 for Al_2O_3. We note that the Al_2O_3 is substantially higher in the fines than the bulk, except for the highland samples where the rocks are aluminum-rich to begin with.

The presence of a highland component (about 20–30%) at the Apollo 11 mare site is a question that begs explanation. It had been stated previously that the lateral mixing is a small-scale effect. Further, the highlands nearest to the Apollo 11 site are some 50 km distant. To deal with this, an interesting suggestion has been made is that this component comes mainly from the excavation of bedrock having a highland composition, which underlies the mare basalts and which has been excavated by meteorite impact cratering. Such a process would require a thin basalt layer. Taylor (8) points out, based on published studies by Rhodes (9) and Hörz (10), that with the exception of Apollo 12, the lunar landings were near the edges of the basins where the basalt flows would be expected to be thin.

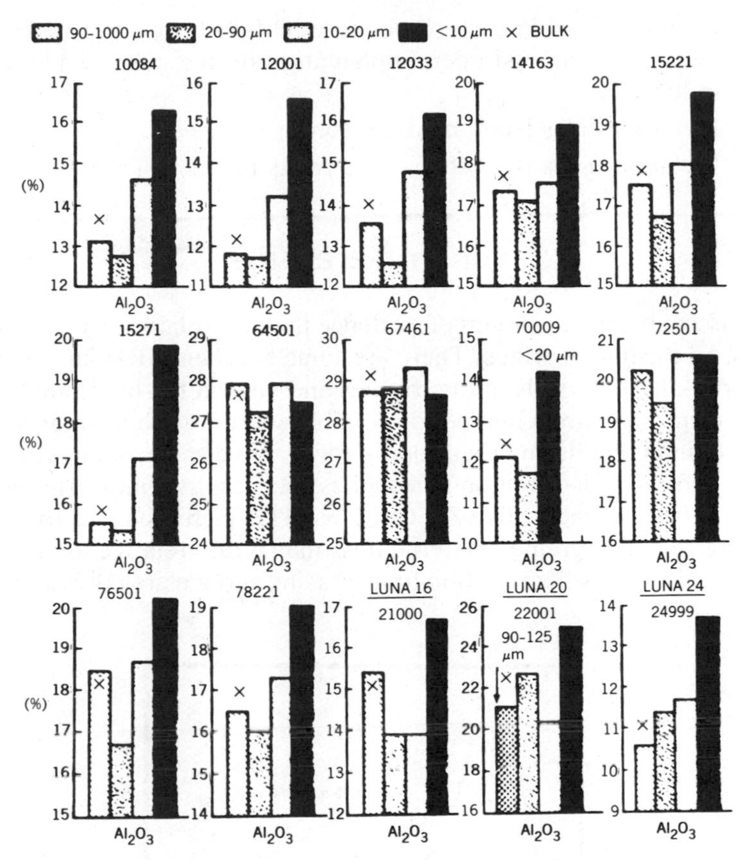

Figure 4.1. Size dependence of Al_2O_3 in bulk, coarse (1000–90, 90–20, and 20–10 μm), and < 10 fine fractions of soils in the reference suite. Source: J. C. Laul et al. *Proc. Lunar Plan. Sci. Conf. 9th* (1978) (6b).

METALLIC IRON

A study of the regolith has disclosed the presence of components, some of which are indigenous and others that have been described as "exotic." One especially interesting element is iron, some of which has been found in the metallic state. The presence of metallic iron has a relationship to the lunar magnetic properties (8). Lunar scientists generally agree on three possible sources:

1. The reduction of ferrous iron in the silicate and oxide due to exposure. Yin and associates (11) have proposed that solar wind reduc-

tion is, at least in part, responsible. Others have suggested mechanisms such as hydrogen implantation by the solar wind, causing reduction.

2. Metal particles from micrometeorites.
3. Metal particles from the source rocks for the soils.

VOLATILE ELEMENTS

There is a substantial amount of evidence that the volatile elements (8) are mobile on the lunar surface. There are some excellent SEM images showing vapor-phase crystals in the cavities and vugs in the brecciated rocks. Here again, there are a number of possible sources such as transport due to fire fountaining during the Moon's early active stage, solar wind sputtering, and the redistribution of volatiles as a result of meteorite impact.

The volatile elements like Zn, Cd, In, and Ga increase by factors of 10–20 in the very fine-grained material (less than 5 μm) relative to the coarse grains. This positive correlation to increasing surface area is evidence for

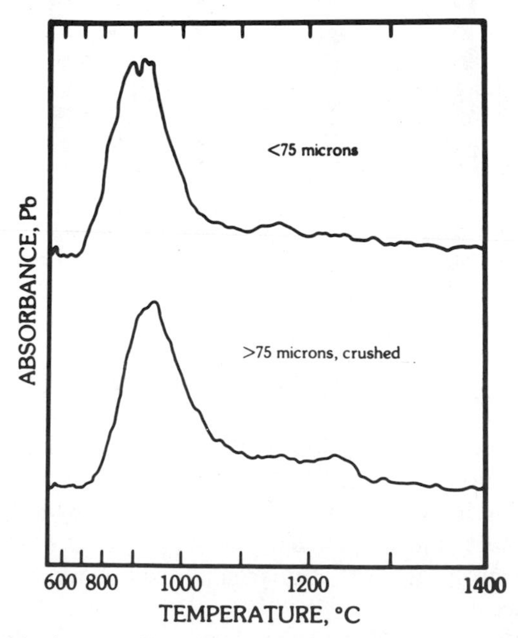

Figure 4.2. Thermal release profiles of Pb in the rock sample 66095. Most of the Pb in both fractions is released below 1000°C, before the surface melts, and is therefore surface Pb. Source: R. S. Taylor (8).

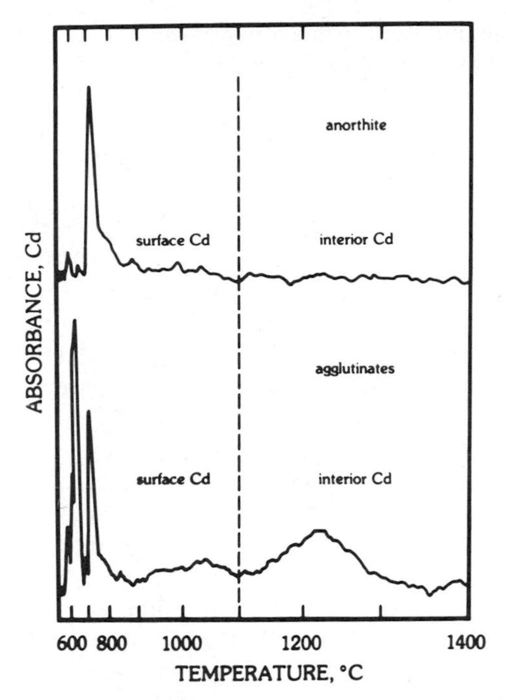

Figure 4.3. Thermal release profiles of Cd in anorthite grains and agglutinates in the fines sample 65701. The Cd in the anorthite fraction is on almost all grain surfaces. A fair amount of interior Cd as well as surface Cd is present in the agglutinates. Source: R. S. Taylor (8).

the selected movement of the volatile elements during regolith evolution. An additional piece of evidence is the concentration of trace siderophile elements in the evolving agglutinates that appear to concentrate the siderophiles due to meteorite bombardment.

Of course, the Moon's environment is somewhat unique. Taylor points out that Hg probably volatilizes during the lunar day when the temperatures climb to 130°C and then condenses out at night when the temperature drops to −170°C. A number of the volatile elements are enriched on the surface of regolith grains. These elements are Au, Bi, Cd, Ga, Ge, Hg, In, Pb. Sb, Te, and Zn. Figure 4.2 shows the method by which it has been determined that the volatiles are on the surface. The curves shown are thermal release profiles for two different fractions. The Pb is released before the sample melts, showing that the Pb resides on the surface. Similar studies have been done on a number of volatile elements. A contrasting example is shown in Figure 4.3. This is a study of Cd release on an Apollo 16 sample 65701. The Cd in the anorthite grain is essentially surficial. The agglutinate, on the other hand, shows not only surface Cd but a substantial amount of interior Cd as well.

THE METEORITIC COMPONENT

Morgan and co-workers (12) have written that it was evident even before the first lunar landing that three types of meteoritic material would be found on the Moon: micrometeorites and small meteorites on the regolith, debris from postmare craters in the ejecta, and ray material and debris from the intense bombardment that occurred early in the Moon's history. While estimates of the amounts of meteoritic material had been made based on crater-scaling laws and the observed fluxes of interplanetary matter, there was no satisfactory procedure for determining the composition, in view of the absence of samples in hand.

The return of lunar samples changes the picture drastically. Suddenly it became possible to study the lunar samples using the most modern of laboratory techniques.

The chemical investigation of the meteorite component was based on two geochemical factors. The first of these was the marked depletion of such siderophile elements as Ni, Ir, Au, and Re in the lunar rocks, a depletion expected on the surface of a differentiated planet. The second factor was the extreme depletion of the Moon in a number of volatile and chalcophile elements such as Ag, Bi, Br, Cd, Ge, Pb, Sb, Se, Te, Tl and Zn. Thus investigators such as Morgan (12), and others were able to use these elements, which are relatively abundant in C1 meteorites, as indicators of a meteoritic component.

Because many of these elements had very low abundances (10^{-9} to 10^{-12} g/g), radiochemical neutron activation analysis was the selected method. The results are shown in Figure 4.4. These represent mature soils with very long exposure ages. Significantly, the soils from all sites gave the same results. The comparison shown in the figure is relative to the crystalline rocks. The net meteoritic component is the result of subtracting the indigenous lunar contribution, which was estimated from the crystalline rocks. The abundance pattern is observed to be relatively flat with the siderophiles and the volatile elements nearly equally abundant. While some variability among the volatile elements was observed, this has been attributed to their mobility in the soils as discussed in the preceding section.

The meteorite component has been found to have a primitive C1 chondritelike composition and to make up about 1–1.5% of the mature soils. To Morgan and co-workers it represents cometary debris.

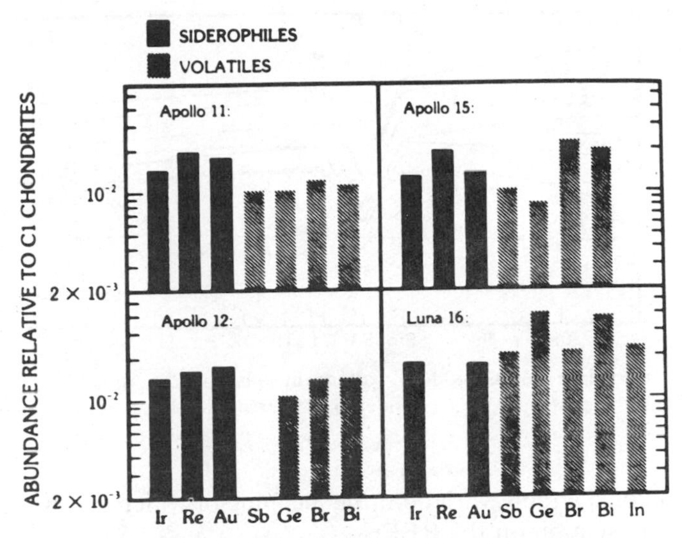

Figure 4.4. All mare soils are enriched in "meteoritic" elements relative to crystalline rocks. The net meteoritic component is obtained by subtracting the indigenous lunar contribution estimated from crystalline rocks. The meteoritic component has a primitive C1 composition. Source: E. Anders (30).

KREEP

KREEP is an acronym for the unusual material first discovered during the Apollo 12 mission: a high potassium (K), high rare earth (REE), and high-phosphorus material (P). As Taylor (8) has stated, it "has beguiled the lunar community since the first discovery during the Appollo 12 mission of a layer in the regolith of light gray fines and the rock 12013, picked up by Pete Conrad on an unscheduled traverse."

Its discovery was a great surprise from a geochemical point of view because of the extreme concentrations of elements typical of the residual liquids produced by fractional crystallization. This was on a Moon thought by some to be a primitive unfractionated body. As an additional surprise, KREEP is now known to be a widespread component of the lunar soil. The fact that it is so widespread makes it unlikely that it is a small or trivial volume of residual melt from the crystallization of a local intrusion.

The REE are very important geochemical elements. The pattern of the REE distribution, based on the data shown in Table 4.2, is shown in Figure 4.5. The values are normalized with respect to the REE in chondrites. The different curves are for the soils from the various Apollo sites.

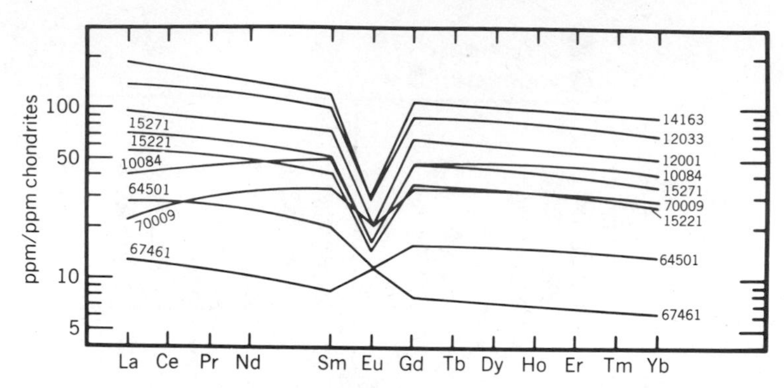

Figure 4.5. Chondrite normalized REE patterns in soils from the Apollo 11, 12, 14, 15, 16, and 17 landing sites. There is a considerable diversity in the patterns. Source: R. S. Taylor (8).

There is a considerable variation, the meaning of which will be discussed below in the section on the REE.

HYDROGEN, CARBON, NITROGEN, AND SULFUR

The presence of these elements in the lunar soil have been discussed by Taylor (8). Hydrogen is interesting for a number of reasons but, in this instance, particularly because it clearly shows the effects of the interaction between the lunar surface and the solar and galactic radiation. For example, the hydrogen in the lunar soil is essentially free of deuterium. Because deuterium is destroyed in nuclear reactions in the Sun, this points very strongly to a solar origin for the hydrogen.

The initial great interest in carbon is easily understood because of the general question of the origin of organic matter and chemical evolution. The hunt for carbon and organic matter has been a disappointing one, however, The overall carbon content of the lunar rocks was about 30 ppm. The soils were relatively richer, containing about 115 ppm. Of this amount, approximately 5–10 ppm was accounted for by meteoritic or cometary sources, and the remainder by solar wind implantation.

The nitrogen in the regolith, like the carbon, proved to be mainly extralunar. The average concentration in the igneous rocks was less than 1 ppm. The concentration in the soil was higher, ranging from 50 to 100 ppm. The low concentration of the nitrogen in the lunar rocks is in accord with the overall depletion of the volatile elements.

Sulfur, by contrast, is relatively abundant in lunar rocks (about 1000 ppm), particularly where it is present as the mineral troilite (FeS). An extralunar component has been identified in the soil, which shows about the same abundance level of the sulfur. Isotopic fractionation of sulfur in the lunar soil has been found, producing a relative enrichment in the heavier isotopes. The proposed mechanism is sputtering by micrometeorite bombardment.

RARE GASES

One of the surprises reported by the Apollo 11 Preliminary Examination Team was the large concentration of the rare gases in the lunar fines. The numbers were of the order of 0.1 to 1.0 cm^3 STP/g. This was an unusually large value. Table 4.3 provides a detailed comparison of the isotopic composition of the rare gases trapped on the lunar regolith against the solar wind and the terrestrial atmosphere. The similarity between the bulk lunar fines and the solar wind is rather striking. Investigations of the nature and location of the trapping of such large volumes of gas have been published by Phakey (13). The gas was found to occur as bubbles, 50–100 Å in diameter in the soil grains under very high pressures (about 5000 atm). The gas content increased as the particle size was reduced. The noble gas was apparently being driven into the grains by the solar wind. Estimates of solar wind penetration were reported as about 1000 Å. More energetic solar wind ions penetrated to millimeter depths.

Based on the above observations, there is no evidence that the rare gases are remains of a primitive atmosphere. The large quantities of rare gas found are thought to be of a secondary origin, coming from a variety of sources such as the solar wind, cosmic-ray interactions, fission products, and radioactive decay.

A comparison of the highland soils with the mare soils shows the latter to be richer in He and Ne. A likely explanation (8) is that the iron and

Table 4.3. Isotopic Composition of Trapped Rare Gases in the Lunar Regolith Compared with Solar Wind and Terrestrial Atmospheric Values

	$^{4}He/^{3}He$	$^{4}He/^{20}Ne$	$^{20}Ne/^{22}Ne$	$^{22}Ne/^{21}Ne$	$^{20}Ne/^{36}Ar$
Bulk lunar fines	2300–2800	96 ± 18	12.4–12.8	31 ± 1.2	
Solar wind	2350 ± 120	570 ± 70	13.7 ± 0.03	30 ± 4	28 ± 9
Terrestrial atmosphere	7×10^5	0.3	9.8 ± 0.08	34.5 ± 1.0	0.5

Sources: D. D. Bogard and L. E. Nyquist, *Proc. Lun. Conf. 2nd,* 1972. J. Geiss et al. *NASA SP-315,* 1972; D. Heyman, *Phys. Chem. Earth,* **10;** 49 (1977); P. Eberhardt et al, *Proc. Lun. Conf. 3rd,* 1972.

titanium minerals found in the maria are more effective traps. Furthermore, the He and Ne are generally depleted, relative to the heavier gases such as Ar, Kr, and Xe, compared to the solar abundances. This is again consistent with the general depletion of the volatile elements.

Among the rare gases one finds an interesting anomaly in the ^{40}Ar in the lunar soils. Heymann and Yaniv (14) have observed that there is an excess of ^{40}Ar in the lunar soils when compared to predicted values by solar wind trapping and ^{40}K decay. Several mechanisms have been proposed to account for this phenomenon. Heymann and Yaniv have suggested that ^{40}Ar produced by ^{40}K decay, escapes from the lunar surface, is ionized by solar radiation, and then is driven back into the lunar surface by the magnetic fields associated with the solar wind. Those ions with impact energies greater than 1 KeV are then trapped in the fines. Those with lower energies are neutralized but not trapped and then are recycled. An alternative explanation is that ^{40}K volatilized during impacts has coated grain surfaces and thus produced the excess ^{40}Ar. Finally, it has been suggested that the Ar was implanted 3–4 aeons ago, and thus calculations based on present day fluxes of solar wind are not applicable. It is the feeling of the lunar community that the earlier suggestions are the most acceptable.

RADIONUCLIDES

The airlessness of the Moon makes it an interesting place for the finding of radionuclides, produced by interactions of solar flare particles and cosmic rays with the surface. Some of these nuclides, such as ^{52}Mn, have relatively short half-lives and, consequently, are of limited geochemical value. Other species, such as ^{53}Mn, have much longer half-lives (3.7 m.y.) and are thus more valuable for geochemical studies.

Because of limited penetration, the solar-flare protons account for most of the production at the surface. Below 1 cm, the predominant species are produced by the secondary galactic cosmic ray particles. Investigations on some of the larger rocks by Reedy and Arnold (15) were successful in separating the contributions of the solar radiation from the galactic cosmic rays as a function of depth. Furthermore, although no plutonium was observed directly, Crozaz and co-workers (16) were able to demonstrate from fission track studies in the uranium-rich minerals that it was there in the past. A list of the radionuclides found in the lunar samples and also in meteorites is given in Table 4.4.

In addition to the above, there were also isotopic anomalies found in elements with high capture cross sections for the low-energy neutrons

Table 4.4. Radionuclides Commonly Detected in Lunar Samples and Meteorites

Radionuclide[a]	Half-life (y)	Targets
^{3}H	12.33	O, Mg, Si
^{10}Be	1.6×10^{6}	O, Mg, Si
^{14}C	5730	O, Mg, Si
^{22}Na	2.60	Mg, Al, Si
^{26}Al	7.3×10^{5}	Al, Si
^{36}Cl	3.0×10^{5}	Ca, Fe
^{37}Ar	0.095	Ca, Fe
^{39}Ar	269	K, Ca, Fe
^{46}Sc	0.23	Ti, Fe
^{48}V	0.044	Ti, Fe
^{53}Mn	3.7×10^{6}	Fe
^{54}Mn	0.86	Fe
^{55}Fe	2.7	Fe
^{56}Co	0.215	Fe
^{59}Ni	8×10^{4}	Fe, Ni
^{60}Co	5.27	Co, Ni
^{81}Kr	2.1×10^{5}	Sr, Y, Zr

[a] ^{10}Be, ^{36}Cl, ^{39}Ar, ^{46}Sc, and ^{60}Co are produced mainly by high-energy galactic cosmic rays: ^{56}Co principally by solar cosmic radiation, the remainder are produced by both type of cosmic rays.
Sources: Reedy, R. C. *Ancient Sun,* 1980, p. 370. Lal, D. *Space Sci. Rev.,* **14,** 25 (1972).

produced by cosmic-ray bombardment. Examples of these are ^{158}Gd from ^{157}Gd, ^{156}Gd from ^{155}Gd, and ^{150}Sm from ^{149}Sm.

RARE EARTH ELEMENTS (REE)

The REE, as we have seen in the discussion of the meteorites and in Figure 4.5, are an especially interesting group of elements. Although their chemistry in nearly constant, they nevertheless will fractionate under certain geological circumstances and are good indicators of such processes.

The geochemical factor, which is largely involved in the behavior of the REE, is the regular decrease in ionic radii from 1.14 Å for La^{3+} to 0.85 Å for Lu^{3+}. The two common cations that are close in size are Na^{+} (0.97 Å) and Ca^{2+} (1.02 Å). While substitution between the trivalent REE and

Na does not occur, it does occur for Ca in many minerals. However, because of the valency difference, the REE tend to concentrate in residual phases during crystallization. In the lunar rocks the REE occur in the interstitial material.

The REE patterns compared to chondritic meteorites is shown in Figure 4.5. One of the unusual features is the Eu depletion. In general, there is a diminishing Eu depletion with decreasing REE abundance, although the effect is not a systematic one. A second feature seen is that the patterns are roughly parallel to each other, although they vary in detail. A marked difference occurs at the large REE end, and the major variation is among the high Ti basalts of the Apollo 11 site, a pattern that is interestingly similar to the low K oceanic tholeiitic basalts. In general, terrestrial basalts show patterns or marked enrichment of the larger REE. The interpretation from these observations is that the lunar patterns demonstrate only a limited amount of fractional crystallization during cooling.

THE EUROPIUM ANOMALY

The following is a quotation from Taylor's *Lunar Science: A Post Apollo View* (17):

> Although many geochemical and geophysical problems have been posed by the study of lunar rocks, the europium anomaly exhibited by nearly all the lunar REE patterns has probably produced most controversy, conflicting explanations and interpretations of the geochemical, mineralogical and experimental petrological results.

The first observation of the unique behavior of Eu came from the analysis of the Apollo 11 rocks. The strong depletion of the Eu relative to the Sm and Gd is shown in Figure 4.5.

Among the REE, Eu is distinct because under reducing conditions the divalent state is stable. The ionic radius of the Eu^{2+} is 1.25 Å, some 28% larger than the 0.98 Å characteristic of the Eu^{3+}. The size and valence of the Eu^{2+} alters its geochemical behavior almost completely. The divalent Eu can enter larger lattice sites. In fact, divalent Eu is quite similar in size and charge to Sr^{2+} (r = 1.18 Å). Thus Eu and Sr show a close geochemical association (coherence) and readily enter Ca, Na, and K sites in the feldspar lattice.

This is a phenomenon of some significance because Sr is such an important geochemical tracer. The coherence between Eu and Sr is clearly seen in the lunar rocks by similar enrichments or depletions relative to the chondrites. The coherence between Sr and Eu depends on the abundance of Eu^{2+} relative to Eu^{3+}. This has been demonstrated to be a function of the temperature and the oxygen fugacity (18).

It is well established that the Eu^{2+} will preferentially enter into the feldspars (characteristic of the highlands), while the Eu^{3+} will remain with the trivalent REE. This is clearly shown in Figure 4.5. The highland samples 64501 and 67461 do not show the negative Eu anomalies found in the maria samples. When the anomalously low Eu was found in the early missions, it was immediately suspected, based on terrestrial experience, that the highland regions would be feldspathic. It can also be added that about 70% of the Eu in the lunar samples was divalent.

Numerous models have been proposed to explain the depletion of Eu in the maria basalts. Of all these, the greatest credence is given to the view that the Eu was concentrated in the highlands and depleted in the maria basalts source regions by early melting and differentiation. The model described by Taylor (17) involves a two-stage process. There was an initial melting at 4.6 by, which formed the highland crust. Europium (and Sr) was incorporated mainly in the plagioclase lattice sites. A complementary iron-rich cumulate underneath the early crust formed the source for the maria basalts.

LUNAR CRUST

Before discussing the chemistry of the highlands and the implications of highland chemistry for an understanding of the origin of the lunar crust, let us see what the highlands are like physically. The surface is covered by a large number of craters, some quite large, so densely distributed that they overlie each other to form a continuously cratered surface. The indications are of a period of intense bombardment by large bodies so that large craters formed, and in the process either eliminating or subduing earlier ones. Visually, all stages and degrees of crater degradation can be seen.

It has been established that the highlands are older than the maria; however, the difference in age is relatively small. Since the oldest maria are about 3.9 b.y. and the Moon's overall age is 4.6 b.y., the highlands appear to have been exposed to bombardment for some 700 m.y. Further examination of the highlands and the maria tells us that the bombardment before the maria formed must have been substantially greater than after. It is obvious that all or nearly all of the original highland morphology has been obliterated.

What is now known about the highlands comes from the highland site visited, from a Surveyor landing, and from orbital data, all of which will be discussed subsequently. The nature of the highland samples from which attempts have been made to reconstruct the early history have been described by Taylor (8) in the following terms: "The complexity of the highland samples constitutes a severe test for the scientific method of inductive reasoning." A detailed look at the various components making up the highlands can be found in many sources and is particularly well summarized by Taylor (8). The operating assumption in studying the chemistry of these components is that the major element chemistry has not been affected by the bombardment, whose dominant effect is melting and mixing.

HIGHLAND CRUST; CHEMICAL COMPOSITION

The existence of the highlands dates back to the early observations of Galileo (18). When observers began to speculate about the chemical composition of the highlands, it was logical to base these speculations on what was known about the terrestrial continents in terms of the density and composition. As Taylor has written, the nature of the large craters was a key to a proper understanding of the highlands—were they, for example, volcanic in origin? The choice of the Apollo 16 landing site was, in large measure, an attempt to resolve the volcanism question. In any event, if the terrestrial analogue was emphasized, then the highlands would be expected to be granitic in composition (19).

One of the earliest in-situ analyses of a highland surface was done by means of a remote analysis experiment—the Alpha Backscatter Experiment—on the Surveyor VII that landed on the rim of Tycho. The composition was reported (20) to be like terrestrial high-Al basalts, quite different from granitic rocks.

Some idea about the mineralogical character of the highlands arose from the discovery of anorthositic (feldspathic) fragments in the Apollo 11 soils. This, Taylor (8) states, suggested that the highlands represented a "sink" for the depleted Eu discussed above.

A second highly significant find was light gray fines (12033) and a rock sample (12013) at the Apollo 12 site. Both of these samples were highly enriched in large-ion lithophile elements. This was evidence of extreme fractionation, and it was strongly enforced by Apollo samples returned from Fra Mauro.

The presence of large areas of K-rich material containing K, REE, P, U, and Th (KREEPUTH) on the lunar surface was subsequently demonstrated by an orbital gamma-ray experiment. All of the above observations were synthesized into new notions or models of the origin of the highland crust. For example, since the Fra Mauro site was on the ejecta blanket from the large and nearby Imbrium basin, the possible source for the above material was either local or from within the Imbrium basin.

Explanation for the chemical composition of the crust with its high relative abundance of refractory elements led to the development of conflicting models involving heterogenous accretion versus homogeneous accretion. The heterogeneous accretion called for the late addition of a layer rich in refractory elements. Ultimately, the homogeneous model, involving homogenization after accretion, became the most popular. The major evidence for this was the determination that the ratios of volatile to refractory elements were similar in mare basalts—from the deep interior and in the highland crust. Here again, Taylor points out that this linking of the highland chemistry and the source regions of the mare basalts provided a good explanation for the Eu anomalies and many other geochemical questions.

As in the case of the regolith, there was the nagging problem of sampling—how representative of the crust were the surface samples, and how could they be related to the chemical results of the orbital experiments (see the section on Orbital Science). These questions will be examined in greater detail below.

ELEMENTAL CORRELATIONS

The significance of element correlations, particularly among the refractory trace elements, has been described by Wanke (21) and Taylor (22). These elemental correlations for many of the distinctive elements in the highland samples are important to the geochemist interested in the formation mechanism of the lunar crust.

The major factor that determines the correlations is similarity of geochemical behavior, which in turn is a function of the similarity in ionic radius and valency or bond type. Some of the element pairs that show a distinct coherence are K/Rb, Th/U, Zr/Hf, the REE (with the exception of Eu), Ba/Rb, Rb/Cs, and Fe/Mn. A second group showing coherence is made up of incompatible elements such as K/U, Zr/Nb, K/Zn, and K/La. This latter group tends to concentrate in residual melts. In addition, this second group's coherence tends to be accidental to some degree.

The strong positive correlation among the elements has been taken as an indication of the uniformity of the crustal processes, and the coherence between the volatile and refractory elements such as K/U, Ba/Rb, K/Zr, and K/La in the mare and highland regions shows that both regions were,

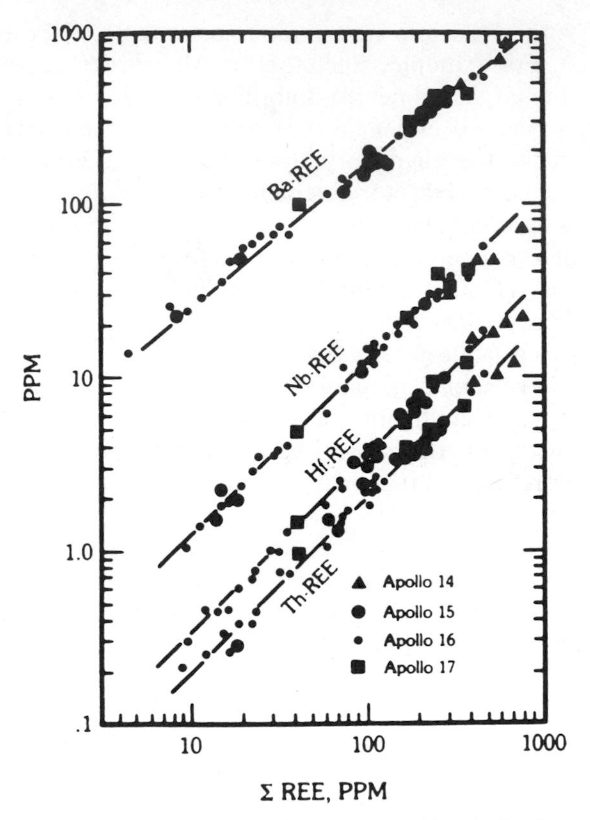

Figure 4.6. Shown is the close correlation between many involatile elements in the highland samples. The abundances are plotted against total REE abundances. The elements chosen here are selected to emphasize correlations between elements of dissimilar chemistry. Source: R. S. Taylor (8).

in the past, homogeneous with respect to the volatile and involatile elements. The coherence described is considered very strong evidence for either homogeneous accretion or homogenization after accretion. Figure 4.6 clearly shows the close correlation observed for some involatile elements. In this instance, the elements were chosen to emphasize the correlations between elements of dissimilar geochemical behavior.

HIGHLAND CRUSTAL ABUNDANCES

The question of representative sampling alluded to previously is obviously a serious problem. How then could the abundances of the highland crust be established in view of the limited sites visited? One of the solutions proposed was that the glass spherules formed by impact would be representative of the principal components of the highland crust. The investigators felt that there might be a group of glass compositions, the granulitic breccias, which would be representative of the average crust. These glasses lent themselves very nicely to electron microprobe analysis.

Table 4.5 shows a highland crustal composition compared with the data obtained from the granulated breccia (78155). The agreement is for the most part excellent, and even surprising, considering the assumptions and manipulation of the data required to produce the table (17).

A brief description of the assumptions used follows: Large-scale use was made of observed interelement ratios. Since there was a relationship between the various ratios, measured ratios could be used to calculate others. Data for large regions of the highlands were obtained from orbital

Table 4.5. Highland Crustal Composition, Compared with Data from Granulitic Breccia 78155

Oxide	Wt.%	78155	Element	ppm	78155	Element	ppm	78155
SiO_2	45	45.6	Cs	0.07	0.11	Eu/Eu	1.4	1.3
TiO_2	0.56	0.3	Rb	1.7	2.1	Y	13.4	
Al_2O_3	24.6	25.9	K	600	650	U	0.24	0.28
FeO	6.6	5.8	Ba	66	59	Th	0.9	1.0
MgO	6.8	6.3	Sr	120	147	Th/U	3.8	3.6
CaO	15.8	15.2	La	5.3	4.0	K/U	2500	2320
Na_2O	0.45	0.3	Ce	12	10.2	Zr	63	54
K_2O	0.075	0.08	Pr	1.6	1.5	Hf	1.4	1.49
Cr_2O_3	0.10	0.10	Nd	7.4	6.3	Zr/Hf	45	36
			Sm	2.0	1.81	Nb	4.5	—
			Eu	1.0	0.87	Zr/Nb	14	—
			Gd	2.3	2.3	Ti	3350	—
			Tb	0.41	0.39			
			Dy	2.6	2.6	Cr	680	680
			Ho	0.53	0.61	V	24	—
			Er	1.51	1.69	Sc	10	13
			Tm	0.22	—	Ni	100	80
			Yb	1.4	1.73	Co	15	14
			Lu	0.21	0.26	Fe%	5.13	4.51
			ΣREE	39.3	34.3	Mg%	4.1	3.8

Source: J. L. Warner et al. *Proc. Lun. Conf. 8th,* 1977; H. Wanke et al. *Proc. Lun. Conf. 7th,* 1976; N. J. Hubbard et al, *Proc. Lun. Conf. 5th,* 1974.

experiments (see the section on Orbital Science). Orbital X-ray fluorescence and gamma-ray data were used to obtain Al/Si, Mg/Si, and Th values for large regions of the highlands (23, 24). The Al/Si and Mg/Si average values were taken as 0.62($\pm$ 0.10) and 0.24($\pm$ 0.05), respectively. The value of 45% for the SiO_2, 45% was considered relatively uniform in the highlands.

Based on such values, the concentrations of the Al_2O_3 and MgO were calculated to be 24.6% and 8.6%, respectively. The value obtained for FeO was 6.6% on the basis of the MgO/FeO ratio in the highland soils and breccias. The Fe/Cr relationship was employed to determine the Cr_2O_3 as 0.10%. Because of the meteoritic component, no estimates were made of the Ni and CO. The Na_2O abundance was taken as typically 0.45%. On this basis the CaO, determined by difference, is 14.2%, a value consistent with the Ca/Al relationship.

The thorium average for the highlands, in the range of 1–2 ppm, comes from the work of Metzger and associates (25). Assuming a ratio of 3.6, the U abundance has been determined as 0.4 ppm. In a similar way, using a K/U value of 1.5, the K value has been calculated to be 600 ppm. From the K/Rb = 350, Rb was calculated at 1.7 ppm and the Rb/^{23}Cs, Cs has been determined as 0.07 ppm. Numerous other determinations have been

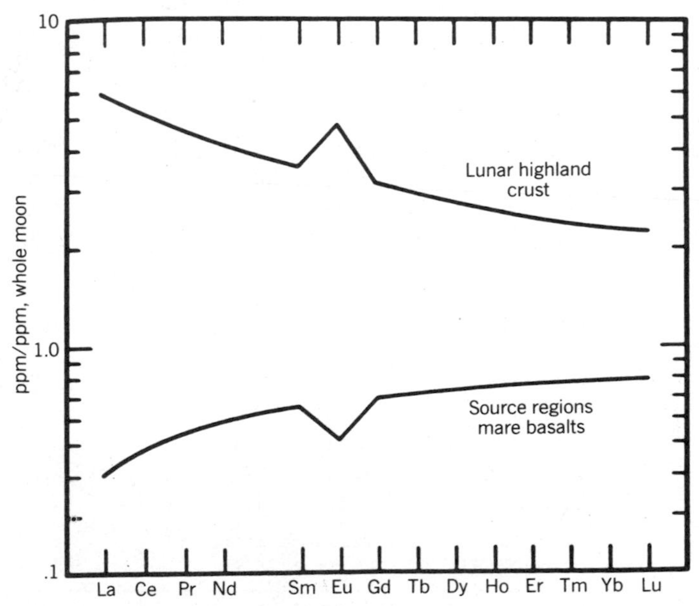

Figure 4.7. Rare earth element abundances in highland rock types. The average highland composition has a positive Eu anomaly. Source: R. S. Taylor (8).

made using the relationships shown in Figure 4.6. The REE abundances in highland rock types, shown in Figure 4.7, demonstrate a positive Eu anomaly. This is in sharp contrast to the different Fra Mauro basalts.

Finally, let us consider the relationship between the highland surface crustal composition shown in Table 4.5 and the composition of the total highland crust. This has been well summarized by Taylor (8). The questions he raises is how representative is this surface composition to that of the bulk highland crust down to depths of about 60–80 km. How reliable are the estimates of crustal thickness? Does the crust vary laterally in chemistry and in thickness? From what depths have materials been brought to the surface by the large cratering events? How much mixing and homogenization has occurred. These are obviously difficult questions to answer. However, the overall evidence is that the compositions determined so far are representative of at least the upper 30 km of the crust.

LUNAR BASALTS

The lunar basalts (26) by their nature, more than any other lunar material, carry the key to an understanding of the source regions in the lunar interior for the lunar crust. Basalts, in general, can be considered as probes of planetary interiors. Further, the data on the major and trace element chemistry provide constraints on the processes producing the basalts. The problem, however, is recognized as being complex.

Basalts are partial melts of planetary interiors, and their compositions are determined by many factors such as the preaccretionary history of the planetary materials, the planetary body's history, the nature of the partial melting and subsequent magmatic evolutionary processes, and the residual mineralogy and phase chemistry of the source. Constraints can be placed on the formation and evolution of a planetary surface by the chemical, isotopic, and experimental studies of the basalts.

Within the above as a background, let us examine the major element compositions of the lunar basalts. To begin with, there is a considerable variety. Taylor (8) writes of 12 distinct compositions, although others have proposed an even greater subdivision. Basalts of different compositions can be found not only at different sites, but even at a single site. As an example, recent studies of the Apollo 11 low-K basalts have shown that in fact there are three distinct low-K basalts. By contrast, the high-K basalts, based on chemistry and isotopic data, seem to have come from a single flow.

The high-Al basalts have been recognized as a class, and samples continue to be found in the lunar sample collection. The newly found samples tend to be small, so there is a question about sampling size; however, the major element compositions all are characteristic of the high-Al basalts. The REE patterns of these basalts do resemble those of the high-Al basalts, but they show such differences in detail that they may come from separate source regions. As a reflection of the complexity of the problem, the Luna 16 basalts from Mare Fecunditatis have high Al, moderate Ti, and high Mg/Fe. They have low abundances of Cr_2O_3 for lunar basalts and are not related to other basalt types. To quote Ma and co-workers (27), "No combination of fractionation of major mineral phases produces the observed compositional characteristics of Luna 16 basalts."

Taylor (8) summarizes the above findings as follows:

> The variations among the major elements and the great variety of mare basalts argue for heterogeneous source regions within the lunar interior. Variations in degree of partial melting of a uniform or primitive unfractionated source region could produce the observed diversity.

The behavior of the large cations is of interest because they nicely demonstrate the effect of geochemical constraints. It has already been indicated that the concentration of K is low in contrast to most terrestrial surface rocks. In fact, the concentration of the K is low enough to make it a minor or trace element. The closest resemblance to the lunar rocks in this regard is the low-K tholeiites or the chondritic meteorites. The K and other large cations are contained principally in the interstitial material in the lunar rocks (referred to as the mesostasis). They enter the lattice sites in the main rock-forming minerals only to a minor degree.

Ba occurs in high concentrations in the mare basalts as a consequence of the high liquid/crystal distribution coefficients that cause it to enter partial melts. In addition, it does have a high intrinsic lunar abundance. Ba is refractory and K volatile, which is reflected in the high Ba/K ratio, compared with chondritic or terrestrial values.

Strontium, another refractory element, is also enriched strongly (about 15 times) in mare basalts relative to the chondrites. Because it readily substitutes for Ca, it is found mainly in the plagioclase feldspar. When Sr is compared to the Ba, it is less strongly enriched in the lavas than Ba. This follows from the difference in geochemical behavior. Because it readily substitutes for Ca, it enters the main mineral phases in the Moon's interior. Thus it is less readily partitioned than Ba into the liquid phase. The Eu^{2+}, because its radius is similar to the Sr^{2+}, is also concentrated in the plagioclase feldspar.

Na, in contrast to the heavier alkalis, K, Rb, and Cs occurs in relatively uniform abundance levels. Geochemically, because of its smaller size it enters more readily into the plagioclase—something the larger ions can do only with difficulty. However, in comparison to terrestrial basalts it is depleted by a factor of about 5. The overall consensus is that this does not come from volatile loss during extrusion but is, rather, an inherent feature reflecting the overall content of volatile elements in the Moon.

Coherence in the basalts can be studied by examining the ratio of Rb/Ba, respectively, a volatile and involatile element. The ratios are low relative to the carbonaceous chondrites. The present evidence, based on isotopic studies, is that this depletion occurred before the Moon accreted. The relationships are illustrated in Figure 4.8. Two characteristics show clearly: (1) the lunar highland and mare basalts show similar volatile/involatile ratios, and (2) the ratio of involatile/to volatile (Ba/Rb) is considerably higher on the Moon than in the chondrites.

The implication, as previously stated in the discussion on the regolith, is that this rules out a heterogeneous accretion of the highland crust from a source enriched in the refractory elements relative to the lunar material. If, in fact, heterogeneous accretion did occur, then homogenization must have occurred before the formation of the highland crust.

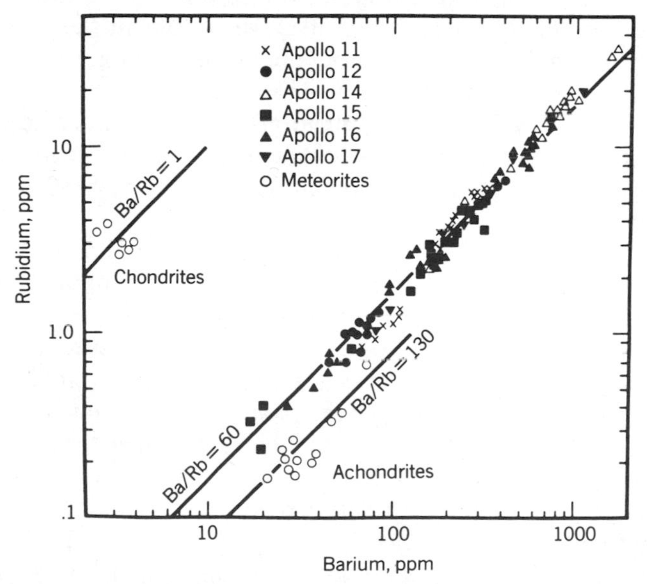

Figure 4.8. Shown is the correlation between the volatile element Rb and the refractory element Ba in mare basalts, highland samples, and meteorites. Note that the highland and the mare samples contain similar volatile/refractory element ratios. The possibilities are that these portions of the Moon have accreted from homogeneous materials or were homogenized following accretion. The meteorites have distinctly different ratios. Source: R. S. Taylor (8).

LARGE HIGH-VALENCE CATIONS

We now know that the Moon as a whole, relative to the Earth, is enriched in refractory elements. Analytical studies have shown that the large, high-valency cations are enriched in the lunar basalts relative to the terrestrial basalts. Not only are these elements refractory, but they are also incompatible, for reasons that will be made clear. For reasons of size and charge, during crystallization of the magma ocean, these elements were excluded from the major mineral phases such as the olivines, the pyroxenes, and the plagioclase. Thus, increasing amounts were trapped in the interstitial liquids or the accessory minerals in the later stages of crystallization.

The concentration of the radioactive elements (K, U, Th) is of great significance because these elements could provide the energy for the partial melting leading to the formation of the mare basalts. There is, in fact, a cumulate model of a mechanism for the concentration of these heat-producing elements. Taylor (8) states that there are adequate amounts of K, U, and Th to provide enough energy for the secondary melting, involving about 0.1% of the lunar volume. Based on such considerations, the source regions of the high-Ti basalts formed late in the magma ocean crystallization and, as a result, contained greater concentrations of the incompatible elements than the low Ti basalts.

KREEP samples from the highland crust contain extreme concentrations of the large cation elements; a probable reason is the invasion of the highlands by the residual melts resulting from the crystallization of the magma ocean.

Thorium and U are especially interesting. They are important for radiometric dating and among the most significant sources of heat. It has been estimated that the abundance of these elements in the lunar rocks would produce sufficient heat to melt the entire Moon. However, in view of the Moon's shape and other geophysical parameters, which indicate that the Moon has probably been rigid for the past 3 b.y., the conclusion is drawn that the U and Th must have been concentrated near the surface.

The K/U ratios for the mare basalts are quite low, about 2500. This is in marked contrast to many terrestrial rocks that have a nearly constant ratio of about 10,000, and the chondrites where the ratio is quite high, about 50,000–80,000. The K/U ratios distinguish the geochemistry of the Moon from other Solar System bodies. In contrast, the Th/U ratios for the Moon, meteorites, and the Earth are quite similar (3.5–4.0). The exceptions, such as the Apollo 17 basalts and some highland feldspathic rocks, have slightly lower ratios of about 3.

The Zr/Nb is constant at about 13–15 in mare basalts as well as the highlands. This strongly emphasizes that for many elements there is a close relationship, a fact that must be accounted for in any theory of lunar origin. However, the high-Ti basalts, found by the Apollo 11 and 17 missions, show exceptions to the constancy of the interelement ratios such as K/Zr and Zr/Nb, which are generally observed in the other lunar samples. The variations show increased amounts of those elements associated with Ti^{4+} such as Zr^{4+}. The Zr/K ratios increase with the Ti content. The evidence is consistent with the derivation of high-Ti basalts from a source already enriched in Ti–Zr–Nb phases. This observation is consistent with at least a two-stage origin for the high-Ti basalts (8).

FERROMAGNESIAN ELEMENTS

The ferromagnesian elements such as Ni, Cu, Ga, V, Cr, Sc, and Ti present interesting problems to the lunar investigators. The above elements show extensive depletion and enrichment compared to either chondritic meteorites or terrestrial basalts. Ni, Cu, Ga, and V are strongly depleted, while Cr, Sc, and Ti are enriched. The depletion in Cu and Ni has been traced in part to special circumstances—the existence of strong reducing conditions during the partial melting and crystallization of the lunar lavas. Under such conditions, the Cu and Ni could be expected to be sequestered by the metallic phases. Additionally, Ni and the other siderophiles were quite probably depleted before accretion. The Ni problem is complicated, as we can see from the argument presented below (8).

Ni is very strongly depleted in the aluminous mare basalts and the high-Ti basalts (1–2 ppm are common). The Fe/Ni ratios are extreme, with values of 15,000–20,000 in comparison to 100–500 in the terrestrial basalts and 20 in the chondritic meteorites. The quartz normative basalts (28) from the source regions, which crystallize before those of the high-Ti basalts, contain 10–20 ppm Ni (very low by terrestrial standards). The olivine normative basalts (from even more primitive source regions) contain 40–50 ppm Ni. The soils, of course, have the highest concentrations of Ni (100–300 ppm), but this reflects the meteoritic component. Thus the lunar samples show both lower Ni and higher Fe/Ni ratios than the terrestrial counterparts.

The major question is: does the above reflect an overall depletion of Ni in the Moon? If the primitive basalts mentioned above come from primitive, unfractionated interiors, then the source regions are highly depleted in Ni compared to the accessible portions of the Earth's mantle.

A second scenario has the mare basalts deriving from differentiated source regions (as proposed in the cumulate model (29). Then the Ni will already be depleted in Ni. The Ni enters the early crystallizing olivine and orthopyroxene lattice sites and thus becomes stored in the early cumulate phases in a crystallizing magma ocean. Taylor and Jakes (29) have developed a model that predicts the existence of large volumes of olivine and orthopyroxene at depths below the source regions of the mare basalts. Thus in their view, whether the Ni content of the bulk Moon is inherently lower than that of the terrestrial mantle depends "on the amount of Ni trapped in the deep and inaccessible olivine and orthopyroxene cumulates."

The Cr story is also rather complex. Compared to the Ni, the Cr is strongly enriched in the mare basalts, relative to the terrestrial basalts by about an order of magnitude. In the section above, the low abundance of Ni was attributed to a combination of an inherently low lunar abundance and the removal of Ni^{2+} in the early crystallizing olivine in the magma ocean.

Any attempt to explain the behavior of the Cr based on a terrestrial analogue would go as follows (8): Cr^{3+} would, like Ni, enter the early crystallizing phases and be retained in residual phases during partial melting. The presence of Cr at levels of 2000–4000 in lunar basalts compared to terrestrial abundances of a few hundred ppm was considered remarkable. There is no known cosmological process known to enrich the bulk Moon in Cr relative to the other ferromagnesian elements. To resolve this paradox, the lunar investigators have invoked the highly reducing conditions in the Moon. A large portion of the Cr in the lunar interior is present as Cr^{2+} (31, 32). It is speculated that because of the large size of the Cr^{2+} ion, it concentrates in the melt relative to the olivine, so that during the crystallization of the magma ocean Cr will be selectively enriched in later pyroxenes. A consequence is the building up of large concentrations in the source regions of the mare basalts. During crystallization of the mare basalts, the Cr behaves mainly like the Cr^{3+}, entering such phases as spinel and chromite. The unusual redox conditions on the Moon thus span the transition from divalent to trivalent Cr and provide a possible solution to the Cr paradox.

Vanadium shows a similar distribution as the Cr, while Sc parallels the Ti content. Once again, the ferromagnesian elements, like some of the others discussed above, show a distribution that is consistent with a derivation of the mare basalts from source regions that have undergone prior crystallization of olivine, pyroxene, and plagioclase.

THE SIDEROPHILES

The siderophiles (the elements that enter the metallic phases) have low abundances in the lunar basalts. Taylor (8) asserts that the basalt data is the key to our understanding of the siderophile elements in the Moon. The lavas come from deep source regions that have escaped the meteorite bombardment to which the surface has been exposed.

Meteoritic bombardment has added a variable siderophile element component to the highland rocks, which makes them uncertain as a source for the establishment of the indigenous lunar abundances for the siderophiles. While the basalt rock samples do not have this problem, they are nevertheless derived from source regions that have a fractionation history. Like a number of elements previously discussed, siderophile elements will also fractionate during the partial melting process and the formation of the metal phases.

Gold and iridium range from 0.1 to 0.001 ppm. Rhenium is more strongly depleted (a factor of 10 lower). Nickel abundances range from less than 2 to 64 ppm, and silver abundances are typically about 1–0.1 ppm. Germanium values are typically about 1–10 ppb.

The relationship of the siderophile element content of the Moon compared to the Earth is an important component in theories of lunar origin. The extreme depletion of the siderophiles in the mare basalt region is once again traced to two causes: (1) a loss of the siderophiles before accretion, and (2) a possible depletion of the siderophiles of the interior by the formation of an Fe or FeS core.

OXYGEN ISOTOPES

The variation in the oxygen isotope oxygen-18 in the mare basalts is very small. Expressed in $\delta^{18}O$ values (30), the values range from +5.4 to +6.8. There is a variation in the minerals ranging from $\delta^{18}O$ of 3.8 to 7.2 in the sequence ilmenite, olivine, clinopyroxene, plagioclase, and silica minerals. This range is the same for the minerals from different rocks. Clayton and associates (28) have stated, based on the study of the crystalline rocks from the Apollo 15 and 16 collections and the clasts in the breccia, that the mineral-pair oxygen isotopic fractionations are consistent with equilibration at temperatures near 1100°C. In general, the dispersion of the $\delta^{18}O$ values for all the lunar rocks is so small as to lead to the conclusion that the oxygen isotopic composition of the lunar interior is now known with great certainty. Furthermore, the isotopic evidence favors a ferromagnesian composition rather than a Ca–Al silicate composition for the interior.

CARBON

The low carbon content in the lunar regolith has been described (8). The carbon found in the lunar basalts from the various sites range from about 10 to 80 ppm. In general, however, these values are much lower than the values found in the soils. It is suspected that the carbon is present in the lunar rocks as carbides, although there is no direct mineralogical evidence of their presence. Their existence is inferred from the release of methane and other gases under acid treatment. The $\delta^{13}C$ values range from -20 to -30 in terrestrial basalts. By contrast, the lunar soils have values ranging from $+10$ to $+20$.

SUMMARY

The important role played by analytical chemistry in providing an understanding of the lunar surface has been introduced. There are now substantial data that strongly reflects the intense effort that has gone into the study of the lunar samples. We have also seen how vital are some of the chemical results in producing a model of the Moon's formation and evolution. It is a problem to synthesize this large amount of data.

Taylor (8) has written that the composition of the bulk Moon is by now as well, if not better, understood than that of the Earth, particularly since the composition of the lower terrestrial mantle is controversial. In order to calculate the overall composition of the Moon, it is necessary to make some assumptions. On the Earth, the crustal composition is secondary to the estimates of the mantle's composition in terms of the major elements. The amounts of Si, Al, Mg, or Ca in the crust can be neglected in a first approximation of the Earth's mantle composition. By contrast, the abundance of the major elements such as Al and Ca in the lunar highland crust are considered critical for evaluating the whole Moon abundances. In addition, the crustal abundances of K, U, Th, Sr and Eu form a large percentage of the Moon's total content. A key element is the thickness and composition of the highland crust.

With regard to the highland crust, it is now established that it contains data showing that the Moon is enriched in refractory elements relative to the primitive solar nebula abundances. The early viewpoint that the highlands are made up of material that was directly accreted as a late-stage accumulation of a heterogeneous Moon has been disproved by the following chemical evidence: (1) there is a reciprocal relationship between the highland and mare basalt abundances derived from the deep interior, (2) there is a similarity in the volatile/involatile element ratios, and (3) the geochemical evidence for crystal–liquid fractionation processes in the evolution of the crust.

We have also seen that the elemental abundances in the mare basalts establish constraints on lunar processes. The basalts are, in fact, the end-stage products of several processes. These have been tabulated (8) and are listed below:

1. Preaccretion fractionation resulted in the loss of the volatile elements and probably of the siderophile elements.
2. Following the accretion of the Moon, widespread melting and fractionation occurred. Shortly after formation at about 4.4 b.y., the melting and fractionation led to zoned structures in which the source regions of the mare basalts were enriched in some elements and depleted in others. A noteworthy example, discussed previously is Eu.
3. Estimates are that between 4.0 and 3.2 b.y., partial melting occurred, fired by radioactive heating, which produced element fractionation in the source regions of the mare basalts. These basalts were then extruded to the surface.
4. During and following extrusion, crystallization made it possible for some phases to be removed, which caused further fractionation.

The importance of the incompatible element abundance was discussed. The proposal is that during the crystallization of the magma ocean and during the partial melting that generated the basalts, the elements became concentrated in the residual liquids or entered the early melts. Because there is a tendency for the ratios of these elements to remain constant, they may reflect the whole Moon abundance ratios. However, as pointed out by Taylor (8), the use of such correlated element ratios for estimating the composition of the Moon should be done only with care, keeping in mind that geochemical considerations, such as ionic radius, distribution coefficients, valency, and crystal field effects, may affect the picture. Correlating the bulk composition is probably better done employing the analyses or recognized, primitive undifferentiated samples, of which the glass, sample number 12001, is an example.

Finally, an interesting comparison has been drawn between the lunar mare basalts and the euchrites, a class of basaltic achondritic meteorites. It has been observed that there is a good major element match between the aluminous mare basalts and the euchrites. The elements that match particularly well are Cr, Na, and K. Such volatile elements as Tl, Bi, and siderophiles like Ir also match well. However, there are some distinctive differences like the refractory Ba, Zr, and Ti, which are consistently higher in the lunar basalts, showing the consistent lunar enrichment. Also different are the oxygen isotopes. While the Moon is not a source for the euchrites, what is interesting is the idea that the euchrites show evidence of similar evolutionary processes and that the same kind of major element chemistry appears to be going on elsewhere.

REFERENCES

1. W. Compston, B. W. Chappell, P. A. Arriens, and M. J. Vernon. *Proc. Apollo 11 Lun. Sci. Conf.*, **2,** 1007 (1970).
2. K. Norrish and B. W. Chappell, X-Ray Fluorescence Spectrography, in J. Zussman, Ed., *Physical Methods in Determinative Mineralogy,* Academic, New York, 1967, p. 161.
3. F. Cuttita, H. J. Rose, C. S. Annel, M. K. Carron, R. P. Christian, E. J. Dwornik, L. P. Greenland, A. W. Helz, and D. T. Ligon, Jr., *Proc. Lun. Sci. Conf. 11th,* **2,** 1217 (1971).
4. A. A. Smales, D. Mapper, M. S. W. Webb, R. K. Webster, and J. D. Wilson, *Proc. Apollo 11 Lun. Sci. Conf.,* **2,** 1253 (1970).
5. K. Norrish and J. T. Hutton, *Geochim. Cosmochim. Acta.,* **33,** 431 (1969).
6. J. C. Laul, *At. Energy Rev.,* **17,** 603 (1979).

6a. J. C. Laul and J. J. Papike, *Proc. Lun. Sci. Conf. 11th,* **2,** 1307 (1980).

6b. J. C. Laul, D. T. Vaniman, J. J. Papike, and S. Simon, *Proc. Lun. Sci. Conf. 9th,* 2065 (1978).

7. J. J. Papike, S. B. Simon, and J. C. Laul, *Rev. Geophys. Space Phys.,* **20** (1982).
8. S. R. Taylor, *Planetary Science: A Lunar Perspective.* Lunar and Planetary Inst., Houston, Texas, 1982.
9. J. M. Rhodes, *Phil. Trans. Roy. Soc.,* **A285,** 293 (1977).
10. F. Horz, *Proc. Lun. Sci. Conf. 9th,* 3311 (1978).
11. L. I. Yin, S. Ghose, and I. Adler, *Appl. Spectrosc.,* **26,** 355 (1972).
12. J. W. Morgan, R. Ganapathy, H. Higuchi, and E. Anders, *The Soviet-American Conf. on Geochemistry of the Moon and Planets,* NASA SP-370 (1977).
13. P. P. Phakey et al., *Proc. Lun. Sci. Conf. 3rd,* 905 (1972).
14. D. Heymann and A. Yaniv, *Proc. Lun. Sci. Conf. 1st,* 1681 (1970).
15. R. C. Reedy and J. R. Arnold, *J. Geophys. Res.,* **77,** 535 (1972).
16. G. Crozaz et al., *Proc. Lun. Sci. Conf. 3rd,* **2,** 1623 (1972).
17. S. R. Taylor, *Lunar Science: A Post-Apollo View,* Pergamon, New York, 1975.
18. G. Galelei, *Nuncius Siderus,* Padua, 1610.
19. J. A. O'Keefe and W. S. Cameron, *Icarus,* **1,** 271 (1962).

20. A. Turkevich, *Science,* **134,** 672 (1961).
21. A. Wanke, H. Baddenhausen, G. Dreibus, E. Jagoutz, H. Kruse, H. Palme, B. Spettel, and F. Teschke, *Proc. Lun. Sci. Conf. 4th,* 1461 (1975).
22. S. R. Taylor, M. P. Gorton, P. Muir, W. Nance, R. Rudowski, and N. Ware, *Lunar Sci. IV (abstracts),* 720 (1973).
23. I. Adler et al., *Preliminary Science Rept.,* NASA SP-289, 1972, p. 17-1; NASA SP-315, 1972, p. 19-1.
24. J. R. Arnold, L. E. Peterson, A. E. Metzger, R. C. Reedy, and J. L. Trombka, Preliminary Science Rept., NASA SP-289, 1972, p. 16-1; NASA SP-315, 1972, p. 18-1.
25. A. E. Metzger, J. I. Trombka, R. C. Reedy, and J. R. Arnold, *Proc. Lun. Sci. Conf. 5th,* 1067 (1974).
26. *Basaltic Vocanism on the Terrestrial Planets,* Pergamon, New York, 1981.
27. M. S. Ma et al., *Geophys. Res. Lett.,* **6,** 909 (1979).
28. R. N. Clayton, J. M. Hurd, and T. K. Mayeda, *Proc. Lun. Sci. Conf. 4th,* **2,** 1535 (1973).
29. S. R. Taylor and P. Jakes, *Lunar Sci. V,* 786 (1973).
30. E. Anders, *The Moon,* **8,** 6 (1973).
31. D. Lal, *Space Sci. Rev.,* **14,** 25 (1972).

CHAPTER

5

ISOTOPES AND COSMOCHRONOLOGY

The developing picture of the origin and evolution of the Solar System is the result, in large part, of the contributions being made by the science of cosmochronology. The subject has been treated in a series of review papers by Wasserburg and associates (1), Lee (2), and Kirsten (3). The use of radiometric dating has made it possible to establish significant milestones along the evolutionary path to the present Solar System and to better understand the significant genetic relationships.

In his review article Kirsten has stated that time is an important parameter in clarifying the physical processes that have occurred in the evolution of matter from the beginning—the nucleosynthesis stage—through the formation of the Solar System to the present. It is time that places boundary conditions on the physicochemical and geological processes that have led to the formation and evolution of the planets—processes such as condensation, accretion, differentiation, and metamorphism. In this chapter we shall examine advances in radiometric age determinations and the measurements of isotopes that provide new clues to Solar System formation.

Among the accomplishments resulting from the impetus supplied by the planetary exploration program have been the development in the methods of absolute age determination. In order to better appreciate the advances in the field, let us begin with a brief review of the radiometric methods.

It has been found that not all nuclei are stable. In general, the lighter, stable nuclei have equal numbers of protons and neutrons. Heavier stable nuclei have somewhat larger numbers of neutrons than protons. It is possible to plot a curve of the number of protons versus neutrons and to define a stability line showing this relationship.

Nuclides falling to the right and left of this line tend to be unstable and are radioactive. It is also true that isotopes with atomic numbers greater than 83 are radioactive. The radioactive isotopes are known to undergo radioactive decay at various rates which are both specific and invariant. This process has become the powerful and precise basis for the dating of rocks.

There are several routes by which the atom decays to achieve stability. It may, for example, emit an alpha particle consisting of two protons and two neutrons. This alpha particle (the nucleus of helium) is generally ejected at high velocity and with specific energy, so that an alpha spectrum is a line spectrum. The process leaves an atom with an atomic number decreased by two and a mass decreased by four. A second form of radioactive decay is by the emission of an electron (beta particle) from the nucleus. In view of the fact that the nucleus does not contain electrons as such, we visualize this as the decay of a neutron into a proton and an electron, so that the original nuclide becomes the next higher one in the periodic table. Because of the interactions of the emerging electrons with the various atoms and the resulting energy losses, the beta particle can have a continuum of energies ranging between zero and the maximum available for the transition.

Still another method for radioactive decay is the process of electron capture. The nucleus in some instances will capture an electron from the innermost shell of the atom (K capture). This process is the reverse of beta emission and results in the decrease in the atomic number by one. A proton combines with an electron to form a neutron. The resulting radiation is a characteristic X ray.

Some nucleii may undergo competing types of radioactive decay, called a branching decay. For example, ^{40}K may decay either by beta emission into ^{40}Ca or by K capture into ^{40}Ar.

It is well known that some nucleii decay into daughter products that are also radioactive. In some instances the number of steps in the decay chain are substantial before the original element ends up as a stable, radiogenic nuclide. This process is known as series decay, and such a chain is said to be in secular equilibrium when all but the last member are present in inverse proportion to their decay constants.

NUCLEAR CLOCKS

There are a variety of nuclear clocks, of which a few illustrative examples will be discussed. Many of these techniques have been applied to extraterrestrial materials to provide a historical background for important events in the history of the Solar System.

DECAY CLOCKS

In principle, when a radioactive nuclide is produced by any of the nuclear reactions operative in the universe and this production continues at a constant rate, the amount of this nuclide builds up to a constant value as the system approaches secular equilibrium. If the radioactive nuclide is removed from the system and is stored, as so often happens in nature, the nuclide will begin to decay in accordance with its decay constant. The amount present after some time t is given by

$$P(t) = P_s e^{-\lambda t}$$

It is possible to solve for t, the time that has elapsed since the sample was set aside by the expression

$$t = \frac{1}{\lambda} \ln \frac{P_s}{P}$$

where λ is the decay constant and P is the number of atoms to which P_s (the original number of atoms) has decayed in time t.

The decay clock has limits because the error in determining P becomes quite large as P_s gets small. Thus such a clock is limited to about 10 half-lives. This is about the time required for the radioactivity to have decayed to about 0.1 the original value at secular equilibrium.

An excellent example of such a decay clock is the ^{14}C clock. The nuclear process producing the ^{14}C clock is a neutron reaction with ^{14}N, a reaction occurring in the upper atmosphere due to cosmic ray interactions. The ^{14}C decays by beta emission to ^{14}N, with a half-life of about 5700 y.

RATIO CLOCKS

Cosmic-ray interactions with intersteller matter will produce a number of radioactive nuclides such as ^{10}Be, ^{26}Al, ^{32}Si, ^{36}Cl, and ^{53}Mn. Such radionuclides have relatively short half-lives but can yield very useful data about the residence time in space—a very useful parameter in dealing with meteorites. Many of these isotopes are to be found in the cosmic dust being swept up by the Earth in its orbit. The estimates run as high as 10,000 tons a day. In view of the extremely low amounts of the radionuclides, their determination requires maximum care and the best analytical techniques available. As we have seen in the chapter on the meteorites, many of these isotopes are now within reach and have been successfully used in cosmochemical studies.

A serious problem complicates the use of radionuclides from secular equilibrium in age determinations. This arises from the difficulty in reconstructing their cosmochemical behavior between the time they were produced and their time of deposition. Because local conditions are likely to affect their concentrations and rates of deposition, the secular equilibrium quantity is likely to vary from place to place. However, if one finds two nuclides that have been deposited side by side in fixed proportions (independent of absolute amounts), and if a sample of material containing two such nuclides is sealed off by some cosmochemical action at some time (t) so that neither nuclide is being added to, then one can write the ratio

$$\frac{N_1}{N_2} = \frac{N_{1(s)}e^{-\lambda_1 t}}{N_{2(s)}e^{-\lambda_2 t}}$$

where $N_{1(s)}/N_{2(s)}$ is the secular equilibrium ratio R_s. Therefore the above equation reduces to

$$R = R_s e^{(\lambda_2 - \lambda_1)t} \quad \text{and} \quad t = \frac{1}{(\lambda_2 - \lambda_1)} \ln \frac{R}{R_s}$$

For the ratio R/R_s to change sufficiently with time to be useful for age determination, the decay constants should differ by at least a factor of 2. R is determined by measurement in the sample, but R_s may have to be assumed in some cases.

ACCUMULATION CLOCKS

A widely used and highly successful method of dating is the accumulation clock, based on the accumulation of a daughter product by the decay of a radioactive parent. The relationship between daughter and parent is given by the following:

$$D = P_0 - P_t$$

P_0 and P_t are the number of atoms of parent nuclide at $t = 0$ and $t = T$, respectively. Two important assumptions are being made here: (1) The system is a closed one (neither parent or daughter is removed by any process other than radioactive decay); and (2) no daughter atoms were present at time $t = 0$. As Faul points out (4), rigorously closed systems probably do not exist in nature; however, a surprisingly large number of minerals and rocks satisfy these conditions well enough to be useful subjects for age determinations.

The problem of the daughter nuclide† having been present at $t = 0$ is dealt with in the following manner:

$$D^* = D_{\text{total}} - D_o$$

where D_o is the original concentration. Then

$$D^* = P_o - P$$

This tells us that D^*, the number of radiogenic daughter atoms, equals the number of parent atoms that have decayed. Since $P = P_o\, e^{-\lambda t}$,

$$D = Pe^{-\lambda t} - P = P(e^{\lambda t} - 1)$$

and therefore

$$t = \frac{1}{\lambda} \ln\left(\frac{D^*}{P + 1}\right)$$

Until very recently, the major radionuclides used in accumulation clock studies were potassium-40, rubidium-87, uranium-238, and uranium-235. Now improved analytical methods have made new radionuclide systems available (e.g., Sm–Nd). These will be discussed later.

GROSS URANIUM–LEAD CLOCK

Historically, the earliest nuclear age determinations were performed on high-grade uranium ore minerals, such as uraninite and pitchblende, using wet chemical methods for the uranium and lead. Not all of the results made sense, and it became obvious that the discrepancies were caused in large part by the relatively high mobility of the uranium and lead in the natural environment. It was realized that not many of the uranium minerals act as closed systems. There were, however, some exceptions.

With the development of isotopic analysis, it became possible to determine if a particular uranium system remained closed. Because uranium has two long-lived isotopes, it became feasible to calculate two entirely independent ages from the accumulation of lead-206 and lead-207. The two parent isotopes were uranium-238 and uranium-235. Both of these nuclides not only have different decay rates but also different intermedi-

† D^* is the radiogenic portion of the daughter nuclide.

ate members in their decay chain. Thus unless concordant ages were determined, it was concluded that the systems had not been closed and neither value represented the true age. The principle of concordancy is an important cornerstone in the science of cosmochronology.

CONCORDANCE

Fortunately, it is often possible to measure the age of a rock by an independent method or methods, using more than one-parent daughter system. The results that agree within experimental error are called concordant. Such agreement gives the investigator great confidence that the determined age is real in terms of actual time.

There are, however, numerous reasons why two or more systems having a common origin will nevertheless show differences in determined ages and be discordant. For the most part, the errors do not come from the laboratory because in a modern, well-run laboratory errors in analysis should be insignificant.

Errors in calculations may arise from an incorrectly chosen decay constant or uncertainty in the correction for some original abundance of a given nuclide. One example cited by Faul (4) is that while most of the common decay constants have assigned errors of 2% or less (std. dev.), the uncertainty in the decay constant of rubidium-87 is much greater, and the two values that were being used at the time of writing of the Faul volume differed by about 6%. Rubidium is mentioned specifically because the Rb–Sr scheme for dating is so widely used. Where analyses and calculations are done with care, the cause of discordance is in the rocks themselves, due to natural geochemical processes. As an example, the daughter products of the radioactive decay will find themselves in a crystal where they do not fit. As a consequence, these nuclides will tend to diffuse out of the crystal. This process can be accelerated by a rise in temperature, and such heating episodes are rather common in nature. The loss of a daughter product is the most frequent source of error.

A particularly graphic example is the loss of argon-40, the daughter product of potassium-40 decay. Such losses from an assortment of minerals have been systematically studied, and the conclusion drawn that the losses would be enhanced by heating and low ambient pressures. Figure 5.1 shows the variation of the apparent ages of different minerals as a function of their distance from a source of heat (an intrusive contact). Both minerals and parent–daughter systems show substantial differences in response to the heating by the intrusive body.

The potassium–argon system also demonstrates the problem of cor-

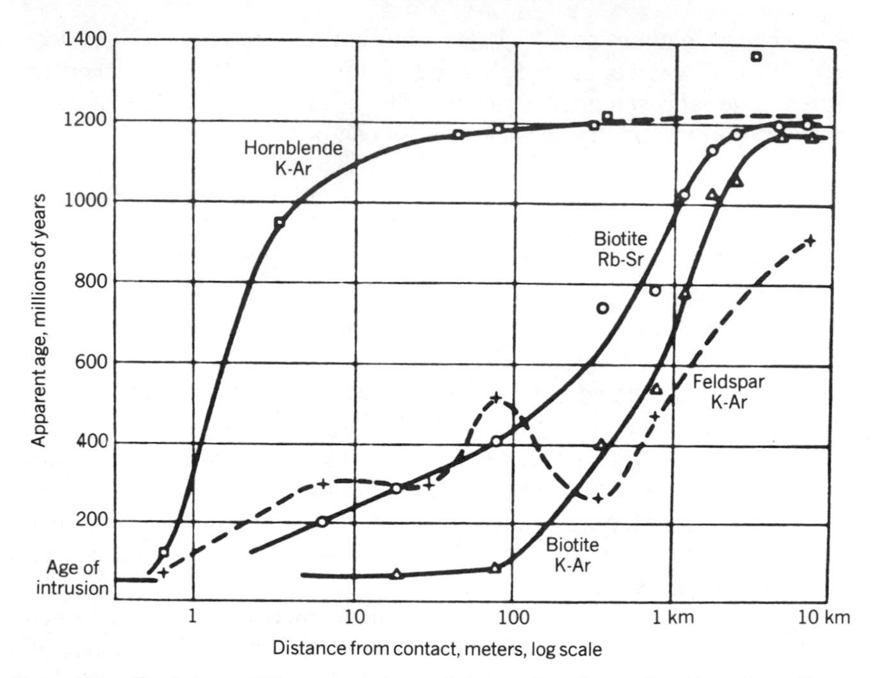

Figure 5.1. Variations of the apparent age of three minerals as a function of the distance from a source of heat—in this instance, an intrusive contact. The different minerals and parent–daughter systems show wide differences due to the heating by the intrusive body. Source: S. R. Hart (5), H. Faul, Ages of Rocks, *Planets and Stars,* 1966; permission, McGraw Hill.

recting for the quantities of original daughter abundances. It is to be expected that magmas will contain some argon-40, depending on the age and the potassium content of the material from which the magma derived, the time through which the magma was molten, the ambient pressure, and all the other factors that affect the outgassing from molten rocks. Thus one would expect a wide range of argon-40 in magmas of diverse origin.

In some instances, minerals have crystalline structures that will accommodate noble gases. Such crystals as they grow would capture argon from the melt. Factors such as these place a burden on the investigator to determine the amount of original argon. These values can be estimated by comparing the measured potassium–argon age with ages determined by other methods. This is a difficult problem because it is necessary to decide whether the problem is due to "original argon" or the recent loss of potassium.

CONCORDIA

Uranium–lead dating is a notable example of observed discordancies. In fact, except for some specific minerals, it is unusual for the uranium-238/lead-206 and the uranium-235/lead-207 ratios to agree to within less than 10%. Such disagreement occurs within rocks that have undergone complete metamorphism, and those rocks that are relatively undisturbed. The mechanisms for this discordancy have been extensively studied.

The stable daughter products of uranium decay are lead-206 and lead-207. Because they are chemically identical and the mass difference is less than 0.5%, even mass-dependent processes would separate them only to a negligible extent in nature. Thus, if lead is lost from a mincral, it is usually lost in the isotopic proportions in which it is present.

The study of the uranium–lead system has contributed greatly to an understanding of discordant systems. The problem was treated by Wetherill (6) in the following manner: The half-life of the uranium-235 to lead-206 decay is small compared to the half-life of the uranium-238 to lead-206 decay. Thus, if a ratio plot is done with $^{206}Pb/^{238}U$ as the ordinate and $^{207}Pb/^{235}U$ as the abscissa, the curve will be shown as in Figure 5.2. In the younger rocks, the proportion of radiogenic ^{207}Pb to ^{206}Pb is high. The basic expressions are

$$D_1/P_1 = e^{\lambda_1 \tau_0} - 1, \qquad D_1/P_1 = {}^{206}\text{Pb}/{}^{238}\text{U}$$

$$D_2/P_2 = e^{\lambda_2 \tau_0} - 1, \qquad D_2/P_2 = {}^{207}\text{Pb}/{}^{235}\text{U}$$

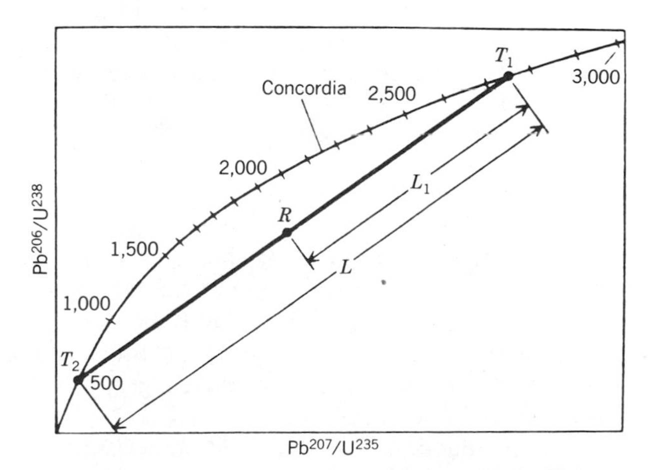

Figure 5.2. The Concordia Curve. Source: Wetherill (6).

Where the ages are concordant, for every age $\tau_0 = T_1 = T_2$ there are unique values of D_1/P_1 and D_2/P_2 as defined by the above equations. The locus of these values for $0 < \tau_0 < \infty$ is the curve named "Concordia" by Wetherill. He has shown that if a mineral of age T_1 underwent a loss of lead by some process at time T_2, then the point obtained would lie somewhere along a chord from T_1 to T_2, the location being determined by the fraction of lead lost. As shown in Figure 5.2, the fractional distance of L_1/L of the point R from T equals the fractional amount of lead lost from the system. Where several events cause the loss of lead, the resulting $^{206}Pb/^{238}U$ divided by the value $^{207}Pb/^{235}U$ fall on a chord that can be extrapolated back to time T_1 when the system originated and T_2 when the lead was lost. In this manner, Wetherill and others have been able to glean important age information from a set of discordant U–Pb data.

It has been subsequently shown by other investigators that the loss of lead need not be episodic. Tilton (7) has proposed that continuous volume diffusion of the lead as a mechanism could also produce the observed irregularities.

MIXED AGES

The situation regarding the ages of rocks has been neatly summarized by Faul (4). He states that

> it is not unusual for an old rock to crystallize, cool and then remain thermally undisturbed through geologic time until the present. More often, the crystalline rock remains cool for some time, then becomes heated again by some geologic process, possibly recrystallizes in part, then cools again, heats again. The process may be repeated several times. It is almost always geologically interesting to try to unravel the history of such a rock and to interpret the discordant sets of ages measured.

Where a rock contains several parent–daughter systems, the effect of some episodic reheating can be quite variable, depending on the resistance of a given system to the redistribution of the parent and daughter. The more resistant systems would tend to give the oldest age, while the least resistant system would yield the age at the time "the clock was reset." It is also conceivable that intermediate ages, which would be called a mixed age, would be observed.

With this as an introduction, let us look at some specific clocks and examine some of the recent advances.

RUBIDIUM–STRONTIUM CLOCKS

Rubidium-87 decays to strontium-87 by beta decay with a half-life of 4.7×10^{10} y. This is so slow a rate that only about 7% of the terrestrial rubidum has converted to strontium since the Earth formed (about 4.6 aeons ago). On the average, strontium is about four times more abundant than rubidium in the Earth's crust. Thus the radiogenic contribution by the rubidium-87 is very small, and the isotopic composition of common strontium has changed very little over geologic time. The variation in strontium-87 is generally reported as the ratio $^{87}Sr/^{86}Sr$.

The modern value is about 0.708 in the ocean today and close to 0.700 in some of the oldest available ocean samples. This represents a change of only about 1% in a period of about 3 aeons. The lowest $^{87}Sr/^{86}Sr$ ratio is a very important value because it is presumably representative of the most primitive material in the Solar System. Such values come from the achondritic meteorites and are considered to be the primordial ratio. The use of this value will be discussed in additional detail below. It is necessary to realize that all the present strontium evolved from the primordial strontium by the addition of the ^{87}Sr from the ^{87}Rb decay process as shown:

$$^{87}\mathrm{Sr} = {}^{87}\mathrm{Rb}(e^{\lambda t} - 1)$$

Since such quantities are usually expressed as a ratio to the constant isotope ^{86}Sr, the expression becomes

$$^{87}\mathrm{Sr}/^{86}\mathrm{Sr} = {}^{87}\mathrm{Rb}/^{86}\mathrm{Sr}\,(e^{\lambda t} - 1)$$

It is customary to plot $^{87}Sr/^{86}Sr$ as a function of $^{87}Rb/^{86}Sr$ in what is known as an isochron diagram or a Rb–Sr evolution diagram. As Faul has pointed out, this is one of the most useful concepts in geochronology. An isochron diagram is shown in Figure 5.3 for four hypothetical minerals. We note the following:

1. At time $t = 0$, all the components of a crystalline rock will have the same $^{87}Sr/^{86}Sr$ ratio, independent of the Rb content. This value plots on a horizontal line, shown in the diagram by the solid circles.
2. At some elapsed time t, proportional amounts of ^{87}Rb will have decayed to ^{87}Sr by beta decay, and now the points plot on a line forming an angle alpha with the horizontal where

$$\tan \alpha = e^{\lambda t} - 1.$$

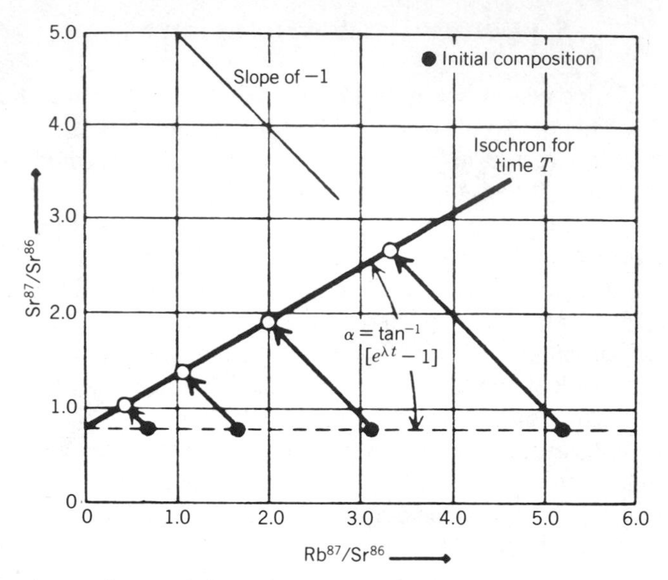

Figure 5.3. A strontium evolution diagram showing the development of the $^{87}Sr/^{86}Sr$ ratio in time T as a consequence of the rubidium decay in four hypothetical minerals of some crystalline rock. Source: Lanphere et al. (8), H. Faul, Ages of Rocks, *Planets and Stars,* 1966; permission, McGraw Hill.

This relationship enables one to calculate the time t from the slope of the line.

3. The zero intercept remains unchanged because there is no addition of ^{87}Sr to a system where ^{87}Rb is absent.

4. Thus, with two Rb/Sr systems of common origin, it is not only possible to calculate an age but also the composition of the original Sr.

An important caveat expressed by Wasserburg and co-workers (1) with regard to the mineral isochron, is the need to establish whether a given linear data array is a true isochron representing several distinct phases rather than a mixture of two end members without a strict time relationship. If the data points do not define a precise line, then the system is complex and neither the age nor the initial Sr. are well defined.

It can and often does happen that following a given time T, after the first crystallization, some geologic event may cause a rock to become hot enough to completely exchange and rehomogenize all the Sr. Such a process can take place under closed-system conditions, nothing being added or removed from the rock. Rubidium will also migrate as a function of the particular equilibrium conditions. As the rock crystallizes, a new set of minerals form. These minerals will once again lie on a horizontal line representing the "new-original" strontium (shown as solid circles in Fig. 5.4) and altered Rb/Sr ratios.

The recrystallization of the rock is equivalent to the clock once being started. After a period of radioactive decay T_2 of the ^{87}Rb to ^{87}Sr, the mineral points will lie on a new line forming an angle beta with the horizontal line where

$$\tan \beta = e^{\lambda T_2} - 1$$

The situation described above not only applies to different mineral phases in a given rock, but to different igneous rock masses formed by differentiation from the same magma, each with a different Rb/Sr ratio. The second melting would be expected to produce a homogenized strontium composition, which are shown as the ordinate intercepts of the internal isochrons as shown in Figure 5.5. In this same figure, we also see that the average samples of each whole rock (the open circles *a, b,* and *c*) fall on a line forming the angle gamma where

$$\tan \gamma = e^{(T_1+T_2)} - 1$$

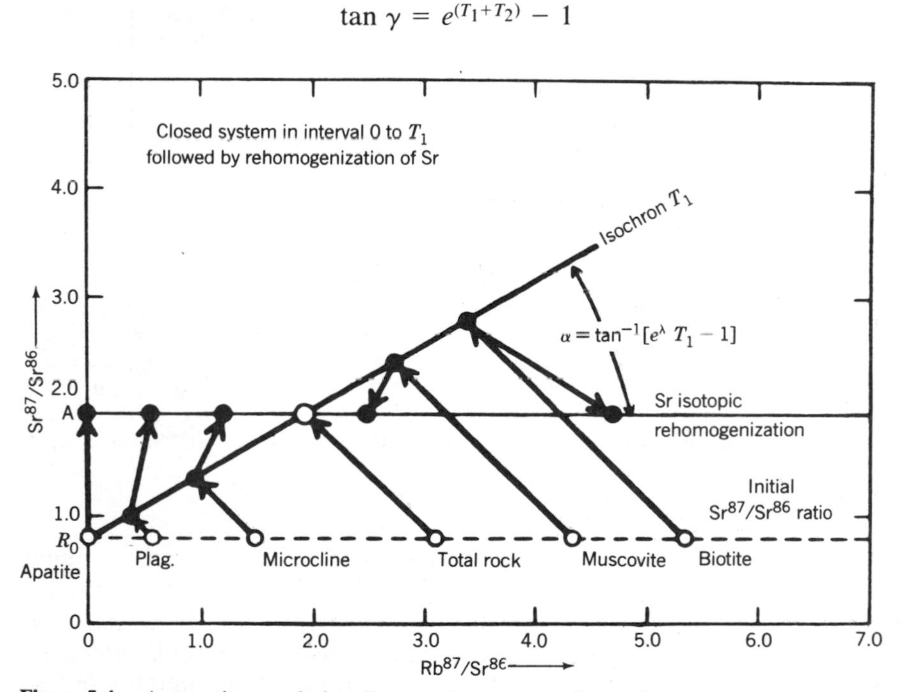

Figure 5.4. A strontium evolution diagram showing the effects of isotopic homogenization for a two-episode model. Source: Lanphere et al. (8), H. Faul, Ages of Rocks, *Planets and Stars,* 1966; permission, McGraw Hill.

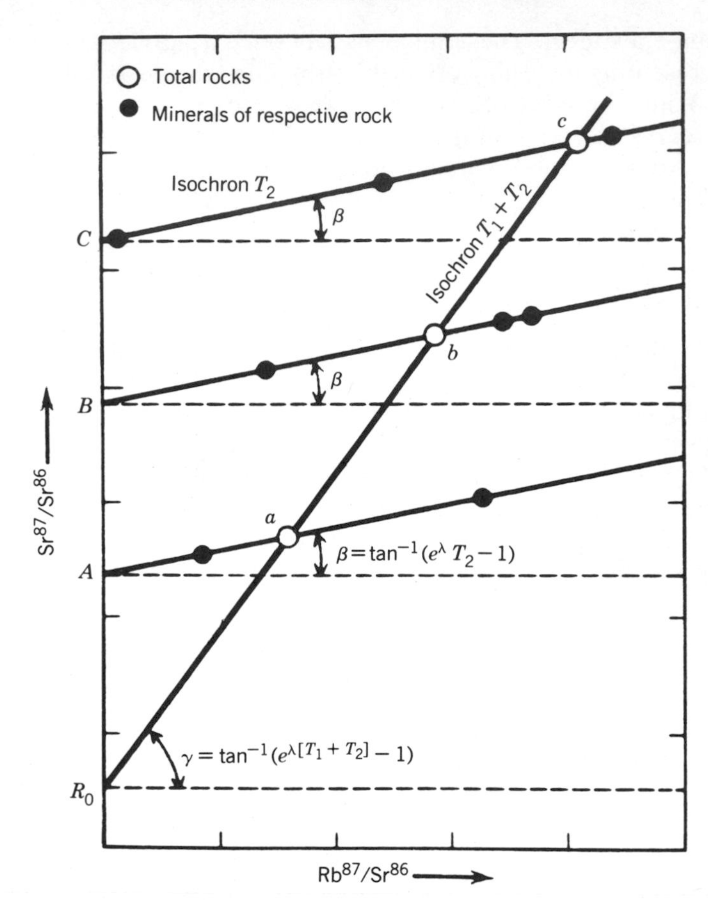

Figure 5.5. A strontium evolution diagram for a two-episode model. In this instance three hypothetical rocks (open circles *a*, *b*, and *c*) were isolated from a common environment $T_1 + T_2$ years ago. The initial $^{87}Sr/^{86}Sr$ ratio was R_0. Each rock taken as a whole, remained a closed system, but T_2 years ago the strontium in each of them was homogenized by some metamorphic event. The mineral isochrons of each rock are drawn through the mineral points (solid circles) and intercept the vertical axis at the homogenized $^{87}Sr/^{86}Sr$ values *A*, *B*, and *C*. Source: Lanphere et al. (8).

The ordinate intercept at *R* gives the initial $^{87}Sr/^{86}Sr$ ratio in the original environment $T_1 + T_2$ years ago.

It is possible to determine a "whole rock age" by using the entire rock. Figure 5.5 shows a whole rock isochron for three different rocks. Because each rock is homogenized, one obtains only a single point in the $^{87}Sr/^{86}Sr$ field. The second point needed to obtain the isochron comes from the "best value" (assumed or determined by other studies) for the initial or primordial isotopic ratio of $^{87}Sr/^{86}Sr$. It is implicitly assumed that this primordial value represents the time that strontium and other elements fractionated in the nebula to form the Solar System of the parent planetary body. Until very recently, the value of the ratio agreed on was 0.69898 ± 0.00003. This value known to the chronologists as BABI (basaltic achondrite best initial) has been derived from numerous analyses of achondritic meteorites, assumed to be samples of the primordial solids that differentiated shortly after planetary accretion. The whole rock age is also known as a model age, an age that defines the origin of the initial rock materials independent of the subsequent history.

The precision with which BABI has been determined and reported is a demonstration of the remarkable quality of the analytical work being performed in some of the dating laboratories.

Wasserburg and associates (1) have recently pointed out that the techniques that permit the precise dating of lunar rocks are the result of a rapid development, over the last decade, in mass spectrometric techniques, refined chemical and mineral separation procedures, and in mineral identification. While these methods have been applied most extensively to lunar materials, in the future they will be employed in a new generation of studies of the meteorites and terrestrial samples.

LUNAR CHRONOLOGY

Some applications of Rb–Sr evolution diagrams to lunar chronology are given below. It has already been stated that it is essential in every case to establish that the data represents a true isochron. It is also important to establish the initial $^{87}Sr/^{86}Sr$ ratio. If the point can be determined when the Moon was first formed, it then becomes possible to calculate model ages of samples relative to this initial lunar value. We have seen that events can occur at some time after the initial formation and produce isotopic homogenization and increased $^{87}Sr/^{86}Sr$ values for the whole rock.

Such increased values over the initial value have provided useful estimates of the time spent in a lunar resevoir prior to crystallization. The calculation of a meaningful model age requires that the rock have a high enough ratio of Rb/Sr so that the enrichment in ^{87}Sr relative to the ^{86}Sr is more than the possible range in the initial lunar values. Where the rock has been a closed system since the Moon was formed, then the age of the model will be that of the Moon.

Figure 5.6 shows an isochron for a typical mare basalt returned by the Apollo 11 mission. It can be seen that the data points lie on a straight line and the slope defines an age of 3.71 aeons—a time substantially lower

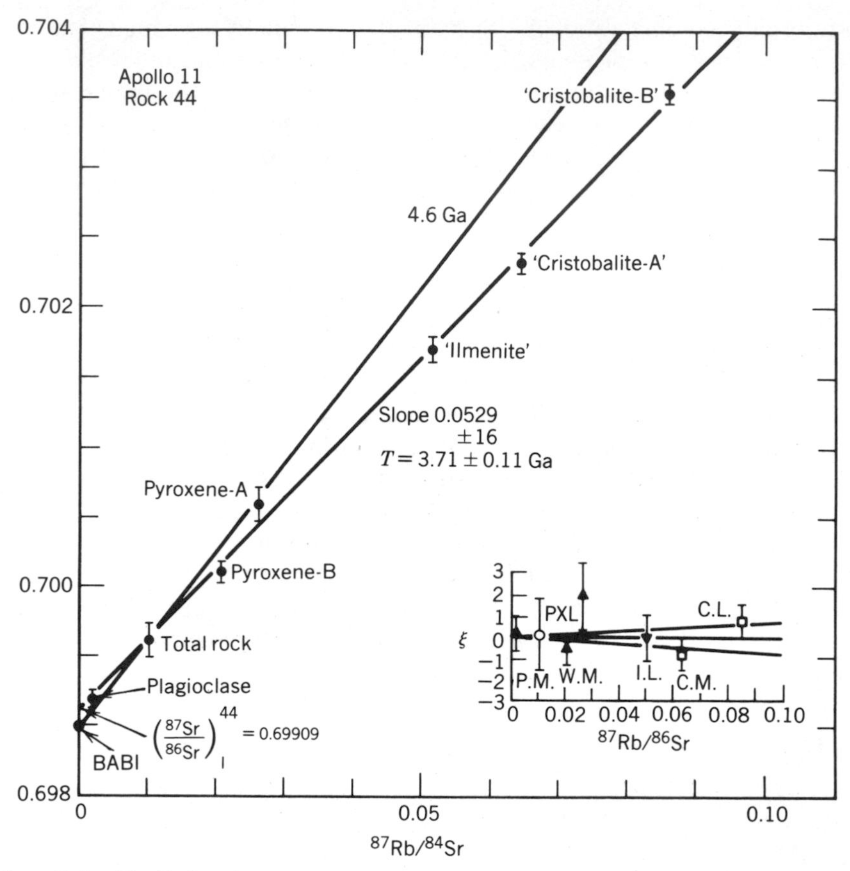

Figure 5.6. Rb–Sr data for a mare basalt from Apollo 11. The initial value is above BABI. A 4.6-b.y. line is drawn through BABI. The fractional deviation in $^{87}Sr/^{86}Sr$ ratio of the data points from the best-fit isochron are shown in the inset in units of 10^{-4}. Source: Wasserburg et al. (1).

than that assigned to the Solar System. The intercept on the ordinate is above the value for BABI and ADOR (Angra dos Reis). Wasserburg et al. (1) attribute the $^{87}Sr/^{86}Sr$ ratio to the fact that the rock came from a magma source region some time after the Moon's formation. The steeper line in the figure is the 4.6-aeon isochron. The insert is described as the fractional deviation in the $^{87}Sr/^{86}Sr$ ratio of the data points from the best fit isochron.

The determination of the ages of the lunar soils has had a considerable impact on the general understanding of the problems inherent in Rb–Sr geochronology. It was observed that the soils were made up of complex mixtures of components of varying ages. This became apparent from the determinations that the mare soils had model Rb–Sr ages around 4.6 aeons, although they were essentially derived from rocks whose crystallization ages from mineral isochron studies and independent $^{40}Ar/^{39}Ar$ data were 3.6–3.8 aeons.

Figure 5.7 is a Rb–Sr evolution diagram for lunar soils. These data are without direct age significance because the soils are complex mixtures of varying ages. Three different lines are shown beginning at BABI and with slopes corresponding to 4.3, 4.6, and 4.9 aeons. There is a strong tendency for the soil data to scatter around the 4.6-aeon line, and it appears that the material had an initial $^{87}Sr/^{86}Sr$ ratio like BABI. From these model ages it appears that the principal fractionation of Rb from Sr occurred at about 4.6 aeons, and that the amount of subsequent fractionation during the partial melting to form the mare basalts was smaller.

Strangely, during later missions it was found that some soils had model ages greater than 4.6 aeons. It has been observed that the apparent age increases with the amount of agglutination of the soil (a maturing process brought about by impacts). In view of the fact that there is such strong evidence that the Moon and Solar System were formed 4.6 aeons ago,

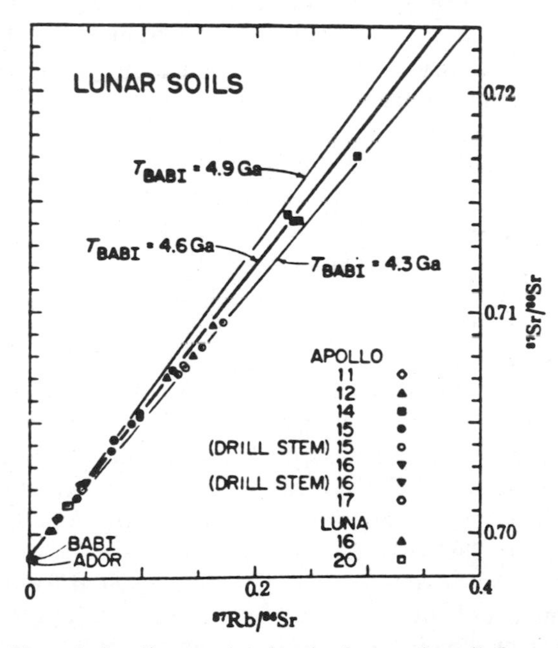

Figure 5.7. Sr–Rb evolution diagram showing the data on the soils from all lunar missions. There is a good linear correlation for these highly varied materials, but not a precise isochron. Source: Wasserburg et al. (1).

explaining the older ages of the soil required an alternate mechanism, particularly since there were no known rock types whose addition could explain the inordinately high model age.

The relationship between agglutination and ages has led to the viewpoint that the "delicate" relationship between the ^{87}Rb and the daughter ^{87}Sr might be occasionally disturbed, resulting in the small loss of the more volatile Rb relative to the Sr during glass formation. Such a phenomenon could cause greater apparent ages.

As Taylor states, "These problems with age dating of lunar soils provide a salutary lesson, not appreciated before the Apollo missions." The experience has made it possible to obtain meaningful ages from such specimens as clasts separated from soils and has provided a valuable lesson to be employed in future sampling missions.

The greatly enhanced interest in extraterrestrial materials experienced during the Apollo era and the development of new analytical techniques, as well as improvement in the existing techniques, has also led to the development of important alternate methods of geochronology. One such technique has already been referred to in the chapter on the meteorites—the Sm–Nd method. This approach to dating was first described by G. W. Lugmair in 1974 (6). The half-life for the decay of ^{147}Sm to ^{144}Nd by alpha decay is 1.06×10^{11} y. In view of the fact that both Nd and Sm are both rare earths and refractory, with similar chemical properties, it was anticipated that the two elements would be less susceptible to chemical fractionation than the Rb–Sr. Thus it was reasonable to assume that the parent–daughter relationship would yield useful radiometric ages, particularly where the Rb–Sr isochrons were ill defined.

In practice, the successful application of the Sm–Nd system did require particularly careful analysis because the variations in Sm/Nd ratio were very small, and the corresponding variations in radiogenic ^{143}Nd was only in parts per thousand. Figure 5.8 shows two internal isochrons determined on a low-potassium basalt. A Rb–Sr isochron is shown in (*A*) and, for comparison, the Sm–Nd isochron in (*B*). The results are in good agreement. In this particular instance, it was felt that the Sm–Nd method yields a more precise age because of the wider range of the $^{143}Nd/^{144}Nd$ ratio compared to the $^{87}Sr/^{86}Sr$ ratio. A third independent method, $^{39}Ar/^{40}Ar$, yielded, within experimental error, further confirmation.

In summary, the last two decades have witnessed a considerable advance in the science of rock dating. This has resulted from advances in mass spectrometry, chemical separation procedures, and clean-room capabilities, much of which has followed from the great interest in Solar System materials. In turn, the results of dating studies is not only making a large contribution to an understanding of the Solar System, but also to the general science of trace analysis.

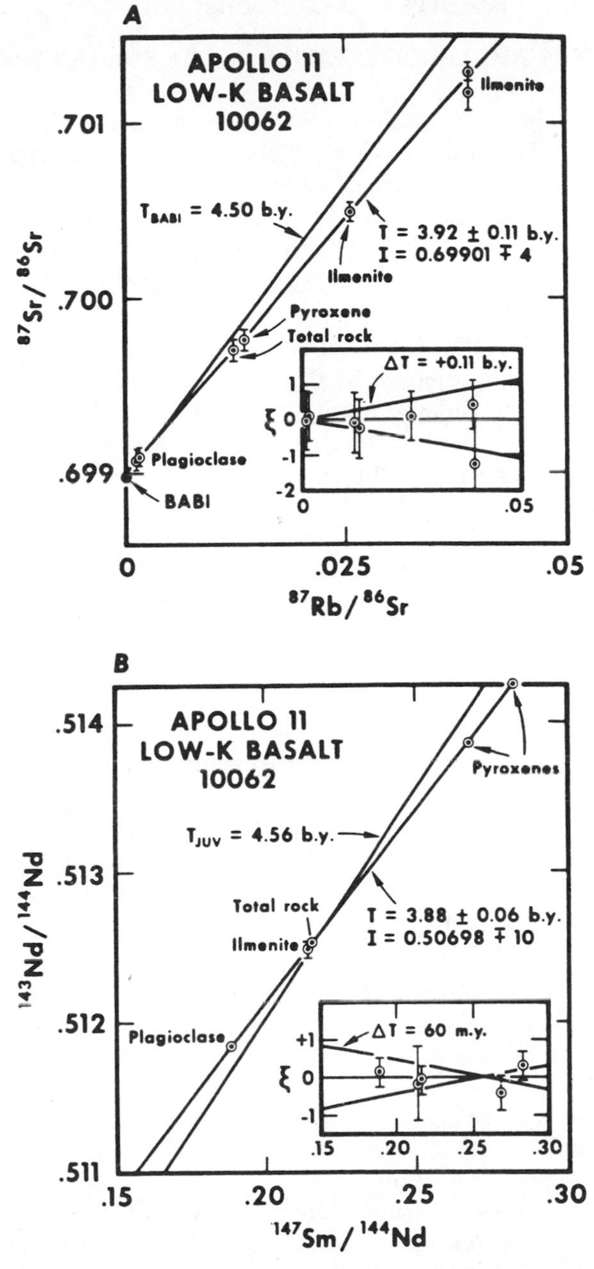

Figure 5.8. Two mineral isochrons for a low-K mare basalt (10062) by the Rb–Sr and Sm–Nd methods. The results from both isochrons are in good agreement. Source: Papanastassiou et al (9) and Guggisberg et al (10).

ISOTOPIC HETEROGENEITIES IN THE SOLAR SYSTEM

We have seen how the radioisotopes have been used in dating Solar System events. The study of isotopic abundances has a broader implication because in the recent years isotopes have revealed new clues to the formation of our Solar System. It appears that nucleosynthetic events occurred almost at the time of Solar System formation. Such information is recorded in unusual isotopic abundances and referred to as isotopic anomalies.

Prior to the discovery of these anomalies, the general impression of the Solar System was that there existed a very high degree of isotopic homogeneity. This was considered to be the situation for many elements measured in a variety of samples from the Earth, Moon, and meteorites. What made it so remarkable was the fact that the cosmic production of the different isotopes of the same element requires markedly different physical conditions and thus different probable sources. The appearance of homogeneity therefore pointed toward an efficient mixing process, which homogenized the different nucleosynthetic components before the planetary bodies formed.

Lee (2) in his review cites the case for barium, where ^{136}Ba is thought to have been produced by slow neutron capture (s process), while most ^{135}Ba is made by the more rapid neutron capture process (r process). The neutron densities required by these processes differ by 10^{20}, which certainly implies production at different sites, and yet it has been observed that the ratio of ^{135}Ba/^{136}Ba is homogeneous to within 0.01%.

There are well-understood variations in isotopic abundances due to modifications in the normal composition by Solar System processes; for example, mass-dependent fractionation such as diffusion, chemical reactions, and so on. Other processes are radioactive decay and nuclear reactions with energetic particles, such as those in the cosmic rays. However, variations that cannot be explained by the above processes are referred to as isotopic anomalies and are attributed to incomplete mixing of nucleosynthetic components.

The discovery of the isotopic anomalies is very recent—1973. One of the very significant observations reported by Clayton (11) concerned anomalies in the oxygen isotopic ratios. It was an example of the fact the isotopic compositions of major elements were not homogeneous in the Solar System. A search for other anomalies provided evidence for the decay of ^{26}Al (half-life about 7.3×10^5 y) into ^{26}Mg in the early Solar System. The relatively short half-life of the ^{26}Al implied that there were nucleosynthetic activities within about 3×10^6 of the formation of the Solar System. Recently, the discovery of ^{107}Ag excesses has been attributed to the decay of ^{107}Pd (half-life about 6.5×10^6 y) strengthens the above conclusion.

Additionally, there is the discovery of a rare class of objects that show correlated isotopic anomalies in many and sometimes all their constituents. This class of material has been given the whimsical acronym of FUN, F standing for large fractionation and UN indicating superposed, unknown nuclear effects. To date O, Mg, Si, Ca, Ba, Sr, Nd, and Sm anomalies have been found in the FUN samples.

There are a number of reasons for the success of the isotopic studies. One very important factor is the present availability of interesting samples, an example of which is the Allende meteorite, a 2-ton sample found in 1969. In particular, the refractory rich inclusions have proved to be a valuable source for study.

A second and very important factor, previously mentioned, is the great improvement in the capability of measuring isotopic compositions with high precision and high sensitivity. Tera and Wasserburg and co-workers (12) have shown that with modern mass spectrometers one can measure isotopic ratios with a precision exceeding 0.005% on samples as small as 10^{-7} g. For elements with high ionization efficiency, (Pb, for example) intermediate precision can be obtained on vanishingly small samples. Thus anomalies too small to be resolved 10 y ago can be detected today. Finally, a vital factor in such studies is the control of laboratory contamination. Such controls have been developed to an extraordinary degree, and some examples will be given below.

Tera and Wasserburg (11) have described a procedure for the precise isotopic analysis of lead in picomole and subpicomole quantities. They were able to obtain high precision isotopic analysis of as little as 8×10^{11} lead atoms. The level of lead contamination was reduced to as little as 4×10^{10} atoms with a controllable blank fluctuation. Where no preliminary chemical separation was necessary, 4×10^{10} atoms were isotopically analyzed with a blank correction of 6×10^9 atoms. In particular, this procedure was applied to a lunar sample of 0.2-mg size that contained 2.5 ppm lead and 1 ppm uranium. Finally, they were able to analyze lead on the surface of individual glass spheres about 150–300 μm in diameter.

In view of the substantial advance that the above work represents in the general area of trace analysis, it is worthwhile looking at some of the details of the procedure. Tera and Wasserburg determined the lead content of all the reagents being used and observed that reagent quantities of the order of 10 g would produce blanks of the order of about 10^{11} atoms. They concluded, therefore, that the only way to decrease the blank was to miniaturize the chemistry in order to minimize the amount of reagents employed.

In the chemical procedures, a study was made of the levels of lead contamination contributed by ion-exchange columns of different sizes. A comparison of the lead contamination contributed by the reagents with that found for the corresponding column procedures showed that the anion exchange resin itself contributed practically no lead blank, while the cation resin contributed a considerable amount of lead. Thus the principal reliance was on the anion resin in the lead separation procedure.

Lead is absorbed from a $CH_3OH–HNO_3$ solution on the anion exchange resin under conditions where major rock-forming elements and U are not absorbed. Small traces of Fe, Ba, Ca, and Ti are retained with the lead and cause serious suppression of the lead signal in the mass spectrometer. The lead is therefore further purified through the use of a microcation column, which also separates the lead from the thorium. Where samples less than 1 mg were used, either the cation or anion separation could be omitted.

More detailed descriptions of the microanion column and microcation column can be found in the original paper. For those practicing mass spectrometry, there is also a detailed description of filament loading. The authors have concluded that using the full microchemistry, good-quality data may be obtained on samples weighing as little as 10^{-4} g and containing 2 ppm of lead and 1 ppm of uranium. The same quality data can be obtained on samples weighing 10^{-7} g and containing 200 ppm of lead and 1000 ppm of uranium. The investigators also state that one can obtain data on a single particle weighing as little as 10^{-11} g.

Kelley, Tera, and Wasserburg (13) have reported on the isotopic determination of silver in picomole quantities by surface ionization mass spectrometry. Following separation from natural samples by anion exchange, they have mass-analyzed the concentrate using surface ionization mass spectrometry and silica gel as an emitter. They have measured isotopic compositions of 50 pmol of silver with a precision of 0.1%, using an electron multiplier detector. By employing stable isotope dilution 0.5 pmol (3×10^{11} atoms) was measured with a precision of better than only 1% and with a 30% blank correction, thus making accurate determinations of silver in natural samples down to 1 pmol/g.

The techniques and procedures touched on here have also been applied to the successful analysis of other heavy metals. Chen and Wasserburg (14) have described the isotopic determination of uranium in picomole and subpicomole quantities. Once again, careful attention was paid to such factors as contamination, chemical yields and purity, ionization efficiency, abundance sensitivity, instrumental fractionation, linearity of the detector, resolution shifts in the ratio of U-238/U-235, and reproducibility.

The message carried by this work is that the determination on ultratrace concentrations of elements can be performed with astonishing precision and sensitivity, provided that extraordinary care is exerted. It is of paramount importance to understand the significance of all of the parameters that bear on the validity of the results.

REFERENCES

1. G. J. Wasserburg, D. A. Papanastassiou, F. Tera, and J. C. Huneke, The Moon, *Royal Soc. London,* **285,** 7 (1977).
2. T. Lee, *Rev. Geophys. Space Phys.,* **17,** 7 (1979).
3. T. Kirsten, in S. F. Dermott, Ed., *The Origin of the Solar System,* Wiley, New York, 1978.
4. H. Faul, *Ages of Rocks, Planets, and Stars,* McGraw-Hill, New York, 1966.
5. S. R. Hart, *J. Geol.,* **72,** (1964).
6. G. W. Lugmair, *Meteoritics,* **9,** 369 (1974).
7. G. R. Tilton, *J. Geophys. Res.,* **65,** 4173 (1960).
8. M. A. Lanphere, G. J. Wasserburg, and A. L. Albee, Redistribution of Sr and Rb isotopes during metamorphism, in H. Craig et al., Eds., *Isotopic and Cosmic Chemistry,* North-Holland, Amsterdam, 1964, p. 269.
9. D. A. Papanastassiou, D. J. Paolo, and G. J. Wasserburg, *Proc. Lunar Sci. Conf. 8th,* 1639 (1977).
10. S. Guggisberg, P. Eberhardt, J. Geiss, N. Grögler, A. Stettler, G. M. Brown, and A. Peckett *Proc. Lunar Sci. Conf., 10th,* 1 (1979).
11. D. D. Clayton, *Icarus,* **32,** 255 (1977).
12. F. Tera and G. J. Wasserburg, *Anal. Chem.,* **47,** 13 (1975).
13. W. R. Kelley, F. Tera, and G. J. Wasserburg, *Anal. Chem.,* **50,** 9 (1978).
14. J. H. Chen and G. J. Wasserburg, *Anal. Chem.,* **53,** 13 (1981).

PART

2

CHAPTER

6

REMOTE ANALYSIS

The first part of this volume was devoted to the terrestrial analysis of extraterrestrial materials that have unambiguously come to us from outer space. On one hand, there are the meteorites, their source or sources the subject of considerable speculation. On the other hand, there are the samples returned from the Moon, an event of major magnitude in the history of science. The study of these samples has advanced our knowledge of the Solar System enormously—in fact, a quantum leap forward. Where the answers continue to elude us, the questions have been brought more sharply into focus and will serve as the basis for more refined investigations.

Further, as we have seen, the study of extraterrestrial materials has led to noteworthy advances in analytical techniques. It is now reasonably obvious that our failure to resolve or solve some of the existing questions is not due to the limitations of analytical methods. The controversies that abound are in the interpretation of the often excellent results. The reconstruction of the past is not a simple matter.

Part 2 of this volume will be devoted to extraterrestrial analysis. It will deal with remote analysis, frequently at enormous distances, as well as in situ analysis by a variety of unmanned probes. Unlike some of the methods described in Part 1, we shall now venture into an area where the methods are far less precise, although remarkably ingenious and representative of extraordinary accomplishments. In some of the illustrations of these methods we shall see that a great deal of circumspection is required in interpreting the results, to avoid pushing the results beyond reasonable constraints.

SOLAR WIND COMPOSITION

The first experiment to be described here may be thought of as a hybrid experiment in that it involved a controlled and systematic collection of extraterrestrial material on the lunar surface, followed by subsequent analyses in terrestrial laboratories. The experiment was called the Solar

Wind Composition Experiment by J. Geiss and co-workers (1). This was a unique experiment, frequently described as simple and elegant, which was deployed for a period of time on the lunar surface to collect and return a sample of the solar wind for subsequent study.

The solar wind has been described as a stream of charged particles, mainly hydrogen, which moves out from the sun with velocities of the order of 300–500 km/sec. In the vicinity of the Moon, the average density is about 10 ions/c. During quiet Sun conditions the energies of these particles is modest (kilovolt range). Geiss has described the background for the experiment. It has been established that the helium ions are present in the solar wind, and the relative abundance of the ions is highly variable. He/H ratios of 0.01 to 0.25 had been observed, but the average He/H ratio in the solar wind was approximately 0.04 to 0.05.

From plasma and magnetic field measurements during Explorer 35, it was established that the Moon behaved as a passive obstacle to the solar wind, and as a consequence, during the normal lunar day the solar wind particles could be expected to reach the lunar surface of the Moon with unchanged energies. It was expected that the grains of the fines on the lunar surface would contain large amounts of the particles, and the analysis of the lunar surface material would yield valuable information on the composition of the solar wind. However, as Geiss pointed out, the dust on the lunar surface would be a solar wind collector of uncertain properties. Accordingly, an experiment was designed to sample the solar wind for a defined period of time. The collector (described below) was a foil with well-understood trapping properties. Excitingly, such an attempt had never been made before.

PRINCIPLES

The experiment to be described is remarkable for its simplicity. A sheet of aluminum foil 30 cm wide and 140 cm long with an approximate area of 4000 cm^2 was carried to the lunar surface by the Apollo 11 crew and there exposed to the solar wind. The foil was placed so that its position was perpendicular to the solar rays. The total exposure time was 77 min, and then the foil was ultimately returned to the Earth. Like other lunar samples, the foil was returned to quarantine in the Lunar Receiving Laboratory before release for study. It was expected that the solar wind particles, arriving with energies of a 1 keV/nucleon, would penetrate only the topmost portion of the foil (1/10,000 cm) and that a large percentage of the particles would be trapped. Upon release from the Lunar Receiving Laboratory, the foil was analyzed for trapped solar-wind noble gas atoms.

Parts of the foil were melted in an ultrahigh vacuum system, and the noble gases of solar wind origin were analyzed by mass spectrometry for both elemental abundance as well as for isotopic composition. Thus, if during the exposure period, the solar wind reached the foil with an intensity comparable to average flux values, the Solar Wind Collector Experiment would allow an assessment of the abundance and isotopic compositions of He, Ne, and Ar, despite the relatively short collection time.

INSTRUMENTATION

Figure 6.1 shows the nature of the hardware that was developed to deploy the foil on the lunar surface. The equipment consisted of a five-section telescopic pole and an aluminum foil screen rolled up on a reel. On the way to the Moon, the reel was stored inside the collapsed pole. The entire weight of the equipment was about 425 g. At deployment, the telescopic pole was extended and the five sections automatically locked. The reel was then pulled out and the foil unrolled and fastened to a hook near the lower end of the pole. For ease in rewinding, the reel was spring-loaded. On deployment, the pole was pressed into the ground as shown in Figure 6.2. During the Apollo mission, the Sun was about 13° above the lunar horizon. Following a collection time of 77 min, the foil was rolled back on the reel. The foil was then detached from the telescopic pole and placed in a Teflon bag for transport back to the Earth in the Apollo lunar sample return container described previously (Chapter 3).

The ever-present concern about the possibility of contamination, in this instance, the lunar samples by the Solar Wind Experiment, was dealt with in the following manner: All the materials used in the construction of the reel and foil assemblies were analyzed for 13 geochemically significant trace elements Li, Be, B, Mg, Rb, Sr, Y, La, Yb, Pb, Th, and U. All concentrations were found to be low enough to be geochemically acceptable. To avoid organic contamination of the lunar surface material in the container, the instrument was subjected to several heating cycles and the degree of contamination checked by mass spectrometry. Finally, prior to shipment to NASA's Kennedy Center for installation in the Lunar Module, the instrument was put into a double bag of Teflon and heat-sterilized.

The details and dimensions of the exposed foil are shown in Figure 6.1. The assembly consisted of a 15-μm-thick aluminum foil. The backside of the foil was anodically covered with approximately 1-μm of aluminum oxide to keep the foil temperature below 100°C during the lunar exposure.

Test pieces 1, 3, and 5 were foils that were bombarded in the laboratory before the mission with a known flux of neon ions having an energy of 15

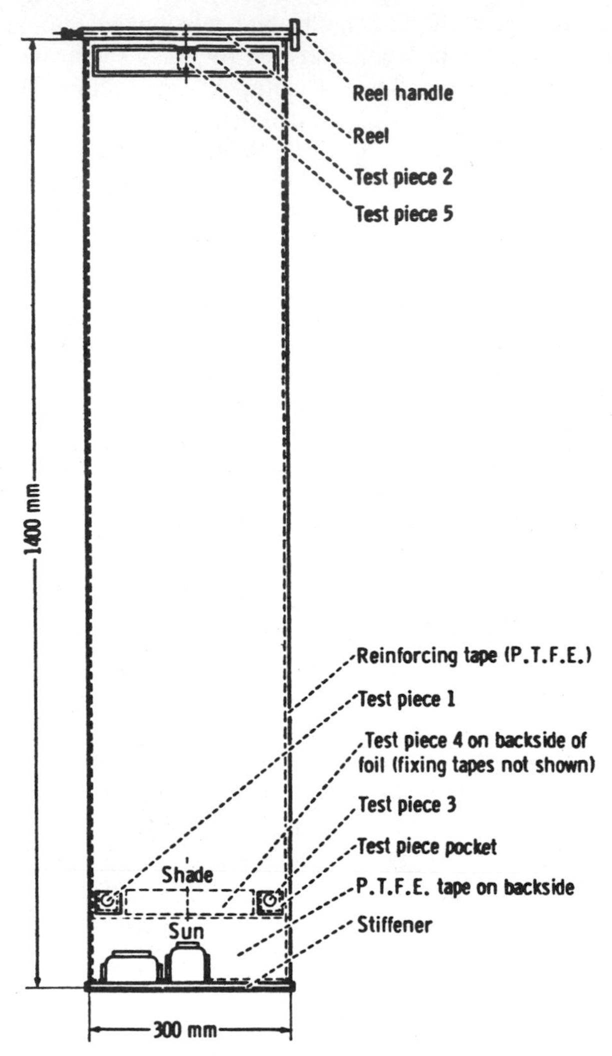

Figure 6.1. Dimensions and details of the exposed Solar Wind Experiment exposed foil assembly. Source: Geiss et al. (1).

keV. The amounts of neon were large by comparison with the expected solar wind neon fluxes. Determination of the amount of neon implanted in these foils were used to determine if solar-wind noble-gas losses occurred during the mission by such processes as unexpected diffusion or surface erosion. Test piece 5 was mounted in a closed pocket and remained shielded during exposure on the lunar surface; test pieces 1 and 3 were

Figure 6.2. The Solar Wind Experiment deployed on the lunar surface. Source: Geiss et al. (1).

exposed to the solar wind. Test piece 4 was a piece of foil taped to the backside of the main foil and used to distinguish a solar wind flux coming from a direction opposite to the Sun. Test piece 2 was mounted in a position that remained shielded from the solar wind.

Before flight, the trapping properties of the foil for kiloelectron-volt ions were investigated in the laboratory. An example of the trapping probabilities were found to be 89% for He (3 keV), 97% for Ne (15 keV), and 100% for Ar (30 keV). The trapping probabilities were found to be independent of foil temperature, and they were not affected by simultaneous bombardment with keV hydrogen ions.

INITIAL RESULTS

The Solar Wind Experiment was kept behind the biological barrier in the Lunar Receiving Laboratory until the quarantine was lifted. However, a portion (1 ft^2) was cut from the midsection of the foil and sterilized inside a high-vacuum container at 125°C for some 39 h. It was then sent to the laboratory of the principal investigator for study. Mass spectrometric analysis of the gas content of the container revealed little air leakage, as shown by the very low Ar content. Small concentrations of He (10^{-16} cc STP) and Ne (10^{-8} cc STP) were found. These were attributed to the lunar dust attached to the foil. While no dust could be seen with the naked eye, microscopic and scanning electron microscopic images did show that a substantial amount of fine dust was added to the procedure for handling the foils. The noble gases extracted from the foil were analyzed in the mass spectrometer. He, Ne, and Ar were found, and the isotopic composition of these elements was determined. A preliminary report stated that the abundances were decidedly nonterrestrial and corresponded to abundance estimates for the Sun.

The Solar Wind Composition Experiment proved so successful that it became a staple of the lunar landing program. A summary was written by Geiss and associates (2) immediately after the Apollo 16 landing at Descartes. The experiment was essentially the same on all the missions, except that on the Apollo 16 flight some platinum test foils were also flown. It had been found that for platinum only a small percentage of the implanted ions would be removed if the foils were washed for several minutes in 2–10% HF. Such treatment would, however, remove a high percentage of implanted noble gas from silicates and other constituents of the lunar fines. This procedure was used to eliminate any possible interference from residual lunar dust adhering to the foils.

SUMMARY

The Solar Wind Composition Experiment, despite its basic simplicity, has provided a great deal of useful information with respect to noble gas abundances in the solar wind—such information as isotopic ratios and some idea about solar behavior. The measurements showed that the He/H of the solar wind (0.04) is much lower than that found in the solar photosphere (about 0.1). It is apparent that some solar wind fractionation occurs with respect to the outer surface of the Sun, but the mechanism is unknown.

A comparison of the solar wind data to the lunar surface and terrestrial atmosphere data is shown in Table 6.1. There is a difference between the solar wind and the lunar fines, which has been attributed to diffusive losses from the soil, although with some retrapping. There is also an observed difference in $^{20}Ne/^{22}Ne$ ratio between the solar wind and the terrestrial atmosphere. Finally, it has been observed that the deuterium/

Table 6.1. Comparison of Weighted Averages of Solar Wind Ion Abundances Obtained from the Solar Wind Collection Experiment[a]

Source	$^4He/^3He$	$^4He/^{20}Ne$	$^{20}Ne/^{22}Ne$	$^{22}Ne/^{21}Ne$	$^{20}Ne/^{36}Ar$
Solar wind (average from 5 SWC experiments)	2350 ± 120	570 ± 70	13.7 ± 0.3	30 ± 4	28 ± 9
Lunar fines 10084	2550 ± 250	96 ± 18	12.65 ± 0.2	31.0 ± 1.2	7 ± 2
Ilmenite from 10084	2720 ± 100	218 ± 8	12.85 ± 0.1	31.1 ± 0.8	27 ± 4
Ilmenite from 12001	2700 ± 80	253 ± 10	12.9 ± 0.1	32.0 ± 0.4	27 ± 5
Ilmenite from breccia 10046	3060 ± 150	231 ± 13	12.65 ± 0.15	31.4 ± 0.4	—
Terrestrial atmosphere	7×10^5	0.3	9.80 ± 0.08	34.5 ± 1.0	0.5

[a] Abundances of surface-correlated gases in lunar fines materials, and a breccia, and in the terrestrial atmosphere. Source: Geiss and associates (1).

hydrogen ratio is quite low. The Sun appears to have very little deuterium. Current steller models have the Sun converting the deuterium to ^3He. There is, however, still a large question in view of the deficiency in the $^3He/^4He$ in the solar wind. The ratio is at least four times less than that indicated for the Earth's deuterium/hydrogen ratio. An explanation offered by Geiss and Reeves (3) is that deuterium may have been enriched in the planets relative to the Sun early in the formation of the Solar System.

REFERENCES

1. J. Geiss, P. Eberhardt, P. Signer, F. Buehler, and J. Meister, *Apollo 11 Preliminary Science Report,* NASA SP-214, 1969, p. 183.
2. J. Geiss, F. Buehler, H. Cerutti, P. Eberhardt, and Ch. Filleux, *Apollo 16 Preliminary Science Report,* NASA SP-315, 1972, p. 14-1.
3. J. Geiss and H. Reeves, *Astron. Astrophys.,* **18,** 126 (1972).

CHAPTER

7

REMOTE SENSING BY REFLECTANCE SPECTROSCOPY

One of the important challenges in the program of planetary exploration was the development of a method for obtaining chemical data from planetary surfaces over great distances. The answer obviously lay in spectral methods. The possibility of employing spectral analysis to obtain mineralogical information from lunar and planetary surfaces had been proposed by a number of investigators such as Lyon (1), Hovis and Lowman (2), and Aronson and associates (3). An extensive study was done for NASA by Lyon (4) at the Stanford Research Institute and published as NASA Tech. Note D-1871. The study included absorption studies of 370 rock and mineral samples and reflection studies of about 80 rocks. The spectral region used was the infrared from 2.5 to 25 μm.

Eventually, Lyon (4) published results for over 330 normal emittance, reflectance, and transmittance spectra of roughened rock and mineral surfaces covering the range from 2.5 to 25 μm. He drew the following conclusions from these studies: Rock and mineral types can be determined from the wavelength position of the minima in the absorption or emission spectra. Further, the details of the shape, intensity, and position of individual maxima or minima could be used to infer differences between mineral groups and mineral assemblages in rocks. Of the various kinds of spectra, absorption spectra had the highest spectral contrast and were the most definitive. On the other hand, he found the emission method most suitable to field operations, particularly from orbit. Aronson and co-workers proposed the use of the near and far infrared (15–500 μm) for the remote sensing of lunar and planetary surfaces. At the time, they attributed the neglect of this spectral region to the inherently poor signal-to-noise and stated that the far infra region is exceedingly rich in information about silicate minerals.

Hovis and Callahan (5) showed that the restrahlen (residual rays) of silicate minerals vary in frequency with the concentration of the silicates, and that the restrahlen peaks vary from 8.5 to 11 μm, depending on the types of rocks. Goetz and Westphal (6) described a method for measuring spectral emissivity differences between two points on the lunar surface in the 8–13-μm region from a terrestrial observatory. They published the results of a lunar survey involving some 22 lunar sites and reported that most of the lunar regions surveyed appeared to be homogeneous in this spectral range. This was attributed to a surface roughened by either micrometeorite bombardment or radiation effects. Only two of these points, Plato and Mare Humorous, showed spectral contrasts that might be due to compositional differences.

A further consideration of spectral techniques led Adams (7) and McCord (8) to propose high-resolution visible and near infrared reflectance spectroscopy, a technique developed prior to the Apollo missions. This approach has proved to be particularly fruitful. The method has evolved steadily, and, with the passage of time, has become more effective, as improved instrumentation and laboratory methods have developed. Both investigators proposed that spectral reflectance data in the 0.3–2.5-μm region would yield geologically significant information. This has proved to be the case, particularly in lunar studies where not only the mineralogy and composition of the surface has been discerned but the effects of the soil-forming processes as well.

The Moon proved unique in this general program because with the Apollo 11 landing and sample return Adams and McCord were able to obtain ground truth data (7,8) and demonstrate that the reflectance spectrum of the soil collected by Armstrong and Aldrin matched the spectra obtained from telescopic observations of the Apollo 11 landing site and several surrounding areas. Adams and McCord's early results showed that there is a lateral compositional homogeneity on a kilometer scale; thus, they were able to extrapolate the conclusions, based on spectra from centimeter-sized samples to unvisited portions of the lunar surface.

The advances in the spectral methods for obtaining data about planetary surfaces, in particular the asteroids, has been summarized in a paper by Gaffey and McCord (9). It was pointed out that in 9 y since the first Tucson Conference on asteroids, there has been a great expansion in the methods for characterizing asteroid surface materials. The data base has expanded in spectral range and resolution and the number of objects observed. They state that techniques and calibrations for interplanetary data have undergone a quantum jump in sophistication.

In 1970, only color comparison photometry based on meteorite samples was available. After 1970, spectral reflectance measurements of laboratory samples with broader spectral coverage and higher resolution became available for meteorites such as basaltic achondrites. In the early 1970s, such investigators as Johnson and Fanale (10), Chapman and Salisbury (11), and Gaffey (12) did a number of reflectance spectral studies on meteorites. A number of investigators such as Adams (13,14), Adams and Goullaud (15), and Bell and Mao (16) related specific spectral properties such as absorption band positions to mineral composition.

There has been a steady growth in the data base dealing with meteorites and other Solar System objects, in addition to the improved understanding of the relationship between spectral observations and mineralogical properties. Particular emphasis was placed on interpreting the data from asteroids. Gaffey and McCord (9) caution that "Care should be exercised in reading the early papers since subsequent improvements in techniques or observational data may have significantly modified their conclusions." There are numerous examples of the early efforts supplied to asteroids that have been summarized by Gaffey and McCord. A few of these are cited below:

McCord, Adams, and Johnson (17) measured the 0.3–1.2-μm spectrum of the asteroid Vesta and identified an absorption feature near 0.9-μm as being due to the mineral pyroxene. They concluded therefore that the surface material was similar to basaltic achondritic meteorites. Chapman and Salisbury (11) found matches in the 0.3–1.1-μm spectra of asteroids and several meteorite types such as enstatite achondrites, a basaltic achondrite, an optically unusual ordinary chondrite, and perhaps a carbonaceous chondrite.

Johnson and Fanale (10) showed that the albedo and spectral characteristics of some asteroids are similar to C1 and C2 carbonaceous chondrites and others to iron meteorites. However, they noted the difficulty of defining a spectral match and raised the question of subtle spectral modification as a consequence of "weathering" of asteroid surfaces. In turn, Salisbury and Hunt (18) considered the effects of terrestrial weathering on meteorite samples and questioned the validity of matches between the spectra of such specimens and the asteroids.

Chapman and associates (19) developed a scheme for classifying asteroids based on spectral data, albedo, and polarization parameters. Most of the bodies fell into two groups, the first having low albedos, strong negative polarizations at small phase angles, and a relatively flat, featureless reflectance curves in the 0.3–1.1-μm region. The second asteroid group had higher albedos, and sometimes the spectral curves showed distinct spectral characteristics. Attempts have been made to relate these to different meteorite types, but such arrangements have led to ambiguities and misunderstandings.

Having discussed, at least briefly, the historical development of reflectance spectroscopy and its use in planetary studies, we shall now examine the principles and applications.

SPECTRAL REFLECTANCE (THEORETICAL BASIS)

The spectral reflectance technique has been defined by McCord (8) as consisting of the measurement of the fraction of incident solar radiation which is reflected from an object as a function of wavelength. The method is based on the wavelength dependence of the spectral reflectance, and not the albedo or absolute amount of reflected radiation. The resultant reflectance spectrum contains optical absorption bands that are diagnostic of surface composition and mineralogy.

It is obvious that when such measurements are made from terrestrial observatories, the useful wavelength range is governed by atmospheric transmission. A second constraint is that the wavelength be in a range where the reflected solar radiation is more intense than the emitted thermal radiation. In addition, there are practical limitations such as those imposed by detector technology and calibration techniques.

Figure 7.1, published by McCord (20), shows the spectral flux from a square kilometer of a lunar mare region. The radiation is shown to consist of two components: the reflected solar radiation and radiation that has been absorbed and reemitted as thermal radiation. The intersection of the two curves and the relative magnitudes of the curves vary for different planetary bodies, depending on their orbits.

McCord states that the material on the surfaces of Solar System objects such as Mercury, the Moon, Mars, and the Asteroids are composed of randomly oriented fragments of crustal material. When these materials are observed in reflected light, the soils return two components of radiation: a specular first surface reflection and a diffuse component, composed of light that has entered at least one grain and then been scattered back toward the observer. It is this diffuse component that contains the most compositional information.

The absorption phenomena involved in the formation of the spectral features that contain the characteristic compositional information being sought have been thoroughly described in a recent paper by McFadden and associates (21). Absorption in the UV–visible region of the spectrum can be ascribed to four different phenomena: color centers, conduction bands, crystal field transitions, and intervalence charge transfer of ions. These will be described here based on McFadden's treatment of the subject.

1. *Color Centers.* Trapped electrons in crystal structure lattice defects produce color centers. Naturally occurring color centers are commonly found in high-symmetry crystals where lattice defects are easily

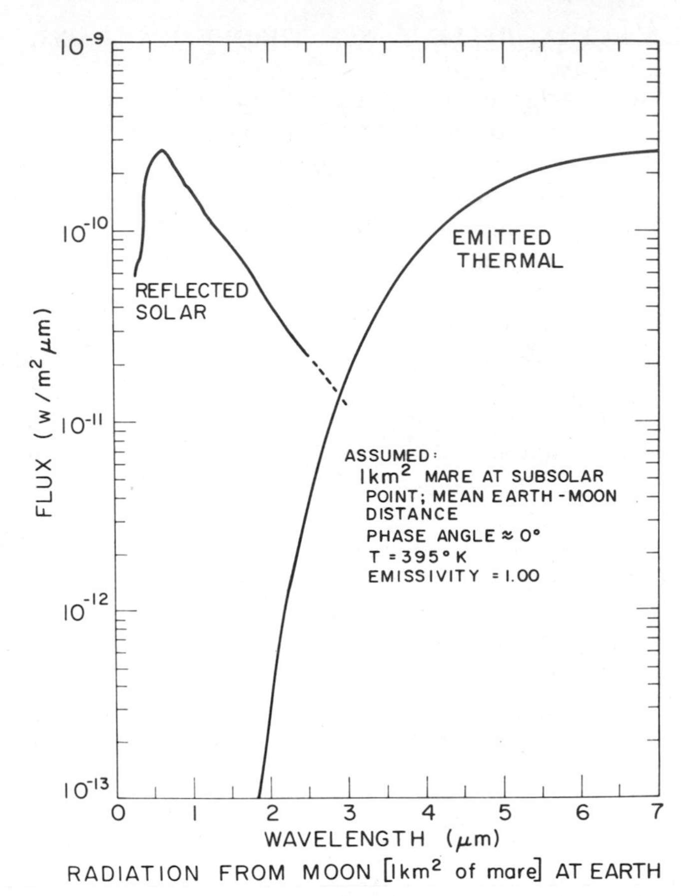

Figure 7.1. Radiation received from a 1-km² mare area at the mean Earth–Moon distance. Source: McCord and Adams (20).

produced. When an impurity is present, which produces a charge imbalance in a crystal lattice, the electrons are bound by the crystal fields of the surrounding ions to preserve electrical neutrality. Figure 7.2*a* (21) is a schematic representation of a point defect at a lattice site, a condition under which color center absorption may occur. In Figure 7.2*b* we see examples of color center absorption for some fluorite specimens. The unambiguous identification of a color center feature can be difficult, particularly in studying a remote object. It requires eliminating other types of absorption and independent evidence that conditions promoting lattice defects exist. It must be added that as yet no color center absorptions have been observed on planetary surfaces.

2. *Conduction Bands.* The effect of the conduction band depends on whether one is dealing with a semiconductor or a conductor. The energy required for transitions between some valence and conduction bands (the forbidden gap) in semiconductors is equivalent to photon energies in the UV–visible region. The wavelength of onset of the band depends on the particular conductor. The onset of the band is quite abrupt for pure semiconductors, but becomes less so with an increase of impurities. Conductivity is proportional to temperature, so that a decrease in temperature leads to a lower conductivity and, consequently, a decrease in band intensity.

In the case of conductors, the valence and conduction bands overlap; thus there is no forbidden gap and absorption is continuous.

Opaque minerals are conductors or semiconductors which totally absorb photons at all the wavelengths observed. The most common Solar System opaques include elemental carbon, magnetite (spinels), and fine-grained metallic particles, which have been reported to have oxide coatings (10, 11). These minerals yield important information about rock formation such as oxidation state.

A mechanism that results in total absorption is the formation of energy bands at ions instead of discrete energy levels. As an example, when anions with a full orbital energy state interact with cations with empty adjacent orbitals, they overlap to form energy bands; therefore photons are absorbed at all optical wavelengths. This phenomenon is common to oxides. The opaque nature of magnetite is due to a highly efficient charge transfer (24).

Figure 7.3 shows the absorption effect of increasing proportions of the opaque mineral magnetite on the crystal field absorption of olivine. Figure 7.4 shows the effect of crystal field splitting of an ion in an octahedral field. The spectrum of elemental carbon and small-to-large grains of magnetite is featureless. Submicron-sized magnetite becomes transparent in the infrared. McFadden states that, unfortunately, existing interpretive techniques of visible and infrared cannot distinguish between opaque types. However, it is important to derive methods of remotely determining the mineralogy and chemistry of the opaques because of their value in determining formation conditions, that is, the oxidation states and chemistry.

It has been found that the addition of small amounts of an opaque component, as little as 1–5%, such as carbon or magnetite to a mafic silicate assemblage drastically reduces the strength of any absorption and its albedo. Such effects can also be caused by shock, but the effects of the opaques is much more marked.

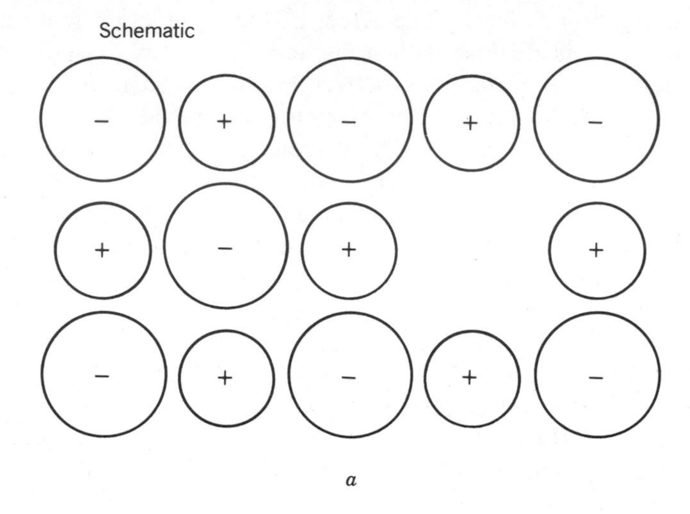

a

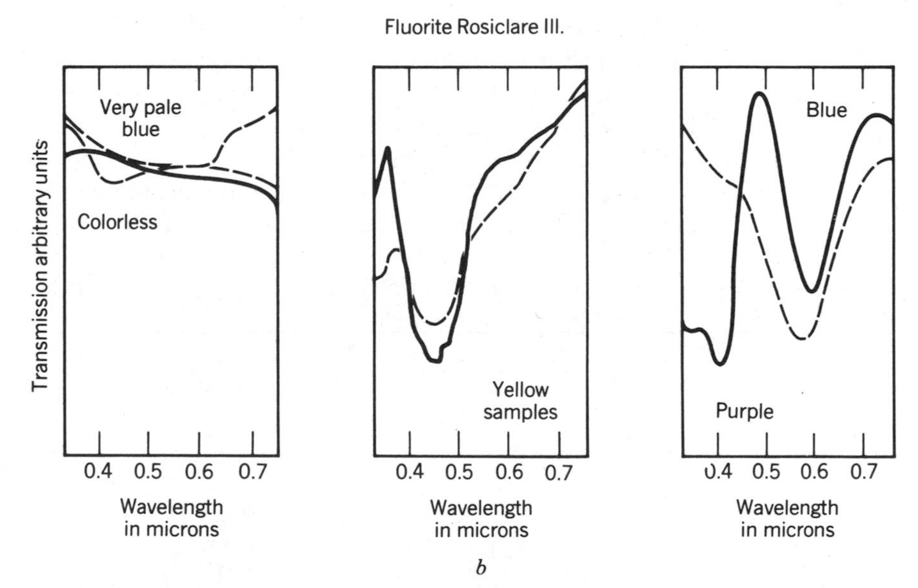

b

Figure 7.2. Schematic representations of a point defect and the situation for a semiconductor. Source: McFadden et al. (21).

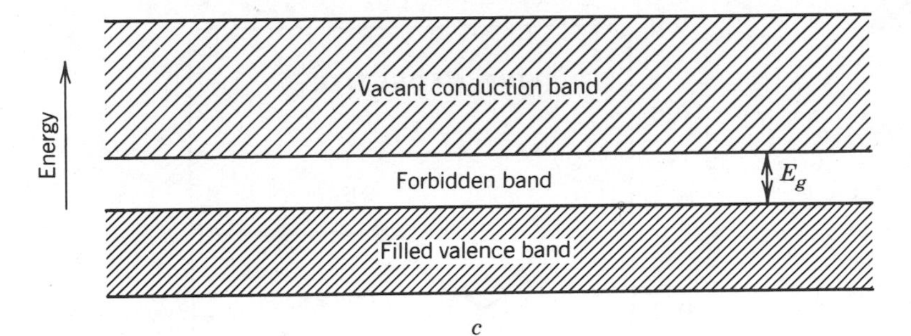

c

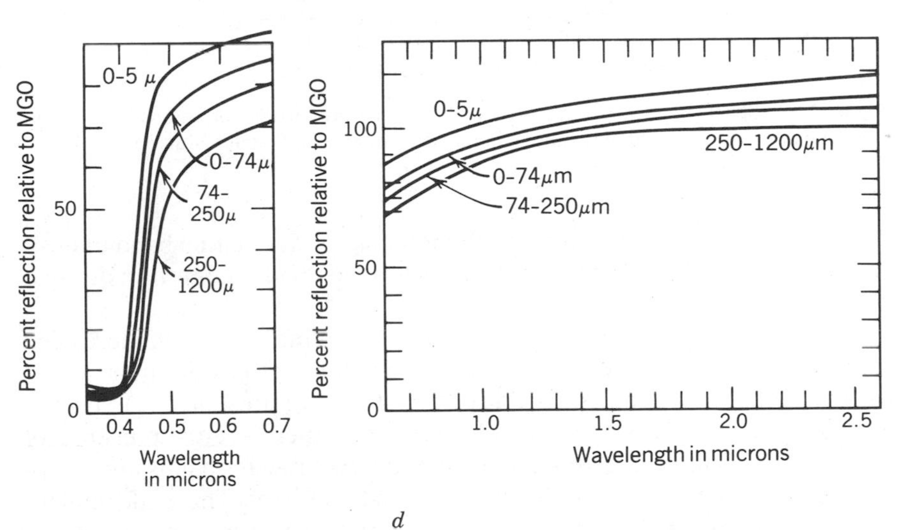

d

Figure 7.2. (*Continued*)

3. *Charge Transfer Absorptions* These absorption features originate from the transfer of electrons between two valence states of adjacent ions. This type of absorption is the strongest and, in certain cases and with additional knowledge of spectral parameters, can be assigned to specific ion interactions. However, because of the short wavelength cutoff of Earth-based measurements, charge transfer absorptions are usually observed as absorption edges, making it difficult or impossible to determine band centers and to assign specific ion interactions.

With a decrease in temperature and increase in pressure, the charge transfer bands increase in intensity. Thus a reflectance spectrometer mounted aboard a spacecraft could measure the same area of a surface at

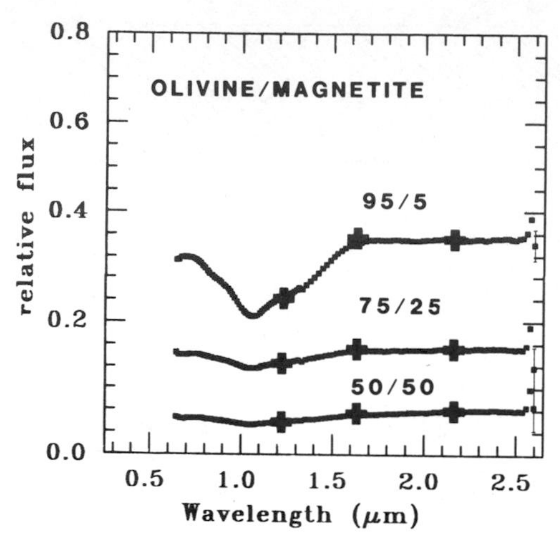

Figure 7.3. Spectra showing the effects of increasing proportions of an opaque material (magnetite) on crystal field absorptions in olivine. Source: McFadden et al. (21).

different times during a diurnal cycle, looking for fluctuating band intensities, and distinguish between conduction bands and charge transfer features.

While the conduction band would increase in intensity with increasing temperature, the charge transfer feature would decrease in intensity. If both a conduction band and a charge transfer were present, the expected intensity variation would not be observed, and therefore the presence of both bands could be inferred. At present, charge transfer bands are identified on the basis of position, bandwidth, and intensity. This is dependent then on mineralogy, temperature, and abundance. The charge transfer band is nonlinear and has a characteristically concave shape.

ABSORPTION BAND STRENGTHS

The proper interpretation of absorption band features depend strongly on an understanding of the factors affecting them. Such an understanding has flowed from laboratory experiments on "common" Solar System materials combined with theoretical modeling based on physical scattering phenomena. The following factors have been observed to be significant: the nature of the crystallographic sites, chemistry (oxidation states and abundance), particle size of the samples, morphology, physical packing, and, finally, the viewing geometry. The alteration of crystal structure or

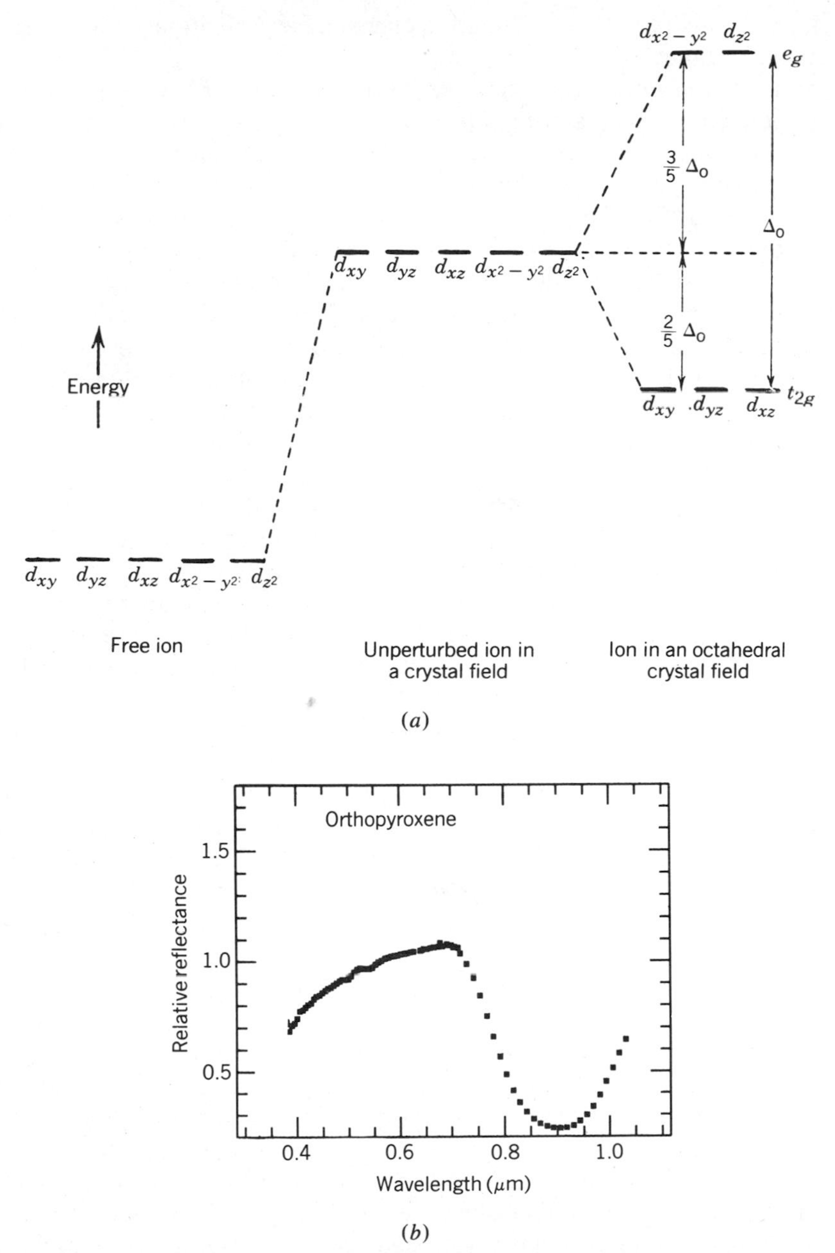

Figure 7.4. (*a*) Schematic representation of crystal field splitting of an ion in an octahedral field. (*b*) Spectrum showing UV charge transfer band and crystal field band at 0.9 μm. Source: McFadden et al. (21).

surface chemistry will be reflected in changed band positions and strength in the reflectance spectrum.

As a case in point, a single spectrum of an asteroid surface will not permit the controlling factors of physical scattering to be determined. The mineralogical composition or physical state must be independently determined. It was stated above that this has been done by laboratory measurements on Solar System materials, their mixtures, and naturally occurring Solar System assemblages such as meteorites (12,22) and lunar samples (23). In her review, McFadden points out that interpretation is not always unique and that the total number of possibilities need to be considered. Independent measurements of albedo, polarization, and radar can occasionally provide important constraints (21). In addition, combinations of data sets from different portions of the spectrum that measure different physical phenomena may serve to provide limits of physical parameters.

Experimental observations have also shown the effects of viewing geometry, packing density, particle size, shock, and temperature. Of the above, the first three are scattering processes. By contrast, temperature and shock produce changes in the structure of the minerals, though some shock effects are related to scatter. Significantly, examination of variation of the UV absorption band strength indicates that the maximum range of variation is no greater than 10%, with particle size and shock two exceptions that are discussed later.

UV band strengths of this magnitude have not been used to date for quantitative constraints on physical parameters of Solar System surfaces. The controlling factors of the UV band depth are multiple, and variations less than 5% cannot be separated from observational errors. Variations greater than 10% are attributed to compositional differences in the abundance of an absorbing species, particle size, or shock effects. These parameters will be discussed in greater detail based on the McFadden review.

The intensity of reflected and absorbed radiation depends on the mean optical path length (mopl) of photons through the sampled material, a value proportional to the probability of photon absorption. Composition, viewing geometry, individual grain morphology, and overall texture affect the value of mopl, which is also wavelength dependent. Almost all materials show an increased reflectance relative to a standard wavelength (about 0.56-μm) with increasing wavelength and phase angle (24). Band depth values relative to 0.56-μm vary nonlinearly with phase angle. To date, phase-angle effects have not been applied to the interpretation of reflectance spectra. However, measured UV absorption coefficients of Io and Europa have been used to place quantitative constraints on their relative porosities.

If a material has a preferred orientation, its particle shape will affect the mean optical path. Such situations are rarely expected in planetary surfaces, because few monomineralic surfaces are known. Effects of particle shape should be randomized when looking at real surfaces.

Effects of particle packing or porosity have been observed for a few materials (25). In the visible portion of the spectrum, a powder's albedo decreases with increased porosity (reduced density), but the opposite effect is seen in the near-infrared where reflectance increases with reduced density. This behavior can be understood in terms of shadowing, but the material must be more transparent in the IR than in the visible. A wide range of materials have not been observed for packing-density variations, due to a large extent to the difficulty in measuring packing density quantitatively.

The particle-size effect is a substantial one, which has been explained in the following terms (22). As the particle size decreases, the number of grain boundaries increase relative to particle volume, and the reflectance increases. For a given mineral, there is a particle-size distribution where spectral contrast is maximum, that is, the specular scattering relative to the diffusely reflected component and wavelength-dependent absorption results in the largest difference between in band absorption and out-of-band reflectance. With increasing particle size beyond the optimum distribution, specular scattering decreases, absorption increases, and the apparent band depth decreases. Figure 7.5*a, b,* and *c* show the effects of particle size on the spectrum of an enstatite pyroxene (22). It can be seen that between <30, 150, and 200-μm particle sizes, the albedo changes by more than 100% at 0.56-μm, and the 0.9-μm relative band depth changes by 100%. The relative strength of the UV absorption changes by 50%.

A functional relationship between albedo and band depth could determine particle size. However, conditions on such bodies as asteroids do not lend themselves to the establishment of these relationships. The combination of more than one mineral component and particle size are not independent functions. Thus an independent method of determining particle is necessary in order to maximize the interpretative potential of the spectral reflectance method. Progress towards developing such a method has been made and described by Clark and Roush (26).

In addition to viewing geometry, the question of gross shape of the body being measured is also of concern. Gradie and Veverka (27) have shown that measurement of integrated reflectance from an asteroid surface depends on shape. Based on calculations of the effect of body shape, it has been shown that the spectral differences are only of the order of about 6% at the extremes of an ellipsoid, with a major to minor axis ratio, $a/b = 3$. It appears that anything but extreme shape differences would not

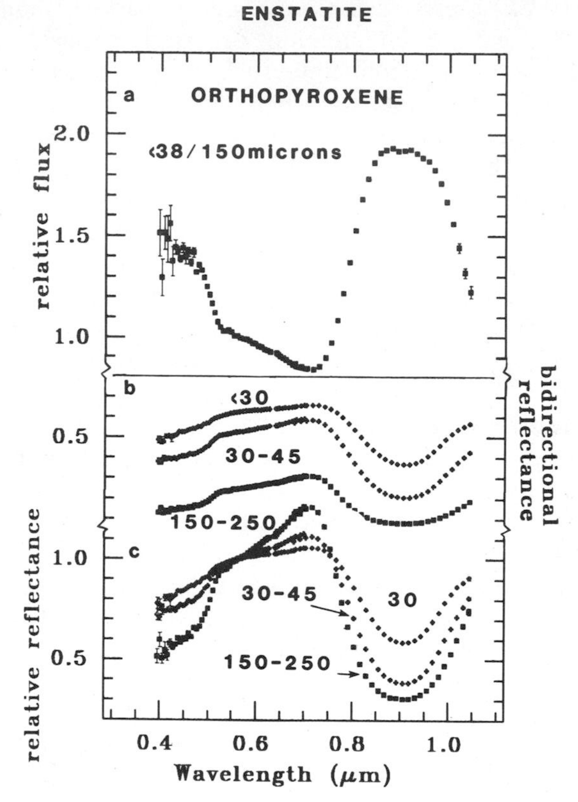

Figure 7.5. (*a*) Curve showing the ratios of the fluxes from extremes of particle size. Seen is the effect on the band strength. (*b*) Changes in albedo and band depth as a function of particle size. (*c*) Reflectance scaled to 1.0 at 0.56 μm. Source: McFadden et al. (21).

be detected in asteroid reflectance spectra. The conclusions drawn are that the interpretation of a spectrum at any rotational phase angle should not be significantly affected by scattering due to body shape, and that the effect is of secondary importance relative to the mineralogical influence on spectral reflectance curves.

Temperature is another factor to be considered. As the temperature rises, the increased thermal motions lead to greater vibrations about the crystallographic sites, producing effectively a larger mean atomic distance. Based on this, one would predict such effects as a larger bandwidth crystal field absorption band, depending on the range of motion of the vibrating metal and oxygen atoms. A shift of band center to longer wavelengths, depending on the range of motion of the vibrating metal and oxygen atoms, would also be predicted.

On the other hand, charge transfer bands would decrease in intensity with increasing temperature. The probability of an electron transfer decreases as the ion's amplitude of vibration increases. Sung and associates (28) and Singer and Roush (29) have observed shifts of the 2.0-μm band position and 1.0- and 2.0-μm bandwidths in pyroxene with temperature. The change in strength of the UV absorption band is negligible, and the position of the 1.0-μm band does not change as a function of temperature for pyroxene. The width of the 1.0-μm band broadens with increasing temperature (see Fig. 7.6). Thus a hot pyroxene spectrum might be incorrectly interpretated as an olivine–pyroxene mixture. Because all of the above is related to temperature, it must be of concern when observing near Earth or inner belt Asteroids, measured up to the 2.5-μm region.

The final parameter, shock, will alter a reflectance spectrum to the extent that a crystal structure is changed. It has been shown experimen-

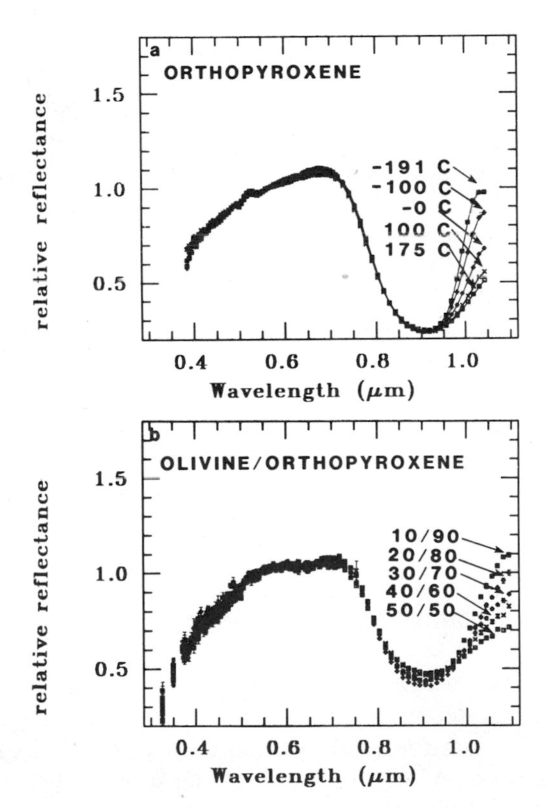

Figure 7.6. (*a*) Spectra of orthopyroxene at different temperatures. (*b*) Spectra of olivine and orthopyroxene of different proportions. Source: McFadden et al. (21).

tally by Adams and associates (30) that only slight changes in the strength of the absorption bands of pyroxene occur after shock pressures of 597 kbar, although the albedo is reduced by 33%. On the whole, the effects of shock are dependent on the strength of the bonds within each particular mineral. In a reflectance spectrum, the presence of the mineral pyroxene will dominate the signature given by shocked mixtures with plagioclase/ and or olivine. There are as yet no detailed studies comparing these two effects. Particle-size spectral variations and shock effects ae generally similar and cannot be differentiated by visible reflectance alone.

With respect to asteroid studies, there has been some speculation that the presence of shocked phases or opaques can be defined by UV spectroscopy. For example, in the vacuum UV, the position of reflectance maxima due to an interband energy gap in mafic silicates shifts to higher energy, perhaps showing a decreasing unit cell volume of shocked crystals. However, the existing vacuum ultraviolet Asteroid spectra do not cover the region of the spectrum containing this diagnostic absorption.

In summary, we see that reflectance spectroscopy in the 0.33–2.5-μm region has been developed as a powerful tool for determining chemical and mineralogical information about distant Solar System bodies. The major effort has involved the Asteroids, although other objects like the Moon have also been studied. It is clear that generally the diagnostic features in the spectra are relatively diffuse in comparison to other types of spectroscopy. As a consequence, the interpretations can be beset by ambiguities, and a special burden is placed on the observer to understand those parameters that add to the uncertainties and to attempt to unfold them from the processed data. This is not always an easy task, and considerable research still needs to be done.

Despite all of this, reflectance spectroscopy offers one of the more powerful analytical methods for the remote analysis of Solar System bodies at great distances. In the section that follows, I shall briefly describe the practice and select some samples of results. A relatively complete survey would be voluminous and beyond the scope of this text.

MEASUREMENT TECHNIQUES

In general, the use of reflectance spectroscopy to study Solar System objects involves telescopic measurements from ground-based observatories, as well as spacecraft measurements. In the former category, spectral reflectance has been and continues to be measured at telescopes using a wide variety of spectrometers and imaging systems. The method of measurement has been described by T. B. McCord (41) and is as follows:

1. The flux from the object of interest is measured and represented as $f_o(\lambda)$.
2. The flux from a nearby star is measured at about the same time and labeled $f_s(\lambda)$.
3. By forming a ratio $f_o(\lambda)/f_x(\lambda)$, instrumental response as well as atmospheric effects are eliminated.
4. The value of $f_x(G)$ is determined as the flux of a standard spectral type G, that is, a solar analogue, is based on the work of Hardorp (31). The reflectance is calculated as

$$R(\lambda) = [f_o(\lambda)/f_s(\lambda)][f_s(\lambda)/f_s(G)(\lambda)] = f_o(\lambda)/f_s(G)(\lambda)$$

assuming $f_s(G)(\lambda) = F_{sun}(\lambda)$. Hardorp has shown this assumption to be valid for certain stars.

The spectral reflectance is displayed as a function scaled to unity at 0.56-μm in order to eliminate albedo effects and to concentrate on differential reflectance spectra. McCord states that this technique has been used to obtain reflected spectra using the light from small areas on the Moon, the integrated light from Mercury, Mars, many Asteroids, and the Galilean satellites.

In order to check the precision of the telescopic spectra, comparisons have been made to returned Apollo samples. Spectral reflectance curves of lunar soils measured in the laboratory have been compared with telescopic data of areas 10–20 km in diameter that contained Apollo and Luna landing sites. Adams and McCord (23) showed the agreement to be within a few percent.

Significantly, the lunar sample spectral curves show the same subtle but important absorption features seen in the telescopic curves. Because the lunar sample studies have made it possible to understand what chemical and mineral properties control the shape of the relectance spectrum, it is possible to obtain information from the entire front surface of the Moon to add to that obtained by direct sampling. The information consists of pyroxene composition of the soil, the titanium dioxide concentration in the lunar glasses, soil maturity, the mare basalt types, and the distribution of the dark mantle deposits.

At this time there have been successful applications of ground-based telescopic observations of reflectance spectra to the determination of the surface composition of Mars (32) and of well over 300 Asteroids (33). On Mercury, an upper limit has been set on the FeO content of its lunarlike soil (34). In addition, a low-Ca pyroxene has been identified.

By far, the bulk of the spectral reflectance data has come from terrestrial observatories. There have been spacecraft measurements of low spectral resolution, and Galileo, an upcoming mission to Jupiter, will carry a near-infrared mapping spectrometer (NIMS). Spectral reflectance curves of Mars (35) and Venus (36) have been determined from Viking Lander and Venera photometric data. The curves are approximate because of the use of broad-band filters; however, it has been possible to make comparisons with terrestrial soil analogues.

RECENT APPLICATIONS

What follows is a brief review of an extensive literature. The illustrations cited are the result of recent efforts and demonstrate the importance of these spectral methods in the remote analysis of the surfaces of Solar System bodies.

At the end of 1979, Singer and co-workers (37) published a summary paper on the composition of Mars by reflectance spectroscopy. They obtained visible and near-infrared reflectance spectra of the martian surface, primarily by Earth-based telescopic observations. In addition, they acquired multispectral images both from spacecraft and Earth-based observations. They observed that all the spectra from Mars show strong Fe^{3+} absorptions from the near-UV to about 0.75-μm.

The darker regions showed the effect to a lesser degree, leading to an interpretation of a lesser degree of oxidation. Dark areas were also observed to have a Fe^{2+} absorption near 1.0-μm, attributed primarily to olivines and pyroxenes.

There was also observational data in the infrared region for highly dessicated mineral hydrates and for H_2O ice and/or absorbed water. Examination of the north polar region showed a strong H_2O-ice spectral signature, but no evidence for CO_2 ice. On the other hand, only CO_2 ice was identified on the south polar cap.

The brightest materials on Mars were found to be widespread, and correlated with aeolian dust. By contrast, the darker materials showed greater mineralogic variability and were thought to be closer in petrology and physical location to their parent rock. The dark materials appeared to fit a model of somewhat oxidized basaltic or ultramafic rock, regionally variable in composition and details of oxidation. The bright materials appeared to be finer-grained assemblages of highly oxygen-sharing dessicated mineral hydrate, some ferric oxides, and other somewhat more minor constituents, including a small amount of relatively unaltered mafic material. The conclusion was that the brightest materials might be primary and/or secondary alteration products of basaltic or ultramafic dark materials.

Figures 7.7–7.9 provide some idea of the nature of the observations. In Figure 7.7, a comparison is made between the Viking Orbiter II and Earth-based reflectance data for a few regions on Mars. All of the spectra have been scaled to a value of 1.0 at 0.56-μm. Figure 7.8 shows a comparison of bright and dark region reflectance spectra scaled to unity at 1.02-μm. Figure 7.9 is a comparison of observed Mars bright and dark region spectra with laboratory spectra of weathered basalt. A single olivine basalt, oxidized in the laboratory, provided the best match for both bright and dark regions in this spectral range. The bright region requires a higher degree of oxidation and a finer grain size than the dark region spectrum.

The detection of water is interesting, particularly the method of unfolding the data. Water was first reported on Mars by Sinton (38) after observing a strong absorption near 3-μm. Pimental and associates (39) observed evidence for small amounts of water in or near the surface using the data acquired by the Mariner 6 and 7 infrared spectrometer in the 3-μm region. McCord and co-workers (40) after analyzing an integral disk spectrum of Mars, showed absorptions in the 1.4–2.9-μm range that was attributable to water in the form of ice plus a highly desiccated mineral hydrate. This was followed by new data showing additional evidence for a mineral hydrate and/or solid H_2O (41). Singer and associates (37) demonstrated that all Martian infrared spectra show a drop in reflectance from 1.3 to 1.4-μm, which is independent of the Mars atmospheric CO_2 absorption. This effect was greater for bright areas than the dark ones.

In an attempt to understand the drop, the investigators approximated the observed Martian reflectance with spectra typical of basalts and their oxidation products. The light areas were modeled with a spectrum consistent with heavy oxidation and the dark areas by a spectrum typical of a basaltic material covered with a thin oxidized layer. The spectra were scaled in the same fashion as the telescopic spectra, and have a smooth reflectance beyond 1.1-μm.

An ice spectrum was added algebraically in small amounts to match the apparent 1.4-μm drop found in the Martian spectra. The ratio of the Martian spectrum to the simulated one shows absorption features that correlate with the expected Martian atmospheric CO_2 bands. Such simulations fit the bright area spectra better than the dark area spectra and show that the relative band intensities are weaker and different for dark areas than the bright ones.

The conclusions drawn are exciting. The results appear to show that water is present in the Martian surface in different forms such as frost ice sheets on the surface and ice mixed in the regolith or bound to it. It has

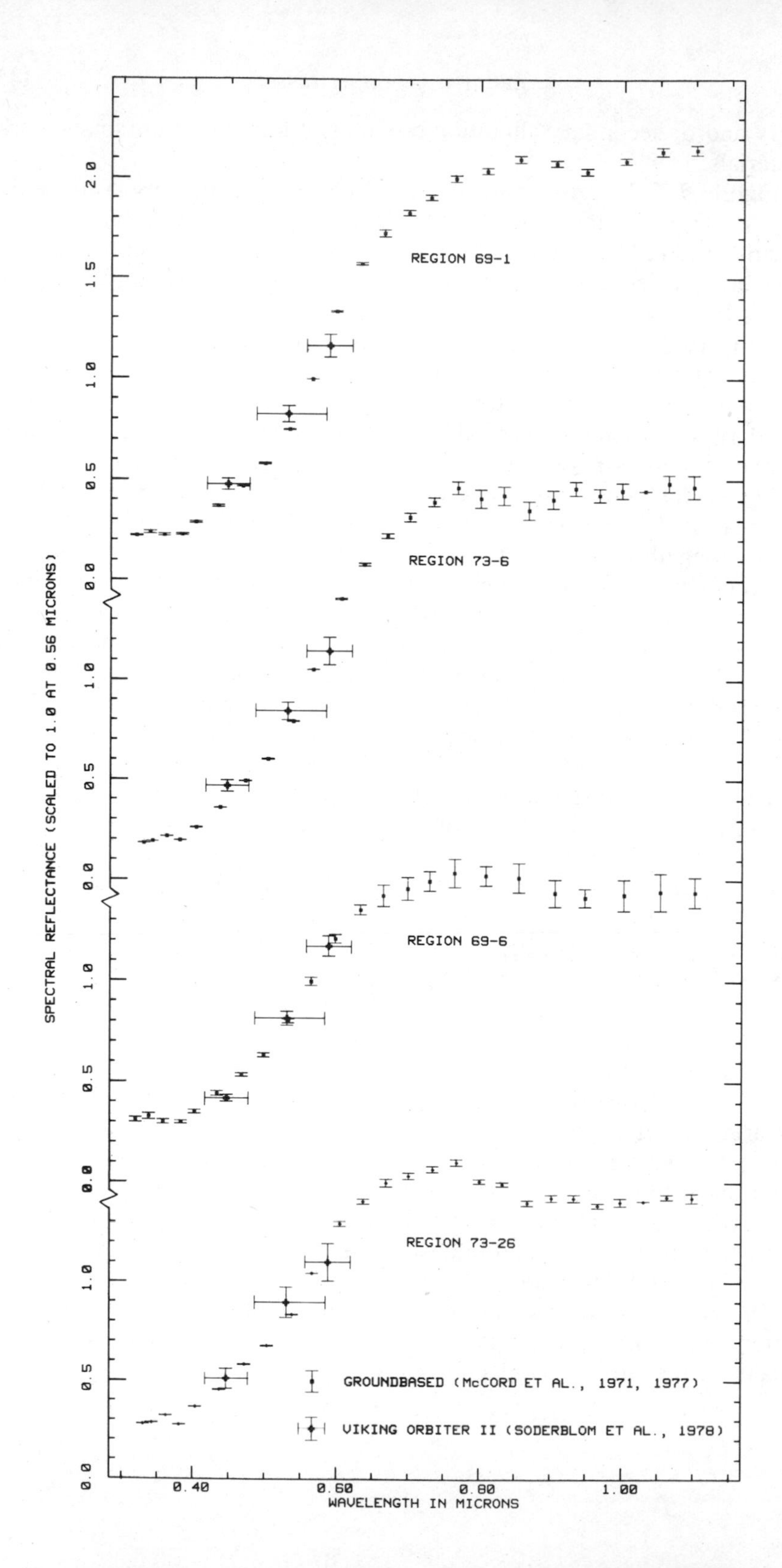

REGION 69-1
REGION 73-6
REGION 69-6
REGION 73-26
GROUNDBASED (McCORD ET AL., 1971, 1977)
VIKING ORBITER II (SODERBLOM ET AL., 1978)
SPECTRAL REFLECTANCE (SCALED TO 1.0 AT 0.56 MICRONS)
WAVELENGTH IN MICRONS
0.40
0.60
0.80
1.00
0.0
0.5
1.0
1.5
2.0

been shown that bound water bands that occur typically at 1.4 and 1.9-μm do not shift appreciably with temperature from 300 to 150°K and that bound water can be distinguished from free ice. The simulation of Martian spectra shows that free water ice is the primary cause of the 1.3–1.4-μm drop in reflectance and that there is more water in the bright areas than the dark. There is evidence at the 1.4- and 1.9-μm spectral region for bound water in the Martian surface. These are, however, difficult regions to observe from the Earth because of uncertainties in telluric water removal and the strong 2-μm Martian atmospheric CO_2 absorption.

Finally, a spectrum of the north polar cap, taken in the northern spring showed very strong water ice bands. McCord, Clark, and associates (41) have successfully modeled this spectrum by assuming that 60% of the radiation is reflected by water ice and 40% by the gray material (flat reflectance at all wavelengths). Singer and associates (37) maintain that the actual amount of water present is difficult to determine because of grain size variations and/or hydration state.

A second review of the application of spectral reflectance in Solar System exploration was published by McCord and co-workers in 1981 (41). Figure 7.10 summarizes the situation in this type of spectroscopy, according to McCord. Prior to 1972, the scatter in the spectral data beyond 10-μm made detailed interpretation very difficult. All that could be said was that the reflectance increases with wavelength throughout the region. In 1970, McCord and Johnson (42a) published spectra for six lunar areas (0.3–2.5-μm), which were relative to an area in Mare Serenatatis. This comparison was able to show important subtle differences in the spectra (Fig. 7.11) and variations from area to area. These early results have been made more exact by improvements in instrumentation and techniques.

An infrared spectrometer has been developed to obtain high precision photometric spectra of planetary surfaces. The details have been described by McCord (42b). In general, the design was dictated by the special problems of looking at an object at great distance and by the instabilities involved in examining Solar System objects. For example, the various components such as the indium antimonide detector and a continuously spinning circular variable filter were cooled to liquid nitrogen temperatures. The spectral region from 0.62 to 2.5-μm was scanned every 10 to 20 s. The object under observation was continuously viewed by the use of a cooled, mirrored aperture plate at 45° to the optical axis

---

Figure 7.7. Comparison of Viking orbiter and ground-based spectrophotometry for identical areas on Mars. The data are scaled to unity at a wavelength of 0.56 μm. Source: Singer et al. (35).

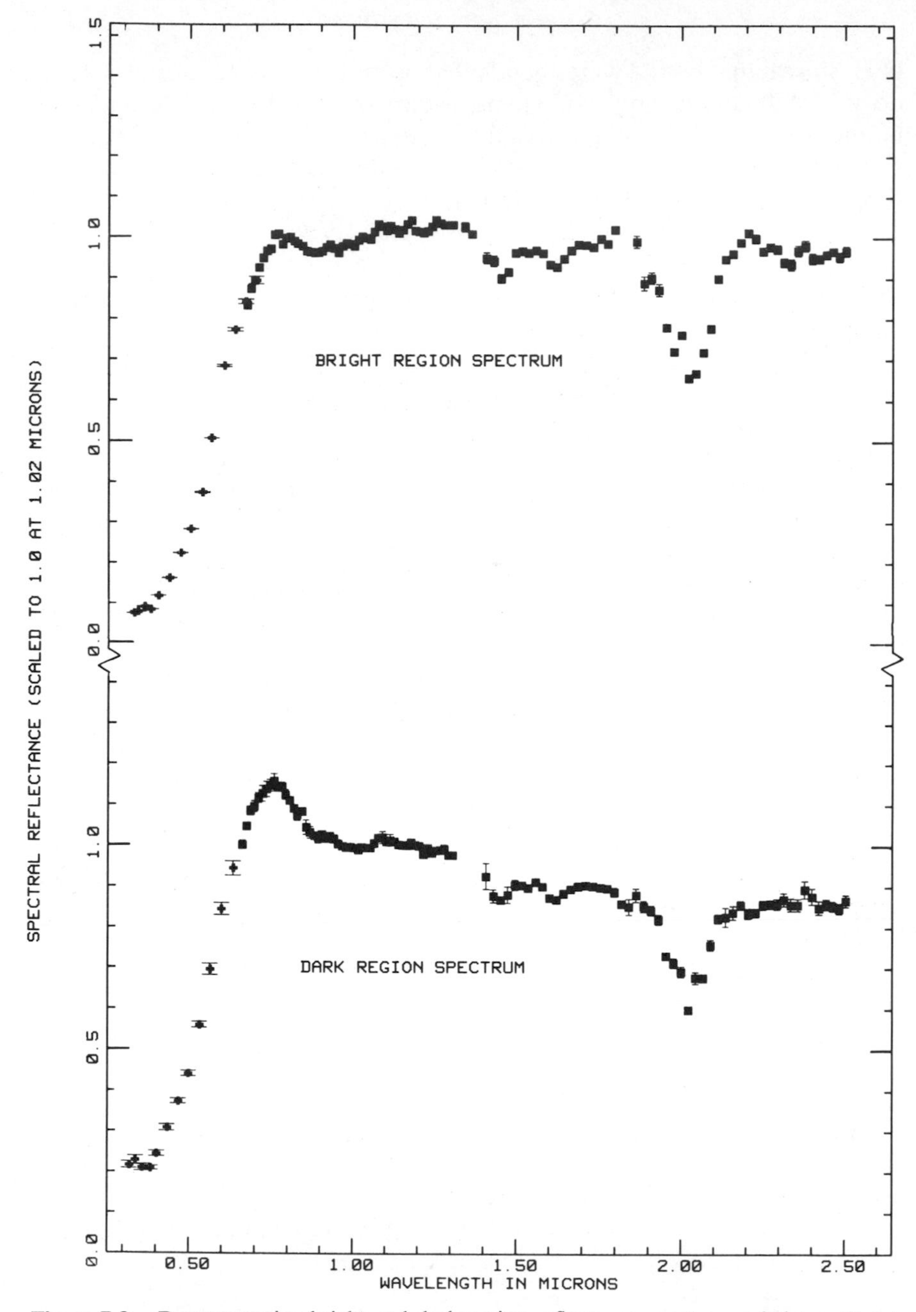

Figure 7.8. Representative bright and dark region reflectance spectra, scaled to unity at 1.02 μm. The bright region spectrum (top) is composed of an average of the brightest areas observed in 1973 (visible) and 1978 (infrared). The dark region spectrum (bottom) is a composite of data from two nearby locations in Iapygia. Source: Singer et al. (35).

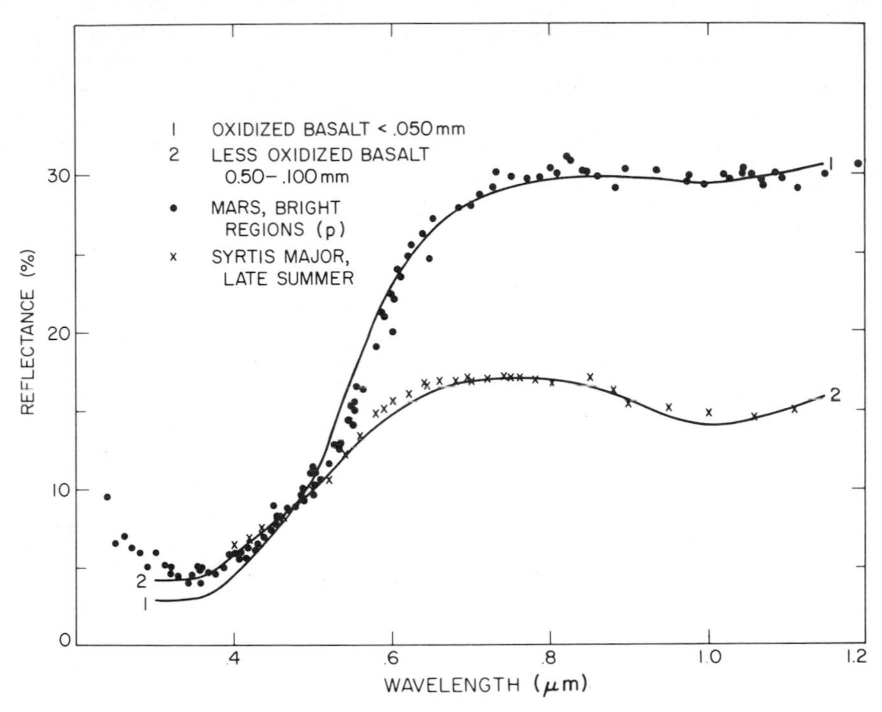

Figure 7.9. Comparison of observed Mars bright and dark region spectra with laboratory spectra of a weathered basalt. Source: Singer (35).

and an image transfer lens. The spectral resolution, $\delta\lambda/\lambda$, was about 1.5% throughout the spectral range. The incoming signal was compared to the signal from a black reference source 240 times per revolution of the continuously variable filter. The signal difference between the black reference and the object was counted, computer, and stored in the instrument data system memory. The 0.6–2.6-μm spectrum was divided into 120 data channels so that the sun of two complete chop cycles correspond to one spectral channel.

The instrument was developed to cover the spectral range of 0.6–2.5-μm, which included the 1.4- and 1.9-μm telluric bands (both quite intense). Thus there was a requirement for a special program and site for observation. The lunar area under observation was compared with a nearby "standard object" (usually a standard lunar area). The time between the target measurement and the standard area was kept small (about < 20 min). Each area was measured for 5–10 revolutions of the circular variable filter, and the data from each revolution was added to the

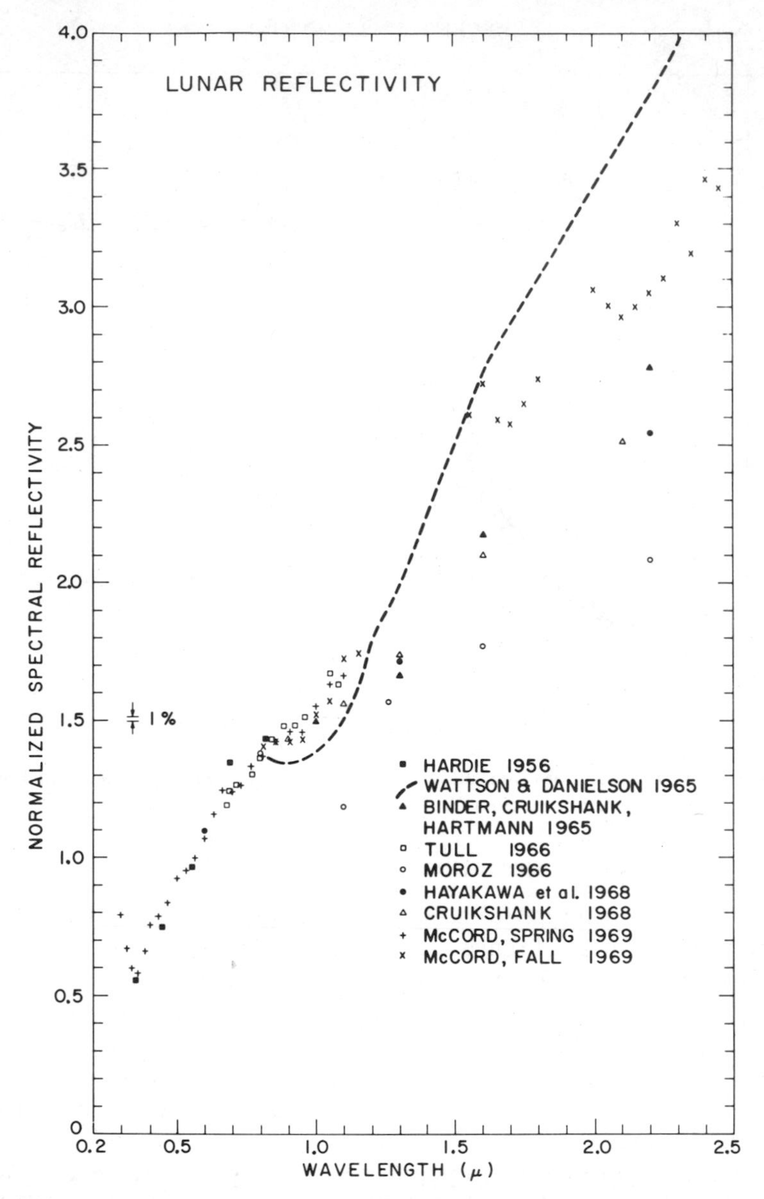

Figure 7.10. The lunar spectral reflectance for all known pre-1969 measurements at infrared wavelengths (areas unspecified). The scatter is great and little can be said with certainty about the data except that they increase toward longer wave-lengths. Source: McCord et al. (41).

Figure 7.11. Lunar spectral reflectance for six small lunar areas plotted as relative spectral reflectance (see text). This relative technique allowed more precise measurement and for the first time showed the presence of absorptions in the lunar spectrum and variations between lunar areas. Source: McCord et al. (41).

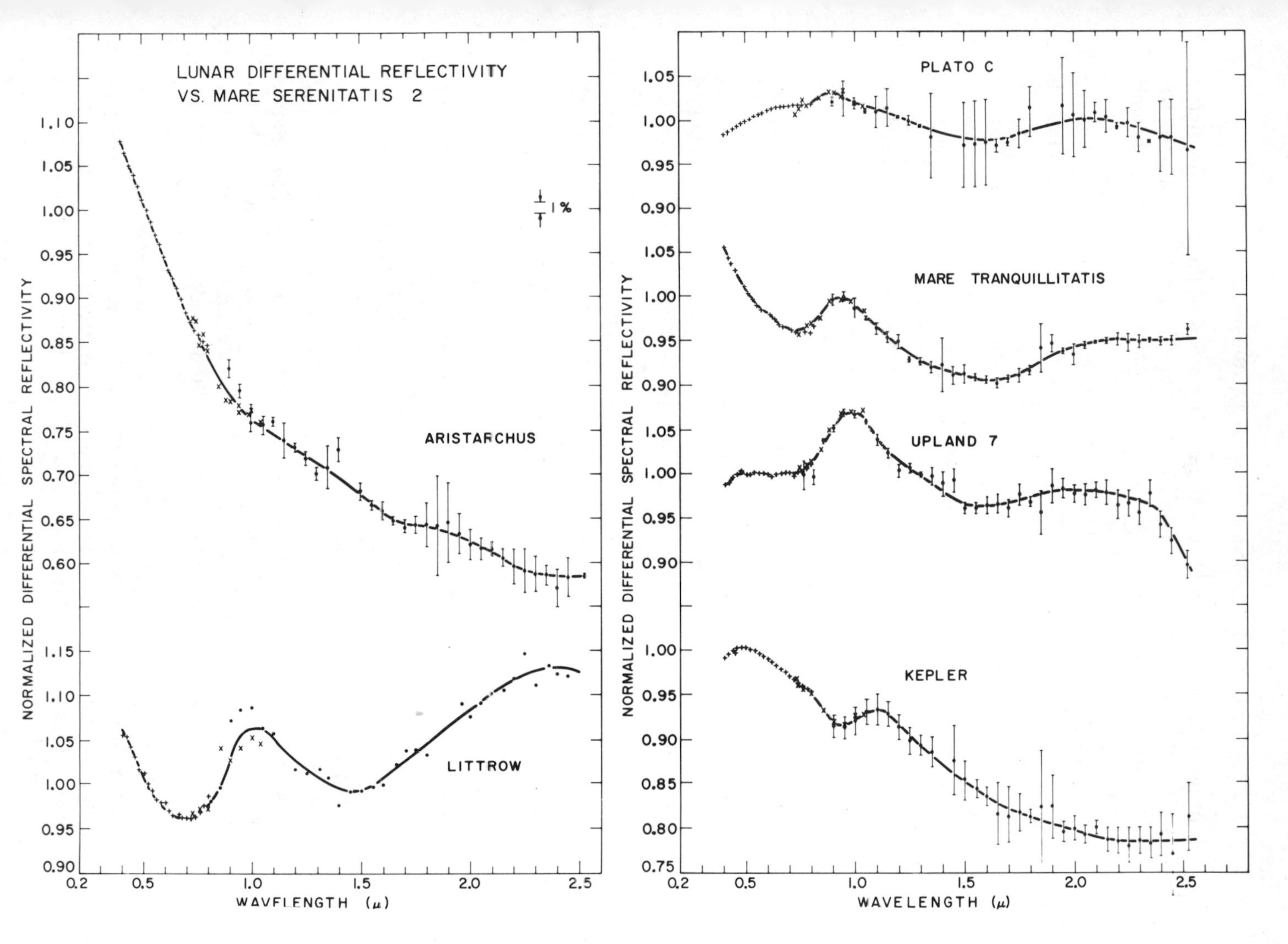
LUNAR DIFFERENTIAL REFLECTIVITY
VS. MARE SERENITATIS 2
1%
ARISTARCHUS
LITTROW
PLATO C
MARE TRANQUILLITATIS
UPLAND 7
KEPLER
NORMALIZED DIFFERENTIAL SPECTRAL REFLECTIVITY
WAVELENGTH (μ)

previous one. The sequence was generally repeated three or more times. Also, the standard was followed for several hours and measured frequently each night so that accurate atmospheric extinction coefficients could be obtained for each spectral channel.

The investigators, McCord and co-workers, (426) agree that while the method was tedious and inefficient (there was a frequent need to reposition the telescope), they obtained high-quality photometric data even in the regions of the telluric water bands. A key element in the measurements was the use of the Mauna Kea observatory in Hawaii, which was particularly well situated because of the high altitude and low water vapor.

It is worthwhile examining the method of lunar data reduction in a little detail. The measurements and spectral reflectance were calibrated using a computer program developed by Clark (43). For each run, a background reading was subtracted for the target area as well as for the standard area. Following this, the standard area measurement was plotted as a function of air mass through which the observation was made, thus calculating the extinction per air mass per spectral channel. The measured extinction was then used to correct all measurements to equal air mass. The calibrated target area measured was divided, spectral channel by spectral channel, by the calibrated standard area measurement, to produce a relative spectrum. However, before the relative spectrum could be determined, it was necessary to remove the thermal flux contribution coming from the hot lunar surface, so that only the reflected radiation remained. The procedure followed was developed by Clark (44).

Figure 7.12 demonstrates this for an Apollo 16 soil sample. The top reflectance spectrum (*a*) is of a soil sample measured in the laboratory and convolved to 1% spectral resolution. The spectrum was fitted with several polynomials. The smoothed spectrum shown is assumed to be the spectrum of the standard reference area near the Apollo 16 landing site used to calibrate all other lunar measurements. The bottom spectrum (*c*) is a relative reflectance spectrum measured at the telescope. It is the flux of the observed standard area in Mare Serenatatis divided by the Apollo 16 standard area flux after removal of the thermal contribution. This relative reflectance spectrum (*c*), when multiplied by the standard Apollo 16 reflectance spectrum (*b*), gives a standard reflectance spectrum (*a*). The error bars are 1 SD of the mean of several independent telescopic observations.

Figure 7.13 shows spectral reflectance for five lunar regions about 10–20 km in diameter: (*a*) Mare Serenetatis, (*b*) J. Herschel, (*c*) Apollo 11 site, (*d*) Apollo 16 (smoothed), and (*e*) Aristarchus. In Herschel, the water absorption bands at 1.4 and 1.9 μm have not been completely removed.

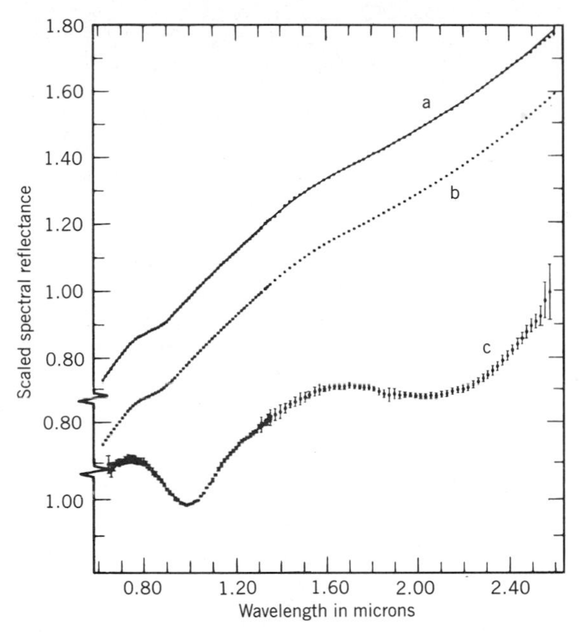

Figure 7.12. The reflectance spectrum of Apollo 16 soil sample is shown as measured in the laboratory (top) and convolved to 1% spectral resolution. The results of this spectrum after fitting with several polynomials to yield a smoothed spectrum is shown in the middle. This smoothed spectrum is assumed to be the spectrum of the standard reference area near the Apollo 16 landing site which is used to calibrate all other lunar spectral measurements. The bottom spectrum is a reflectance spectrum measured at the telescope. It is the flux of the more easily observed standard area in Mare Serenetatis. Source: McCord et al. (50).

The absorption features shown in Figure 7.13 have been analyzed in the following fashion (45,46). First the continuum was estimated and removed, leaving only the absorption bands. The residual absorption bands were then fit simultaneously by mathematical Gaussian functions to specify their characteristics quantitatively. An example of this is shown in Figure 7.14 for three areas.

Based on the techniques described, it has been possible to establish the identification of the minerals olivine, plagioclase, and several types of pyroxene at several locations on the lunar surface. The crater Aristarchus was found to have an average pyroxene composition of augite, and plagioclase was present. A dark mantle deposit in the crater J. Herschel was at least in part composed of a mixture of 70% olivine and 30% pyroxene.

All of the above identifications became possible because the reflectance spectra for 10–20-km diameter lunar features were observed for the first time in the infrared spectral region (0.65–2.5 μm) with sufficient resolution and precision to define mineral electronic bands. Absorption

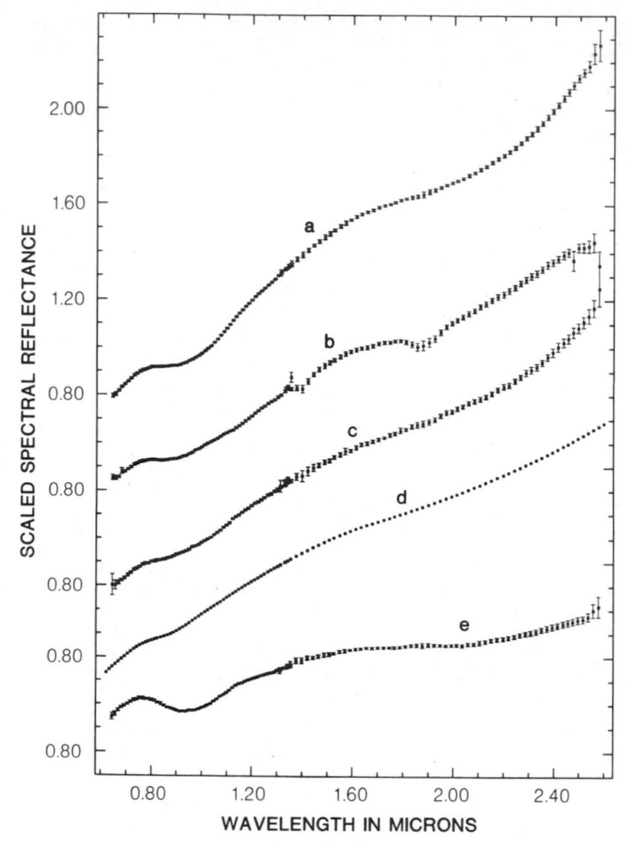

Figure 7.13. The spectral reflectance for five lunar regions approximately 10–20 km in diameter; *a*, Mare Serenetatis; *b*, J. Herschel; *c*, Apollo 11 site; *d*, Apollo 16 (smoothed); *e*, Aristarchus. Terrestrial atmospheric water absorptions not completely removed from the telescopic measurements are evident for Herschel near 1.9 and 1.4 μm and at shorter wavelengths. Source: McCord et al. (42a,b).

features in some spectra were quantitatively analyzed using newly developed computer processing techniques, including thermal flux removal and absorption band fitting making mineral identification more accessible.

A very interesting application of reflectance spectroscopy has recently been reported by C. M. Pieters (47). The study involves the central peak of the crater Copernicus. Originally, there were Apollo plans to visit this particular feature because of the possibility of finding deep-seated lunar materials. However, the Apollo program came to a halt before such a mission could be mounted. As Pieters has pointed out, the only way then

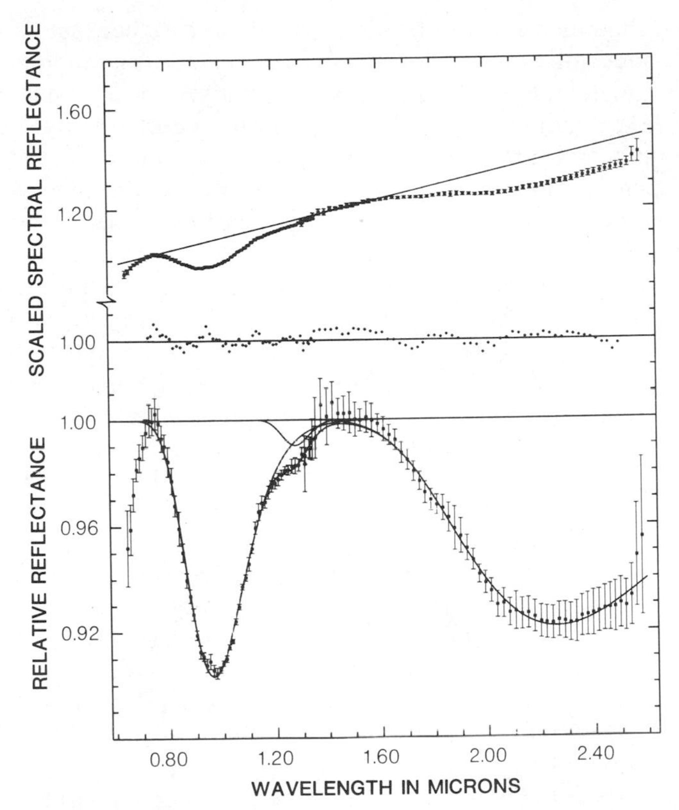

Figure 7.14. Part of the treatment of the spectral reflectance curves used to extract quantitative information is shown here for the crater Aristarchus. The top spectrum is shown corrected for thermal flux. The straight line through the top spectrum approximates the continuum. The bottom plot is the upper spectrum with straight line continuum removed and fitted by Gaussian functions. The middle plot is the difference between the Gaussian function and the fitted spectrum. Source: McCord et al. (42a,b).

to obtain compositional information was through remote sensing techniques. Although there were a number of well-tested orbital techniques available, after 1972 the principal technique was the use of Earth-based telescopic measurements.

The measurements showed that olivine was the major mafic mineral in the central peak. This information was obtained by the use of the near-infrared reflectance for the region from 0.7 to 2.5 μm. The composition of the deep-seated materials comprising the central peak of Copernicus was unique among the various measured areas.

Another particularly interesting study was to attempt to find the lunar source area of the Antarctic meteorite ALHA81005. This meteorite has been described as a regolith breccia, which appears because of physical characteristics and composition to have originated from an impact event in the lunar highlands. In a paper published in 1983, Pieters and associates (48), the investigators employing near-infrared spectroscopy, claimed a nearside limb or a farside source area—a surface unit on the Moon that had not been previously sampled by any U.S. or Soviet mission.

The final examples to be given here are extraordinary because of the great distances over which compositional information has been obtained. R. N. Clark (49) has measured the reflectance spectra from the Galilean satellites Ganymede, Europa, and Callisto, and Saturn's rings. Based on laboratory studies of water frost, water ice, and water and mineral mixtures, he concluded that the spectra of the icy Galilean satellites are characteristic of water ice (e.g., ice blocks or, conceivably, crystals larger than about 1 cm, or frost on ice, rather than pure water frost). The decrease in reflectance at visible wavelengths is caused by mineral grains on the surface. The spectra of Saturn's rings were more characteristic of water frost, with mineral grains mixed in, but not on the surface.

The impurities on these objects are not in spectrally isolated patches but, rather, appear to be intimately mixed with the water. The impurity grains seemed to have reflectance spectra typical of minerals containing Fe^{3+}. Clark states that some carbonaceous chondrite meteorite spectra show the necessary spectral shape. With regard to the Jovian moons, Ganymede was found to have more water ice on the surface than previously supposed (about 90 wt.%). A similar finding was made about Callisto, where the estimates are about 30–90%. Europa's surface is a vast frozen water surface, with only a small percentage of impurities.

The reflectance spectra indicated that Saturn's rings also have only a small percentage of impurities. Estimes of the bound water or bound OH for these objects was about 5–15%, averaged over the entire surface. A new absorption feature was identified in the spectra from Ganymede, Callisto, and Europa at 1.15 μm. This feature has also been seen in the spectra of Io, but not in Saturn's rings. However, this feature has not been seen in the laboratory and, as a consequence, its cause is unknown.

The final illustration of the determination of chemical information at extreme distances has been given by Cruikshank (50). The data involves the study on Triton, the largest of the two Neptunian moons, by reflectance spectroscopy in the near infrared. They have reported that the near-infrared spectrum of Triton is characterized by strong absorption bands of methane, probably in the solid state. They also found an additional absorption band at 2.16 μm, which was tentatively identified as the density-induced band of molecular nitrogen in the liquid state.

The investigators, however, express some reservation about not being able to observe the fundamental and additional overtones of this band system because of limitations due to the terrestrial atmosphere or the low precision of measurements of the spectral signal.

Based on the absorption coefficients for this band, derived from laboratory studies and from the literature, they have calculated that Triton has a layer of nitrogen at least 10 cm deep over most of its surface, a quantity they consider plausible in terms of the cosmic abundance of nitrogen and by comparison with Saturn's moon Titan, where a massive nitrogen atmosphere exists. The investigators modeled the Triton spectrum combining liquid nitrogen and solid methane and found that, by adding a spectral component corresponding to fine-grained water frost, they were able to account for the shape of the continuum in two spectral regions. They have also speculated that still another component, a dark solid, a photochemical derivative of methane may occur as a contaminant of the surface material.

In summary, we have seen some interesting illustrations from analytical chemistry over vast distances and under difficult conditions. The application of the technique of near-infrared reflectance spectroscopy has required careful laboratory calibrations to produce the necessary matches to the observed spectra. It is obvious that there are pitfalls to be guarded against and that considerable investigative work still needs to be done; however, the results to date are quite remarkable.

REFERENCES

1. R. P. Lyon, NASA Contractor Report, *C.R. 100* (1964).
2. W. A. Hovis and P. D. Lowman, private communication (1968).
3. J. A. Aronson, A. G. Emslie, R. V. Allen, and H. G. Linden, NASA Contract Report, *NAS-8-20122* (1966).
4. R. J. Lyon, *Econ. Geol.,* **60,** 715 (1965).
5. W. A. Hovis and W. R. Callahan, *J. Opt. Soc. Am.,* **56,** 639 (1966).
6. A. F. H. Goetz and J. A. Westphal, *Appl. Opt.,* **6,** 1981 (1967).
7. J. B. Adams, *Science,* **159,** 1453 (1968).
8. T. B. McCord, Color Differences on Lunar Samples, PhD. Dissertation, California Institute of Technology (1968).
9. M. J. Gaffey and T. B. McCord, *Space Sci. Rev.,* **21,** 555 (1978).
10. T. V. Johnson and F. P. Fanale, *J. Geophys. Res.,* **78,** 8507 (1973).
11. C. R. Chapman and J. W. Salisbury, *Icarus,* **19,** 507 (1973).
12. M. J. Gaffey, *J. Geophys. Res.,* **81,** 9059 (1976).
13. J. B. Adams, *J. Geophys. Res.,* **79,** 4829 (1974).

14. J. B. Adams, in C. Karr, Ed., *Infrared and Raman Spectroscopy of Lunar and Terrestrial Minerals,* Academic, New York, 1975.
15. J. B. Adams and L. H. Goullaud, *Proc. Lunar Planet. Sci. Conf. 9th,* 2901 (1978).
16. P. M. Bell and H. K. Mao, *Geochim. Cosmochim. Acta,* **37,** 731 (1973).
17. T. B. McCord, J. B. Adams, and T. V. Johnson, *Science,* **168,** 1445 (1970).
18. J. W. Salisbury and G. R. Hunt, *J. Geophys. Res.,* **79,** 4439 (1974).
19. C. R. Chapman, D. Morrison, and B. Zellner, *Icarus,* **25,** 104 (1975).
20. T. B. McCord and J. B. Adams, *The Moon,* (3/4) 251 (1973).
21. L. A. McFadden, M. J. Gaffey, H. Takeda, T. L. Jacowski, and K. L. Reed, *Proc. 7th Symp. Antarctic Meteorites,* National Institute of Polar Research, Tokyo, 1982.
22. L. A. McFadden, T. V. V. King, and M. J. Gaffey, *Icarus,* **59,** 25 (1984).
23. J. B. Adams and T. B. McCord, *Proc. Lun. Sci. Conf. 1st,* 1937 (1970).
24. J. Gradie and B. Zellner, *Science,* **197,** 254 (1977).
25. J. B. Adams and A. L. Felice, *J. Geophys. Res.,* **72,** 5725 (1967).
26. R. N. Clark and T. L. Roush, *J. Geophys. Res.,* **89,** 6329 (1984).
27. J. Gradie and J. Veverka, *Proc. Lun. Planet. Sci. 12th,* 1769 (1981).
28. C. Sung, R. B. Singer, K. M. Parkin, R. G. Burns, and M. Osburne, *Proc. Lun. Sci. Conf. 8th,* 1063 (1977).
29. R. B. Singer and T. L. Roush, *B.A.A.S.,* **14,** 727 (1982).
30. J. B. Adams, F. Horz, and R. V. Gibbons, *Lunar Planet. Sci. Conf. 10th,* 1 (1975).
31. J. Hardorp, *Astron. Astrophys.,* **105,** 120 (1981).
32. R. B. Singer, *J. Geophys. Res.,* **87,** 7967 (1982).
33. C. R. Chapman and M. J. Gaffey, in T. Gehrels, Ed., *Asteroids,* Univ. Arizona Press, Tucson, 1979, p. 655.
34. T. B. McCord and R. N. Clark, *J. Geophys. Res.,* **84,** 7664 (1979).
35. R. B. Singer, *J. Geophys. Res.,* **87,** 10159 (1982).
36. Yu. M. Golovin, B. Ye. Moshkin, and E. Konomov, in D. M. Hunten et al., Ed., *Venus,* Univ. of Arizona Press, Tucson, 1986, p. 131.
37. R. B. Singer, T. B. McCord, R. N. Clark, J. B. Adams, and R. L. Hugenin, *J. Geophys. Res.,* **84,** 7967 (1975).
38. W. M. Sinton, *Icarus,* **6,** 222 (1967).
39. G. C. Pimental, P. B. Forney, and K. C. Kerr, *J. Geophys. Res.,* **79,** 1623 (1979).
40. T. B. McCord, R. N. Clark, and R. L. Huguenin, *J. Geophys. Res.,* **83,** 5433 (1978).
41. T. B. McCord, R. N. Clark, B. R. Hawke, L. A. McFadden, P. D. Owensby, C. M. Pieters, and J. B. Adams, *J. Geophys. Res.,* **86,** 10, 883 (1981).

42a. T. B. McCord and T. V. Johnson, *Science,* **169,** 855 (1970).

42b. T. B. McCord, *Applied Optics,* **7,** 475 (1968).

43. R. N. Clark, *Publ. Astron. Sci. Pac.,* **92,** 221 (1980).

44. R. N. Clark, *Icarus,* **40,** 44 (1979).

45. T. G. Farr, B. A. Bates, R. L. Ralph, and J. B. Adams, *Proc. Lun. Planet. Sci. Conf. 11th (abstracts),* 276 (1980).

46. R. N. Clark, *J. Geophys. Res.,* **86,** 3074 (1981).

47. C. M. Pieters, *Science,* **215,** 59 (1982).

48. C. M. Pieters, B. R. Hawke, M. Gaffey, and L. A. McFadden, *Geophys. Res. Lett.,* **10,** 813 (1983).

49. R. N. Clark, *Icarus,* **44,** 388 (1980).

50. D. P. Cruikshank, submitted for publication.

CHAPTER

8

SURFACE ANALYSIS FROM ORBITING SPACECRAFT

The orbital experiments and measurements described in this chapter were designed around the concept that an understanding of the origin and development of a planetary body requires a knowledge of the overall chemical composition and the chemical distribution through that body. As already pointed out, it is generally accepted that planetary composition is related to the particular mechanisms of condensation and accretion, as well as the nature of the chemical abundances that existed in the primordial clouds of gas and dust. There is also a consensus that, subsequent to formation, large chemical and physical processes exerted a considerable influence on the distribution of the elements in the various bodies of the Solar System.

The focus in this chapter will be largely on the Moon, because in every sense the effort was so large and so highly successful, and because the results have produced an important precedent for future exploration efforts. Our knowledge about the Moon prior to 1960 came from Earth-based telescopic studies. Attempts were made to infer something about the Moon's surface chemistry from observed optical albedos, light scattering, and infrared measurements. Lunar studies began to pick up momentum in the last decade under the impetus supplied by the space programs initiated by the United States and the Soviet Union, all of which provided a great deal of new information. The programs were varied, including Ranger flybys, Surveyor landers, Lunas, Orbiters, and, most recently, Apollo and Lunakhod flights.

The earliest in situ chemical analysis of lunar surface materials was done using instrumented Surveyor soft landers (1) (see Chapter 9). Chemical analysis was performed by means of an alpha backscatter experiment deployed to the lunar surface from the Surveyor spacecraft. An impressive number of sites were visited. Surveyor V landed at Mare Tranquillitatis, Surveyor VI at Sinus Medii, and Surveyor VII on the rim of the crater Tycho. The reported chemical results were subsequently and spectacularly confirmed by the analysis of the returned lunar samples.

The first successful attempt at orbital analysis came with the Russian Luna 10 flight (2), which orbited the Moon carrying a gamma-ray experiment. Analysis of the gamma-ray spectra from the lunar surface convinced the investigators that concentrations of K, Th, and U were comparable to those found in terrestrial basalts and inconsistent with granites, ultrabasic rocks, or chondrites. The results were in fact consistent with reports published by the Surveyor investigators that the maria were similar but not identical in composition, and the highlands, though related, clearly contained less iron and more aluminum, that is, were more feldspathic.

The first lunar landings began with Apollo 11 to Mare Tranquillitatis and resulted in the return of a substantial variety of materials consisting of crystalline rocks, breccias, and soils (see Chapter 4). The samples were essentially basalts, although small fragments of rock were found that were obviously not indigenous to the mare site. Rather, they appeared to come from elsewhere, perhaps the highlands. The fragments were feldspathic in character, ranging from gabbroic anorthosites through anorthositic gabbros to anorthosites. These, interestingly, appeared similar in composition to the analysis reported by the investigators for the alpha backscatter measurements for the Surveyor VII landing in the Tycho highlands.

The Apollo 12 mission proved exciting for a number of reasons, one of which was the return of a type of rock identified as KREEP a rock type enriched in potassium (K) rare earths (REE), phosphorus (P), thorium, and uranium.

With Apollo 14 came an attempt to sample rocks associated with a major cratering event—the production of the enormous Imbrium basin. The Apollo 14 rocks proved to be mainly breccias, with compositions closely allied to the KREEP-type rocks.

The orbital science experiments that will be described below began with the Apollo 15 mission and continued through Apollo 17, when the program was halted. The value of an integrated experiment for performing a remote geochemical survey of the lunar surface was recognized in a published report by Carpenter and associates in 1968 (3). The central idea was that some atomic and nuclear radiation, characteristic of the elements comprising the surface layer, is naturally emitted from the Moon, and that these radiations could be used to determine the chemical composition of the lunar surface. Further, by means of an integrated experiment, one could hope to obtain complementary information that would yield a more comprehensive picture of the distribution of the key elements.

A picture of the radiation environment at the lunar surface is shown in Figure 8.1. From such a model, it appears that the Moon is uniquely suited for remote chemical surveying. Examining the figure, we see that

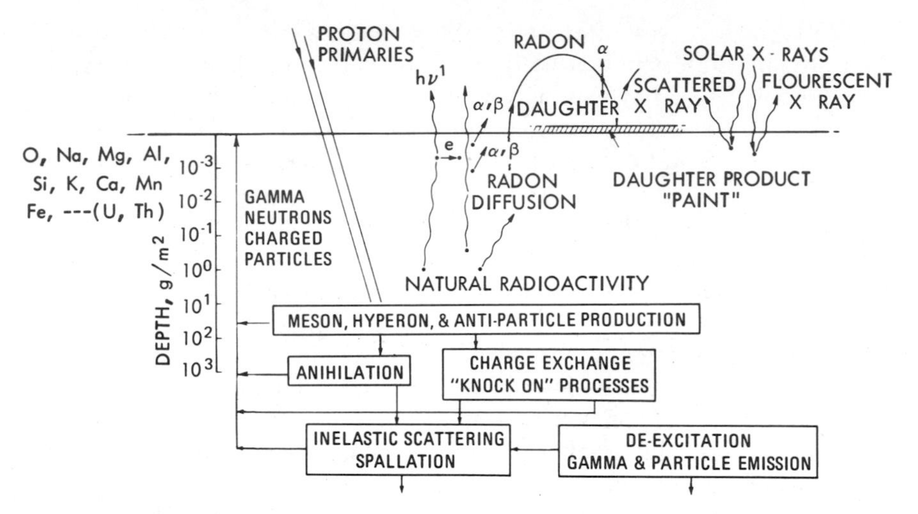

Figure 8.1. The radiation environment at the lunar surface.

both natural and induced sources of radiation are present. The principal, naturally occurring radioactive constituents are the long-lived nuclides ^{40}K, ^{238}U, and ^{232}Th, as well as their decay products. The presence of these elements was suspected before the active exploration of the Moon occurred, this was subsequently borne out by the Russian orbiters and the returned lunar samples. These elements emit alpha, beta, and gamma radiation of various energies. As a further consequence of the existence of these elements in the Moon, one might also expect the appearance of special processes such as radon diffusion (4).

Because of the Moon's special environment (absence of an atmosphere), it was reasonable to expect that cosmic rays and energetic solar protons bombarding the lunar surface would also produce a variety of shorter-lived nuclides, resulting in induced radioactivity with the production of gamma rays. In addition, one might also expect the prompt emission of charged particles, neutrons, and protons.

A number of studies of the Sun had shown it to emit copious quantities of soft X rays even under quiet Sun conditions. These were expected to produce fluorescent X rays characteristic of some of the elements making up the lunar surface, the yield depending on the intensity and spectral character of the solar flux as well as the relative abundance of the elements in the lunar surface.

Finally, as we see in the figure, radon and thoron diffusion through the surface—should it occur—would be expected to produce alpha particles characteristic of the various decay processes. The lunar ambient just described led to the development of a series of experiments built on the utilization of the various types of radiation. These are described below.

LUNAR X-RAY FLUORESCENCE EXPERIMENT

As we have seen, the lunar surface is bombarded by substantial fluxes of soft X rays even under "quiet" Sun conditions. Further, numerous calculations had shown that a typical solar X-ray spectrum was energetically capable of producing measureable amounts of characteristic X rays from all the abundant elements with atomic numbers of about 14 (Si) or smaller. Observations had shown that during enhanced solar activity, such as flare conditions, the intensity of the X-ray output increases, and the spectrum hardens (becomes more energetic), so that one might also expect to see characteristic X rays from some higher atomic number elements.

A typical X-ray output from a quiet Sun is shown in Figure 8.2. The energy distribution decreases markedly with increasing X-ray energy and becomes quite small at energies of about 3 keV. Based on a strictly thermal mechanism of production, the observed spectrum leads to calculated coronal temperatures that range between 10^6 and 10^7 K. Variations of temperature produce changes in fluxes and spectral composition. These changes are reflected in variations of fluorescent intensities from the lunar surface, as well as the relative intensities from the various elements in the lunar surface. As an example, a hardened solar spectrum would produce an enhancement of the X-ray intensities from the heavier elements relative to the lighter ones. To deal with this problem, an X-ray solar monitor was employed during the mission to follow the variations in solar X-ray intensities and the spectral output. In addition, simultaneous measurements of the solar X-ray flux were obtained during the mission from various Explorer satellites in flight at the time.

The spectral distribution shown in Figure 8.2 was based on satellite observations. This is a calculated distribution based on a combined coronal temperature of 1.5×10^6K, with a hot-spot temperature of 3×10^6 K in proportions determined from Solrad data using a model developed by Tucker and Koren (5). The K shell absorption edges for Na, Mg, Al, Si, and K are superimposed on this curve along the energy axis. Only those X rays with greater energy than the absorption edge energies are capable of exciting these elements. The intensities of the fluorescent X rays would thus depend on the incident flux and the ionization cross sections. It is obvious that under quiet Sun conditions the solar flux is most suitable for exciting the light elements, in particular the major rock-forming elements such as Si, Al, and Mg.

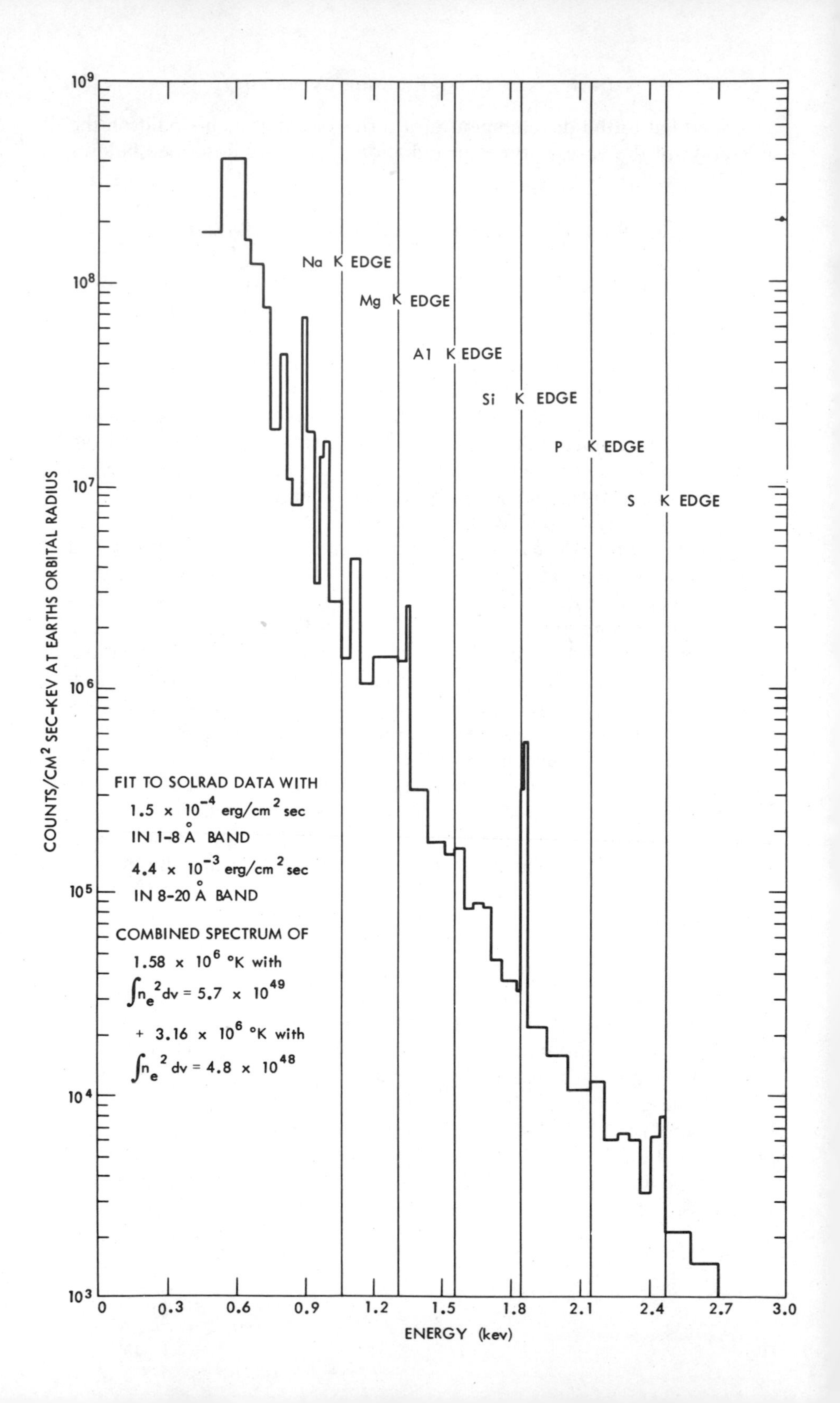

COUNTS/CM² SEC-KEV AT EARTHS ORBITAL RADIUS
ENERGY (kev)
Na K EDGE
Mg K EDGE
Al K EDGE
Si K EDGE
P K EDGE
S K EDGE
FIT TO SOLRAD DATA WITH
1.5 x 10^-4 erg/cm^2 sec
IN 1-8 Å BAND
4.4 x 10^-3 erg/cm^2 sec
IN 8-20 Å BAND
COMBINED SPECTRUM OF
1.58 x 10^6 °K with
∫n_e^2 dv = 5.7 x 10^49
+ 3.16 x 10^6 °K with
∫n_e^2 dv = 4.8 x 10^48

By using the calculated solar X-ray fluxes and spectral distributions, and assuming some typical terrestrial rock compositions, such as those corresponding to basalts and granites, it was possible to calculate the expected yields of fluorescent X rays. Such calculations were done by Gorenstein and co-workers, (6) and Eller (7) using an assumed effective temperature of 4×10^6 K, and a region from 1 to 8 Å.

The numbers determined corresponded to a low lunar X-ray brightness, thus making it obvious that high-resolution devices such as crystal spectrometers with their low inherent efficiency were not feasible. Accordingly, the X-ray experiment was designed around large area proportional counters with thin, high-transmitting windows. The sensing assembly consisted of three gas-filled proportional counters, mechanical collimators, calibration sources for in-flight calibration, a temperature monitor, and the associated electronics.

Figure 8.3 shows a close-up of the arrangement of the detectors, calibration rod, and the collimators. Each individual detector had an approximate window area of 30 cm^2 and a window thickness of 3.8×10^{-2} mm beryllium. The gas fill was 90% Ar, 9.5% CO_2, and 0.5% He. The collimators were designed to provide a nominal $\pm$ 30° field of view. The calibration sources, two for each detector, emitted characteristic Mn and Mg K radiation in order to cover the spectral range of interest. These calibration sources, one set for each detector, were programmed to periodically rotate toward the detectors during flight, while the spacecraft electronics provided a signal indicating that the experiment was in the calibrate mode.

Two methods of X-ray energy resolution were used. The energy output of the detectors was analyzed by eight discriminator channels, covering 0.5–2.75 keV in equal intervals. One of the subsystems involving a bare detector (Fig. 8.3) was programmed to change automatically and periodically to a low-gain mode, thus covering a spectral range of 1.5–5.5 keV (to look for heavier elements should the Sun flare). As an additional method of energy resolution, one of the detectors was operated bare, and the other two with selected X-ray filters. A magnesium foil covered one window and an aluminum foil covered the other. The magnesium foil filter preferentially filtered aluminum and silicon K radiation, while the aluminum filter was more selective for silicon radiation.

The X-ray detector was designed of a nominal 60° field of view, but in actual practice it proved to be closer to 50°. At the mission altitudes of about 110 km, the instantaneous area was approximately 100 km on edge.

Figure 8.2. Calculated typical X-ray output from a "quiet" Sun.

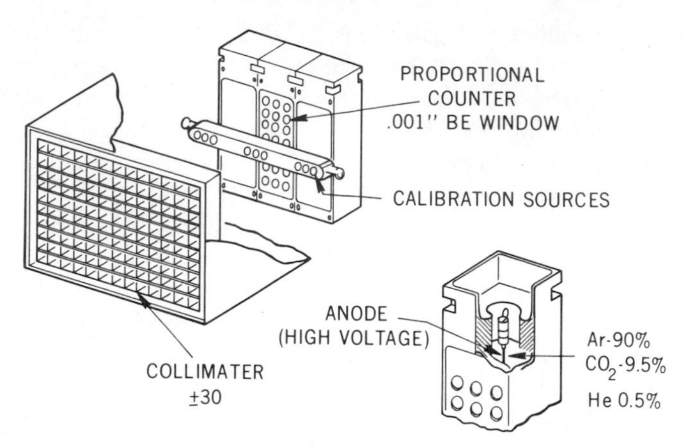

Figure 8.3. Functional configuration of the X-ray spectrometer flown on the Apollo 15 and 16 missions.

The X-ray processing assembly was designed to both sort energies and to discriminate against unwanted pulse shapes, thus rejecting non-X-ray events that would contribute background noise. This mode of operation produced a marked improvement of signal-to-noise ratios under circumstances where gamma rays and cosmic rays produce unwanted backgrounds.

An important component of the X-ray fluorescence experiment was the solar X-ray monitor, a small proportional counter detector mounted on the opposite side of the spacecraft. This detector continuously provided data about the Sun's X-ray flux falling on the lunar surface. The data obtained from the solar monitor was important in interpreting the X-ray fluorescence data.

GAMMA-RAY EXPERIMENT

The importance of gamma-ray measurements to an understanding of the Moon's evolution was appreciated from the beginning of the planning for lunar exploration. It was assumed that K, Th, and U would be among the key elements. In terrestrial processes, the radioactive decay of these elements is considered as one of the primary sources of energy leading to volcanism and magmatic differentiation. In turn, the distribution of these elements is used as a measure of the degree of chemical differentiation; K, Th, and U tend to concentrate in the late stage crystalline rocks such as the granites.

The chemical information about these elements is shown in the gamma-ray spectra by discrete lines that have energies characteristic of the individual elements. Two groups of such lines exist.

The first group results from natural radioactivity due to the spontaneous decay of ^{40}K and the radioactivity products of Th and U. Examples of these are the 1.46-MeV line emitted by the ^{40}K decaying to ^{40}Ar and the 2.62-MeV line of ^{208}Tl, a daughter product of Th. The second group of lines results from galactic cosmic ray interaction with the lunar surface, producing secondary particles and excited nuclei.

Typically, these cosmic ray particles are protons with energies of the order of 10^9 eV. They react with the lunar surface to produce a cascade of lower-energy particles—most importantly, neutrons. These neutrons, in turn, produce excited nuclei, which in turn emit radiation.

Three processes are involved, all of which in turn lead to characteristic gamma rays. Neutrons may scatter inelastically, leaving the target nucleus in an excited state. The deexcitation process yields gamma ray lines characteristic of the elements. One such example is Fe, which under these conditions produces a 0.84-MeV line. Another major process is neutron capture. Neutrons emitted in the initial interactions lose energy by successive collisions. Some neutrons escape, while others are captured. The binding energy of the neutron, approximately 8 MeV, is emitted from the product nucleus by a complex decay scheme, leading to a spectrum that may contain a few strong lines. An example of this is the Fe line at 6.4 MeV. The third process, which is generally a less important one, is the formation of radioactive nuclides by nuclear reactions, which leads to radioactive decay, again with emission of gamma rays. An example of this is ^{26}Al, produced from Al and Si.

A less common source of interaction involves the solar cosmic rays and the lunar surface. During major solar flares, the Sun becomes a source of particles, with energy in the range of 10–100 MeV. Most of these particles lose energy by ionization, but some undergo nuclear reactions. During quiet Sun conditions, these solar particles are of little consequence.

Importantly, the expected intensities of the gamma-ray lines as a function of concentration can be calculated from a knowledge of the physical processes involved. For the natural radioactivities, the calculations are simple and straightforward. The situation is, however, more complex for the cosmic-ray-induced lines, requiring a knowledge of the fluxes, cross-sections and backgrounds, although the problems have been solved in a satisfactory manner.

For the Apollo 15 and 16 missions, the gamma-ray detector consisted of a 7 × 7-cm, right-cylindrical NaI (Tl) crystal detector surrounded by a plastic anticoincidence mantle for suppressing events due to charged particles (see Fig. 8.4). The electronic processor consisted of a 512-channel pulse-height analyzer, including an amplifier but no memory. The information was transmitted to ground control, channel by channel, in real time. When away from a direct communication link, the information was stored on tape for subsequent transmission. Spectra were obtained by sorting the received pulses for various time periods. Additional instrumental details may be found in papers by Arnold and co-workers (8) and Harrington and associates (9). The detector was deployed at the end of a boom some 8.25 m from the spacecraft in order to minimize the background gamma radiation produced in the spacecraft by cosmic-ray interactions. The arrangement of the gamma-ray instrument and the X-ray instrument described above are shown in Figure 8.5, mounted in the Science Instrument Module (SIM).

Two types of data analysis were performed on the measured gamma rays. Integral counts in the 0.6–2.7-MeV region were used to determine the variation of the natural emitters, K, Th, and U. Detailed spectral analysis was also performed, making it possible to obtain quantitative analysis for Th, K, Fe, Ti, Mg, Si, and O. The methods and models employed have been described by Reedy and co-workers (10). The results of the integral method showed that about 90% of the counts in the discrete spectrum in the 0.6–2.7-MeV region were due to gamma-ray emission from the K, Th, and U. This integral count procedure made it possible to study the distribution of the natural radioactive elements all along the flight path.

Quantitative analyses were performed using an interactive matrix inversion procedure, which has been detailed by Reedy and others (10). The most difficult part of the analysis was the derivation of the lunar gamma-ray continuum, the portion of the flux in the 0.5–10-MeV region that does not contain characteristic lines. This was a very significant part of the data reduction because the continuum was found to produce about 85% of the counts in the detector. Further, the shape as well as the magnitude was not constant and had to be derived from the data below 3 MeV. It was also found that the lines of the K, Th, and U made a significant contribution to the continuum by Compton scattering in the lunar surface.

To treat the continuum problem, it was necessary to separate it into two components, one portion relatively constant over the entire lunar surface and the second a variable, as a function of the K, Th, and U. The constant part of the background was derived from an analysis of those regions of the Moon where the overall natural radioactivity was a minimum. Empirical methods were developed for determining the magnitude

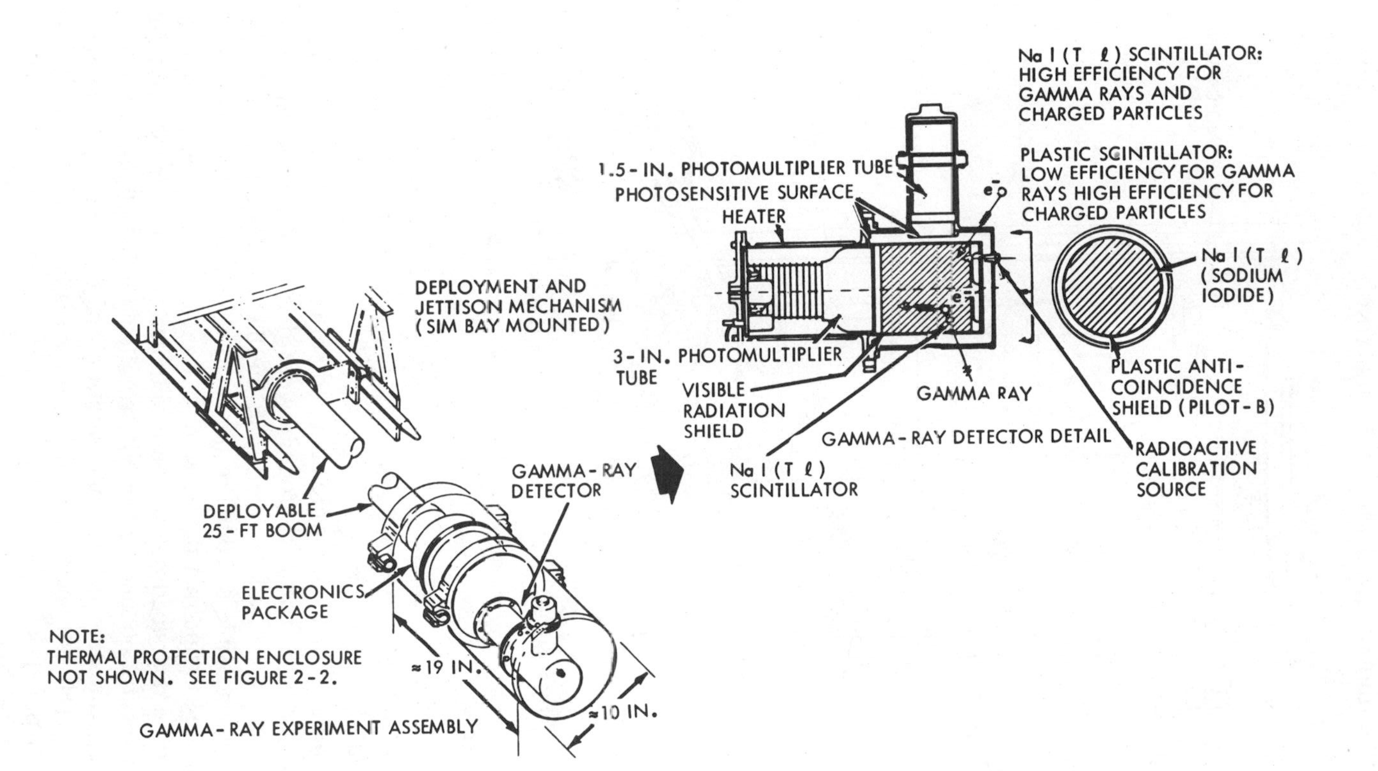

Figure 8.4. Components of the geochemical gamma-ray experiment flown on Apollo 15 and 16.

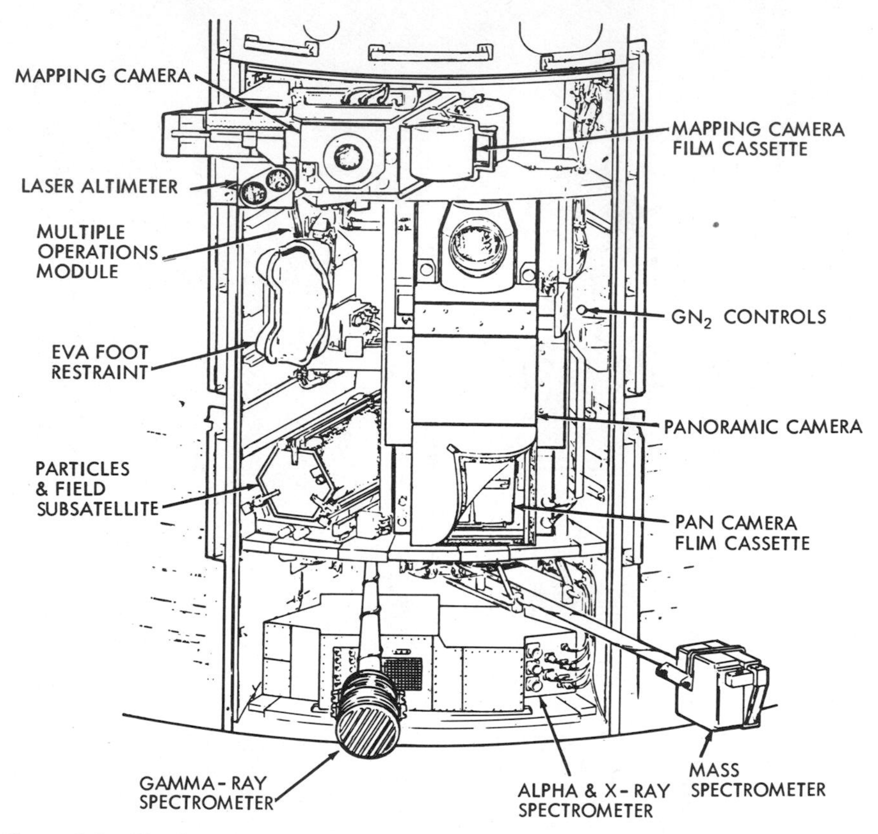

Figure 8.5. The instruments mounted on the Science Instrument Module (SIM) of the Apollo Spacecraft.

and shape of the scatter build-up as a function of the Th, U, and K concentrations.

Some very useful results were obtained both from the Apollo 15 and 16 flights (Table 8.1). The backgrounds were derived from the Apollo 15 data. Because there was an overlap in the flight paths during both missions, it was possible to compare the results for the overlap regions. It was gratifying to find that the agreement was good. The results, some of which will be described below, were obtained from the analyses of discrete line spectra obtained during the Apollo 15 and 16 flights.

The determination of the concentrations of the individual elements from the discrete line spectra was done sequentially. In the first stage, the spectrum from the 5–9-MeV region was analyzed. This is the region where only the Fe, O, Si, and Ti contribute significantly. The component

Table 8.1. Lunar Regions Analyzed

No.	Assigned Name	Coordinates Boundary Regions
	Apollo 15	
34	Van de Graaff region	168°W–168°E
35	Highland east farside	168°E–82°E (south of 10°S) 168°E–88°E (north of 10°S)
36	Highland east nearside	From (Region 35) to 60°E
37	Mare Fecunditatis	60°E–42°E
38	Mare Tranquillitatis	42°E–21°E
39	Mare Serenitatis	21°E–6°E
40	Archimedes region	6°E–15°W
41	Mare Imbrium	15°W–39°W
42	Aristarchus region	39°W–54°W
43	Oceanus Procellarum (north)	54°W–81°W
44	Highland west farside	81°W–168°W
	Apollo 16	
26	Highland east farside	180°E–142°E
22	Mendeleev	142°E–138°E
28	Highland east limb	138°E–92°E
20	Mare Smythii	92°E–83°E
19	Highland east nearside	83°E–55°E
17	Mare Fecunditatis	55°E–44°E
14	Intermediate Mare-Eastern Highlands	44°E–21°E
12	Highland nearside center	21°E–5°E
10	Ptolemaeus–Albategnius	5°E–5°W
23	Fra Mauro region	5°W–20°W
8	Mare Cognitum	20°W–30°W
5	Oceanus Procellarum (south)	30°W–50°W
3	Grimaldi region	50°W–76°W
2	Orientale rings	76°W–105°W
29	Highland west limb	105°W–119°W
1	Hertzsprung region	119°W–136°W
27	Highland west farside	136°W–180°W

Source: I. Adler and J. I. Trombka (10).

intensities were derived by a matrix inversion procedure, and were then subtracted from the spectrum. The second step involved the analysis of the difference spectra, which then determined the remaining components in the low-energy region.

The iron proved to be a particularly interesting element. It has a line spectrum in the 1-MeV region due to the inelastic scatter of neutrons (*n*, *n'*, gamma). In addition, iron emits high-energy gamma rays (>5 MeV) as a result of prompt neutron capture (*n*, gamma). Each of these reactions is differently dependent on the energy distribution of the neutron flux. The (*n*, gamma) ray flux is for the most part proportional to the thermal neutron flux, while the (*n*, *n'*, gamma) is proportional to the fast neutron flux. Thus ratio of the (*n*, gamma) to the (*n*, *n'*, gamma) can be used as a measure of the presence of strong thermal neutron absorbers, such as Gd or other rare earths. Their presence will markedly attenuate the flux of the thermal neutrons and produce a change in intensity of the corresponding spectral lines. Such effects have been observed.

Where possible, data was normalized using "ground truth," conforming the orbital data to returned lunar data at one or more points. The two sites selected as particularly useful were Mare Tranquillitatis and the Apollo 16 region near Descartes.

In the main, the sources of error were statistical uncertainties associated with the low count rate, the continuum correction, the correlation between library components, and uncertainties in "ground truth" normalizations.

GAMMA-RAY RESULTS

The orbital science program proved to be remarkably successful. As an example, the gamma-ray and X-ray experiments yielded many hours of useful data from a large number of orbits. The lunar surface coverage of the Apollo 15 flight was greater than the Apollo 16 because of the higher orbital inclination. The X-ray experiment continuously monitored the Sun's X-ray output, and for most part the sun proved to be a relatively stable excitation source. The initial results of the X-ray fluorescence experiment were used to produce relatively coarse maps showing the gross chemical variations for Al/Si and Mg/Si ratios across the lunar surface. The first results available from the gamma-ray experiment gave the distribution of the natural radioactivity (K, Th, and U as a group) (11). Because of satisfactory statistics, it was possible to map the radiation distribution on a relatively fine scale, 2° × 2°, corresponding to a square area on the lunar surface about 60 km on edge. The results of the analysis is shown in

Figure 8.6*a* and *b*. Based on these early measurements, the investigators were able to state the following:

1. All the 5° regions within and bounding the western maria showed a higher level of radioactivity than any other regions on the lunar surface covered by the Apollo track. There was a striking contrast between the western maria and the rest of the Moon, particularly the eastern maria. An inference was drawn that those portions of the western mare regions not overflown were also highly radioactive in comparison with other parts of the Moon.

2. There is a detailed structure in the distribution of radioactivity within the highly radioactive regions mentioned above. The highest concentrations observed were in the Aristarchus region in the high ground west of the Apollo landing site and south of Archimedes, and in the areas south of the Fra Mauro crater. The Fra Mauro area overflown was about 7° south of the Apollo 14 landing site. The soil from the landing site

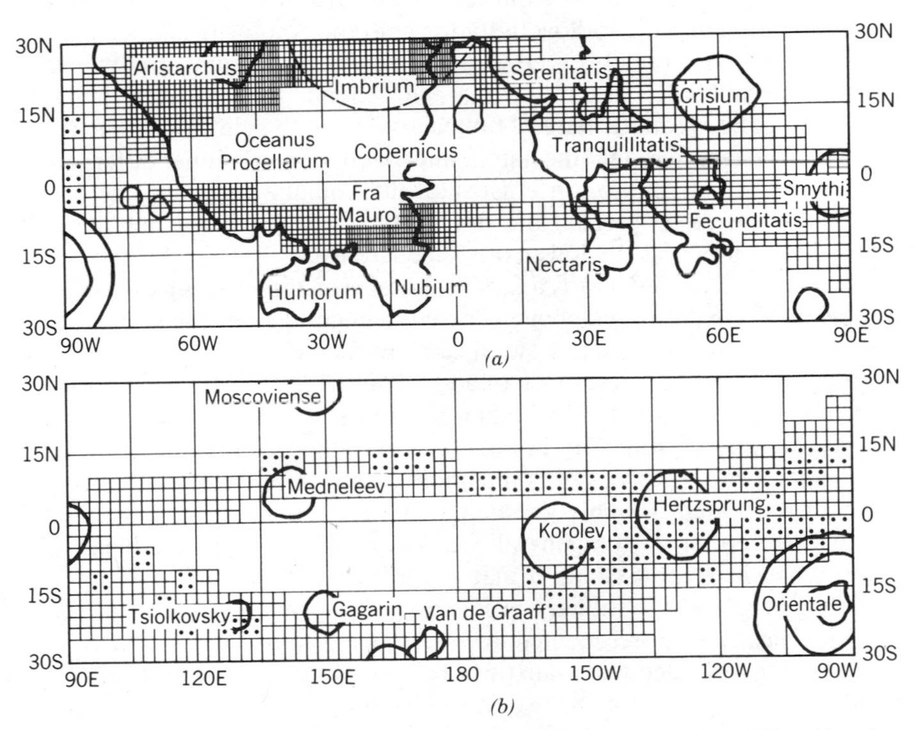

Figure 8.6. The integrated intensities from the K, Th, and U plotted on a 2° × 2° scale. Source: I. Adler and J. I. Trombka (24).

showed levels comparable to the orbital data. The data from this area indicated that the Fra Mauro area is surficially related to the western maria, rather than the adjacent highlands, as has sometimes been inferred from the albedo and topography.

3. The eastern maria were found to show evidence of local enhancement, although the overall radioactivity was lower than the western maria. Higher intensities relative to the surrounding highlands were visible in Mare Tranquillitatis, Mare Fecunditatis, Mare Smythii, and Mare Crisium.

4. The highland regions were observed to show relatively low radioactivity, except on the borders of the western maria where lateral mixing seems to have occurred. The eastern farside highlands (180–90° E) are measureably more radioactive than the western highlands (90–180° W). A small maximum in activity was found near the feature Van de Graaff, where a major magnetic anomaly was also found.

A correlative study that proved fruitful was a comparison of the laser altimeter profiles, obtained from an instrument mounted on the Apollo spacecraft, against the radioactivity for both the Apollo 15 and 16 flights. By superimposing the observed values for the natural radioactivity on the altimeter profiles, it was found that, in general, there was an inverse relationship between the natural radioactivity and the elevation. The relationship appears to reflect the nature and extent of major lunar differentiation processes. If the Moon is isostatically equilibrated, then the more extensive the early anorthositic differentiation was, the higher the expected elevation and the lower the radioactivity.

The crater Van de Graaff, a farside crater, is an interesting example of the relationship described above. There is a major depression in the vicinity of Van de Graaff, which shows the sharpest contrast in elevation to the adjacent highlands, a difference of about 8 km. This same area is also the site of the only major, farside enhancement in natural radioactivity. The depression covers some 30° in longitude on either side of 180° lattitude. The gamma-ray feature is of comparable extent. The gamma-ray investigators have stated that this Van de Graaff area is notable and, to date, a singular exception to the general conditions prevailing on the lunar farside. When it is understood, it may contribute significantly to an understanding of lunar evolutionary processes.

With regard to more detailed chemical analysis, a major problem was the obtaining of adequate statistics. To deal with this, the ground track was divided into a series of regions, chosen by considering major topographic boundaries, the density of high-quality data in a given region, and the contrasts observed in the radioactive maps.

The actual regions are shown in outline form in Figure 8.7. The smallest area, such as Mendeleev (region 22), was selected in order to determine the minimum area for which useful data could be derived. However, the errors in these regions were so large that no statistically valid or meaningful concentrations could be derived. In the version of the analysis being described, only data with sufficient total counts above background giving significantly low statistical errors were used. Complex border regions were assigned to either the mare or highland regions, unquestionably reducing the real chemical contrasts in some regions.

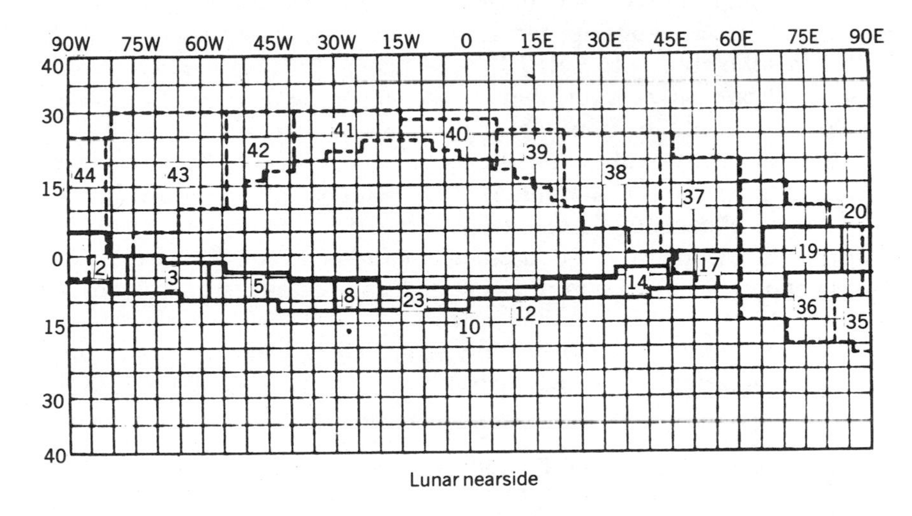

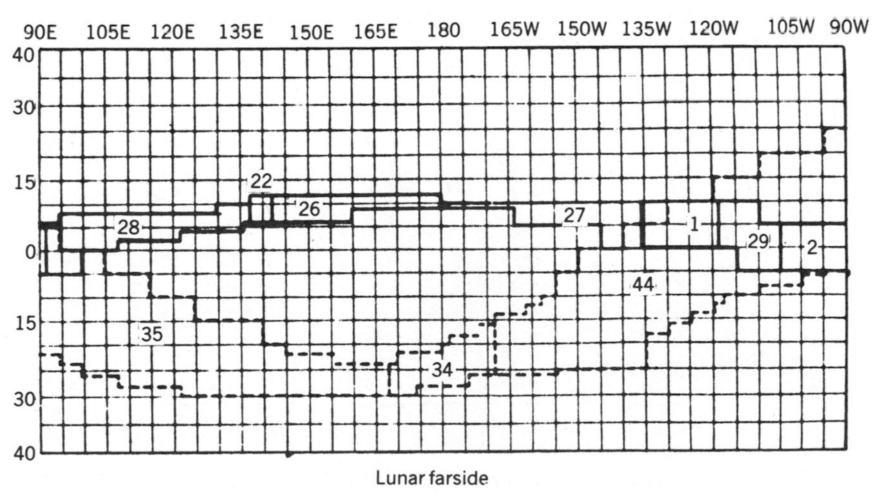

Figure 8.7. Outline of the regions described in Table 8.1.

The results of the early analysis are shown in Tables 8.2 and 8.3. As described above, the Fe, Mg, and Ti values were normalized by using "ground truth" values (Table 8.4), the conforming of the orbital data to returned surface samples at one or more points. A second method was the comparison of results from the Apollo 15 and 16 missions in the area of overlap. The values in parentheses are the ground truth data used to normalize concentrations for these elements in all the regions. The values for Fe are weighted means for the two modes of analysis. Also identified are regions where the values determined from neutron capture (low-energy neutrons) are well below those obtained from inelastic scatter (MeV neutrons). These regions appear to have high concentrations of K, REE, and P, presumably due to KREEP (see Chapter 3). It is expected that the presence of rare earths would depress the flux of the thermal neutrons and thus lead to results for iron, which may be slightly low.

In a similar way, Table 8.3, shows some Apollo 16 results, again by region. The Apollo 15 continuum was used for this work, with good consistency, except for Ti results, which are omitted. Here again, Fra Mauro gives evidence of a depressed, thermal neutron flux. There was also an apparent depression in a few highland regions, which is not yet understood.

Some instructive comparisons are presented in Table 8.4. for regions where the flight paths crossed each other. The regions analyzed on the

Table 8.2. Apollo 15: Element Concentrations by Regions

Region	Fe (%)	Mg (%)	Ti (%)	Th (ppm)	K (ppm)
34	7.7	3.8	0.1	2.3	1600
35	6.5	4.5	1.3	1.0	940
36	9.3	5.7	0.8	1.4	1200
37	11.3	7.0	2.2	1.2	1400
38[a]	(12.1)	(4.8)	(2.9)	1.7	1200
39	10.7	6.6	2.6	2.3	1700
40	8.4[b]	6.3	0.8	6.8	3100
41	13.6	6.2	1.4	5.8	1700
42	9.6[b]	4.9	2.2	6.9	2500
43	10.5	4.6	2.0	3.9	1700
44	5.7	3.5	1.5	0.4	950
All data	8.7	4.8	1.45	2.2	1230

[a] Values in parentheses are ground-truth data used to normalize concentrations for these elements in all regions.
[b] Region of apparent depressed thermal neutron flux (see text).
Source: I. Adler and J. I. Trombka (24).

Table 8.3. Apollo 16: Element Concentrations by Regions

Region	Fe (%)	Mg (%)	Th (ppm)	K (ppm)
26	6.2	3.4	0.6	920
22	4.3	3.0	0.5	960
28	7.2	2.9	0.5	840
20	8.8	2.8	1.3	1900
19	8.6	4.5	1.3	980
17	9.0	4.6	2.1	1100
14	9.0	3.5	1.5	1300
12	5.9	4.0	2.1	1400
10	4.8[a]	5.2	5.0	2700
23	9.7[a]	5.7	10.5	3900
8	12.1	4.9	8.4	3600
5	12.2	5.0	5.0	2300
3	4.4[a]	3.6	2.4	1100
2	4.7	2.9	0.8	1000
29	3.5[a]	2.1	0.4	1200
1	3.6[a]	3.6	0.6	720
27	5.2	2.7	0.5	730
All data	7.2	3.6	2.1	1300

[a] Region of apparent depressed thermal neutron flux (see text).
Source: I. Adler and J. I. Trombka (24).

two missions were roughly the same and the comparisons were expected to give good agreement, which in fact was observed. Also shown are comparisons to the ground truth numbers.

There were two landing sites that were not overflown, where a reasonable comparison was possible (Table 8.4). Apollo 14 observed the northern end of the Fra Mauro formation, while Apollo 16 overflew the southern end. Whatever the origin of this feature, similarities were expected throughout. Apollo 12 sampled an area of Oceanus Procellarum not far from, and topographically similar to, the area overflown by Apollo 16. These comparisons are shown in parts (*c*) and (*d*) of the table. In the cross-over region near the east limb, the regions analyzed on the two missions are roughly the same, and the comparison should give good agreement. The highland region from 5 to 21° east longitude, which includes the Descartes region, is compared with the soil analysis. The deviations seen are in the expected direction if more mare and KREEP material are mixed in near the margins of the region.

All of the comparisons discussed seem generally satisfactory and confirm the validity of the analysis. While further work is underway, no

Table 8.4. Comparisons of Orbital Data with Returned Lunar Samples

	Some Comparisons			
	(*a*) East Crossover		(*b*) Ground Truth	
	Apollo 15 Region 36	Apollo 16 Region 19	Apollo 16 Region 12	Descartes Soil[a]
Fe (%)	9.3	8.6	5.9	4.0
Mg (%)	5.7	4.5	4.0	3.3
Th (ppm)	1.4	1.3	2.1	2.0
K (ppm)	1200	980	1400	940
	Ground Analogies			
	(*c*) Fra Mauro		(*d*) Oceanus Procellarum	
	Apollo 16 Region 23	Apollo 14 Soil[b]	Apollo 16 Region 5	Apollo 12 Soil[c]
Fe (%)	9.7	8.1	12.2	12.5
Mg (%)	5.7	5.6	5.0	6.2
Th (ppm)	10.5	11.6	5.0	7.6
K (ppm)	3900	4400	2300	2600

[a] Fe and Mg: average of soil analysis, taken from five papers, *Third Lunar Sci. Conf.* and PET report. K and Th from Eldridge et al. (1973).
[b] Fe and Mg: average of soil analysis, taken from five papers, *Third Lunar Sci. Conf.* and PET report. K and Th from Eldridge et al. (1972).
[c] Fe and Mg: average of analyses of bulk soil 12070 and related samples in five papers, *Second Lunar Sci. Conf.* K and Th from O'Kelley et al. (1971).
Source: I. Adler and J. I. Trombka (24).

substantial modification of the values reported is expected. An important implication stemming from the results is that the contrasts between the mare and highlands are as expected nearly everywhere.

Iron, and less obviously magnesium, are relatively enriched in the maria. The radioactive elements show the same pattern with one important exception: the backside feature, the Van de Graaff region already mentioned. It shows a chemical composition different from any thus far observed on the Moon. The major elements are like those of the highland, though the iron is a little high. The concentration of K and Th are very similar to Mare Tranquillitatis and typical for a mare. The interpretation is that such a composition could not be formed by mixing of major components observed elsewhere.

X-RAY FLUORESCENCE EXPERIMENT—RESULTS

The X-ray fluorescence experiment, like the gamma-ray experiment, proved to be equally successful and, in fact, complementary in providing an integrated picture of the chemistry of the lunar surface. The results have been summarized by Adler and co-workers in a number of publications (12–16). The compositional information was limited, supplying data on Si, Al and Mg; however, in the case of the Moon these proved to be highly significant elements. The data compared to the gamma-ray information were more detailed in a spatial sense because the X-ray spectrometer had superior spatial resolution. Figure 8.8 shows an example of an Al/Si and Mg/Si profile made during the Apollo 15 flight. Similar profiles were made during the Apollo 16 flight.

The method of calculation of the Al/Si and Mg/Si concentration ratios was previously discussed. The results are shown in Tables 8.5, 8.6, and 8.7. Table 8.5 shows the consistency in measurements between both Apollo missions. Tables 8.6 and 8.7 give results for a number of specific lunar features. It should be pointed out that there were a number of fortuitous circumstances that operated to make the X-ray experiment successful. The first factor was the constancy of the Si concentration on a Moonwide basis, which made it possible to normalize the Al and Mg data. The second favorable factor was that all three elements are adjacent to each other in the periodic chart, which makes their response to the solar flux nearly independent of the nature of the matrix. Although their excita-

Table 8.5. Overlap in Chemistry between the Apollo 15 and 16 Ground Tracks

	Apollo 16 Concentration Ratios		Apollo 15 Concentration Ratios	
Feature[a]	Al/Si $\pm$ 1σ	Mg/Si $\pm$ 1σ	Al/Si $\pm$ 1σ	Mg/Si $\pm$ 1σ
Mare Fecunditatis	0.41 $\pm$ 0.05	0.26 $\pm$ 0.05	0.36 $\pm$ 0.06	0.25 $\pm$ 0.03
Mare Smythii	0.45 $\pm$ 0.08	0.25 $\pm$ 0.05	0.45 $\pm$ 0.06	0.27 $\pm$ 0.06
Langrenus area	0.48 $\pm$ 0.07	0.27 $\pm$ 0.06	0.48 $\pm$ 0.11	0.24 $\pm$ 0.06
Highlands west of Smythii	0.57 $\pm$ 0.07	0.21 $\pm$ 0.03	0.55 $\pm$ 0.06	0.22 $\pm$ 0.03
Western border of Smythii	0.58 $\pm$ 0.08	0.22 $\pm$ 0.04	0.52 $\pm$ 0.06	0.22 $\pm$ 0.06
Eastern border of Smythii	0.61 $\pm$ 0.09	0.20 $\pm$ 0.06	0.60 $\pm$ 0.10	0.21 $\pm$ 0.03

[a] The overlap between corresponding areas of the *Apollo* 16 and 15 ground tracks is not exact, so that differences for the same area may be real.
Source: I. Adler and J. I. Trombka (24).

Table 8.6. Concentration Ratios of Al/Si and Mg/Si for the Features Overflown during the Apollo 15 Flight

Feature	Concentration Ratios	
	Al/Si ± 1σ	Mg/Si ± 1σ
West of Diophantus and Delisle	0.26 ± 0.13	0.21 ± 0.06
North-east of Schröters Valley	0.28 ± 0.08	0.21 ± 0.06
Mare Serenitatis	0.29 ± 0.10	0.26 ± 0.07
Diophantus and Delisle area	0.29 ± 0.08	0.19 ± 0.05
Archimedes Rille area	0.30 ± 0.10	0.21 ± 0.06
Mare Imbrium	0.34 ± 0.06	0.25 ± 0.04
Mare Tranquillitatis	0.35 ± 0.08	0.22 ± 0.04
Mare east of Littrow (Maraldi)	0.35 ± 0.09	0.30 ± 0.03
Faius Putredinus	0.36 ± 0.06	0.25 ± 0.03
Mare Fecunditatis	0.36 ± 0.09	0.23 ± 0.05
Apennine Mountains		
Haemus Mountains, west border of	0.38 ± 0.10	0.25 ± 0.05
Serenitatis	0.39 ± 0.08	0.26 ± 0.05
Mare Crisium	0.39 ± 0.11	0.18 ± 0.02
Tsiolkovsky		
Haemus Mountains, south-	0.40 ± 0.07	0.26 ± 0.04
southwest of Serenitatis	0.42 ± 0.10	0.25 ± 0.06
Littrow area	0.45 ± 0.06	0.27 ± 0.06
Mare Smythii	0.45 ± 0.07	0.26 ± 0.02
Taruntius area, between Tranquillitatis and Fecunditatis		
Langrenus area, east of Fecunditatis to 62.5°E	0.48 ± 0.11	0.24 ± 0.06
Highlands between Crisium and Smythii (Mare Spumans and Mare Undarum area)	0.51 ± 0.06	0.22 ± 0.05
Highlands east of Fecunditatis,	0.51 ± 0.10	0.22 ± 0.05
Kapteyn area 68–73°E 7.5–15°S	0.51 ± 0.10	0.23 ± 0.05
Highlands west of Crisium	0.52 ± 0.10	0.22 ± 0.05
Highlands east of Fecunditatis 62.5–68°E 4–12.5°S		
West border of Smythii to 4–5° out from Rim	0.52 ± 0.06	0.22 ± 0.03
South of Crisium, Apollonius area, to Fecunditatis, 50–60°E	0.53 ± 0.06	0.23 ± 0.03
East border of Crisium out to 6°	0.54 ± 0.09	0.22 ± 0.04
from Rim	0.54 ± 0.12	0.16 ± 0.02
Tsiolkovsky—Rim	0.55 ± 0.06	0.22 ± 0.03
Highlands between Crisium and Smythii, 2.5°S 69°E, 5°S 76°E, 12°N 80°E, 10°N 83°E		

Table 8.6. (*Continued*)

Feature	Concentration Ratios	
	Al/Si ± 1σ	Mg/Si ± 1σ
Highlands west of Tsiolkovsky, 110–124°E to 9–21°S	0.57 ± 0.11	0.19 ± 0.04
Highland east of Fecunditatis 73–85E 10–19°S	0.58 ± 0.13	0.21 ± 0.05
South and southwest of Sklodowska,	0.59 ± 0.14	0.19 ± 0.07
86–101°E 18–23°S	0.59 ± 0.15	0.16 ± 0.05
Pirquet, 135–145°E 18–23°S	0.60 ± 0.10	0.21 ± 0.03
East border of Smythii, out to 4–5°	0.60 ± 0.10	0.18 ± 0.04
Pasteur Hilbert highlands area 101.5–110°E 7–18°S		
Hirayama, highlands east of	0.62 ± 0.07	0.19 ± 0.04
Smythii, 89°E 12°S, 100°E 15°S,		
98°E 2°S, 103°E 5°S	0.62 ± 0.12	0.15 ± 0.06
Highlands around Tsiolkovsky	0.65 ± 0.24	0.14 ± 0.05
South part of Gagarin, 144–153°E 21–23°S		

Source: I. Adler and J. I. Trombka (24).

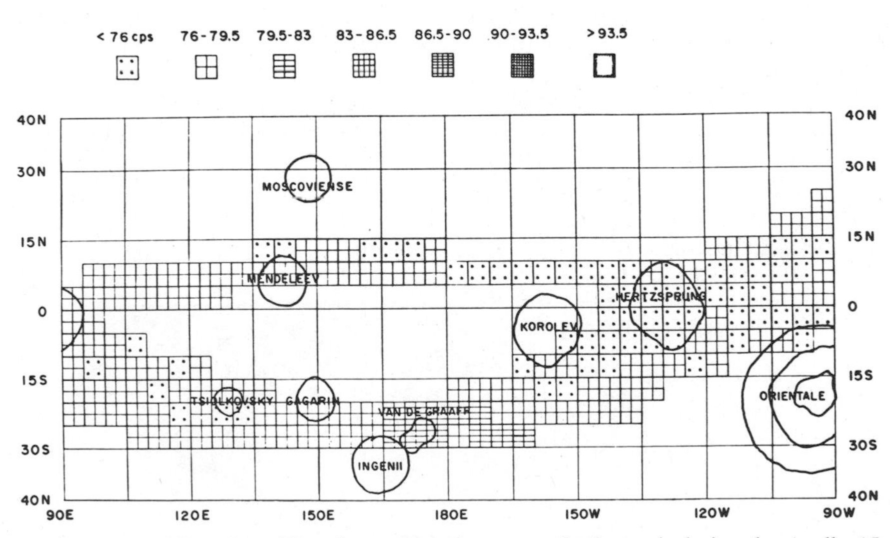

Figure 8.8. Al/Si and Mg/Si ratio profiles along a northerly track during the Apollo 15 flight. Source: I. Adler and J. I. Trombka (24).

Table 8.7. Concentration Ratios of Al/Si and Mg/Si for the Sites Overflown during the Apollo 16 Flight

Feature	Concentration Ratios	
	Al/Si ± 1σ	Mg/Si ± 1σ
Mare Cognitum	0.38 ± 0.11	0.40 ± 0.29
Upper part of Sea of Clouds (9°-13°W)	0.39 ± 0.12	0.20 ± 0.05
Mare Fecunditatis (42°-57°E)	0.41 ± 0.05	0.26 ± 0.05
South of Fra Mauro (13°–19°W)	0.45 ± 0.07	0.26 ± 0.04
Mare Smythii (82°–92.5°E)	0.45 ± 0.08	0.25 ± 0.05
Southern edge of Mare Tranquillitatis, Torricelli area (26°–30°E)	0.47 ± 0.09	0.23 ± 0.05
Eastern edge of Fecunditatis, Langrenus area (57°–64°E)	0.48 ± 0.07	0.27 ± 0.06
Ptolemaeus (4°W–0.5°E)	0.51 ± 0.07	0.21 ± 0.04
Highlands west of Ptolemaeus to Mare Nubium (4°–9°W)	0.51 ± 0.11	0.25 ± 0.12
Highlands west of Mare Fecunditatis (37.5°–42°E)	0.52 ± 0.07	0.24 ± 0.05
Highlands west of Smythii (72°–77°E)	0.57 ± 0.07	0.21 ± 0.03
Western border of Smythii (77°–82°E)	0.58 ± 0.08	0.22 ± 0.04
Highlands east of Descartes (20.5°–26°E)	0.58 ± 0.07	0.21 ± 0.04
South of Mare Spumans (64°–72°E)	0.58 ± 0.07	0.25 ± 0.04
Isidorus and Capella (30°–37.5°E)	0.59 ± 0.11	0.21 ± 0.05
Highlands west of Descartes (3°-14°E)	0.59 ± 0.11	0.21 ± 0.05
Eastern borders of Mare Smythii (92.5°–97.5°E)	0.61 ± 0.09	0.20 ± 0.06
Farside highlands (106°-118°E)	0.63 ± 0.08	0.16 ± 0.05
Descartes area, highlands, *Apollo* 16 landing site (14°-20.5°E)	0.67 ± 0.11	0.19 ± 0.05
East of Ptolemaeus (0.5°–3°E)	0.68 ± 0.14	0.28 ± 0.09
Highlands (97.5°–106°E)	0.68 ± 0.11	0.21 ± 0.05
Farside highlands west of Mendeleev (118°-141°E)	0.71 ± 0.11	0.16 ± 0.04

Source: I. Adler and J. I. Trombka (24).

tion was not completely insensitive to variations in the solar flux, it was sufficient to provide good quantitative estimates of the ratios.

The final results shown in the tables were obtained by integrating the spectra from a number of passes over the same feature. The following picture emerged from the X-ray measurements:

The Al/Si ratios were highest in the eastern limb highlands, and considerably lower in the mare areas. The extreme variation was a factor of 2, the lowest value occurring in the Imbrium basin region. The Mg/Si concentration ratios generally showed an inverse relationship to the Al/Si. If the gamma ray results for the Fe concentrations were considered, it was found that Fe is high in the low Al regions and low where the Al is high. The Al/Si and Mg/Si chemical ratios for the highlands corresponded to that for anothositic gabbro through gabbroic anorthosites or feldspathic basalts. By contrast, the chemical ratios for the mare areas corresponded to mare basalts. This result was consistent with the gamma-ray results on Fe, Th, U, and K previously described.

Early reports to the Apollo 16 astronauts (while the mission was in progress) of very high Al/Si ratios in the Descartes area, obtained by the orbital experiment, were subsequently confirmed by the analysis of the lunar samples returned from the site. It appeared from the data that some material samples at Descartes were similar to that of the eastern limb and far side highlands. This conclusion was further justified by the fact that the Mg/Si concentration ratio was about 0.18 for the returned material, close to the 0.19 reported by the orbital X-ray measurements. The Eastern limb highlands and the far side highlands were about 0.16–0.21.

In both the Apollo 15 and 16 the Al and Mg values derived showed an inverse relationship to each other in most instances, a finding consistent with the mineralogy. The inverse relationship between the Al and Fe, obtained from both the X-ray and gamma ray measurements was even more striking.

There were distinct chemical contrasts between such features as the small mare basins and the highland rims (note for example the crater Tsiolkovsky in Fig. 8.8).

A fruitful use of the X-ray fluorescence data was in comparing the Al/Si ratios against measurements of the optical albedo. These observations were particularly significant in view of the long-standing discussions about whether these albedo differences represent solely topographic differences or also compositional differences among the surface materials.

Early workers such as Whitaker (17) and others recognized convincing evidence for compositional changes where sharp contrasts in albedo occurred. However, it remained for the later Surveyor, Apollo, Lunar, and Lunakhod missions to provide quantitative compositional data. Chemical differences related to albedo variations were first confirmed by the alpha backscattering experiment carried on the Surveyors V, VI, and VII (18).

The Surveyor V and VI experimental data were used to analyze widely separated mare sites and chemically similar surface materials were reported for each site. The Surveyor VII experiment, on the other hand, observed a highland site, and data analysis showed a significant chemical difference between that region and the two mare locations. The Surveyor results and the analysis of returned lunar samples confirmed that albedo is indeed affected by composition as well as topographic features. The X-ray experiment on Apollo 15 and 16 provided data for the correlation of regional albedo with surface composition on essentially a global scale—at least for those elements determined by the X-ray experiment.

The data from both Apollo flights exhibited an excellent correspondence between Al/Si and optical albedo values. An example from the Apollo 16 flights is shown in Figure 8.9. There is a positive correlation

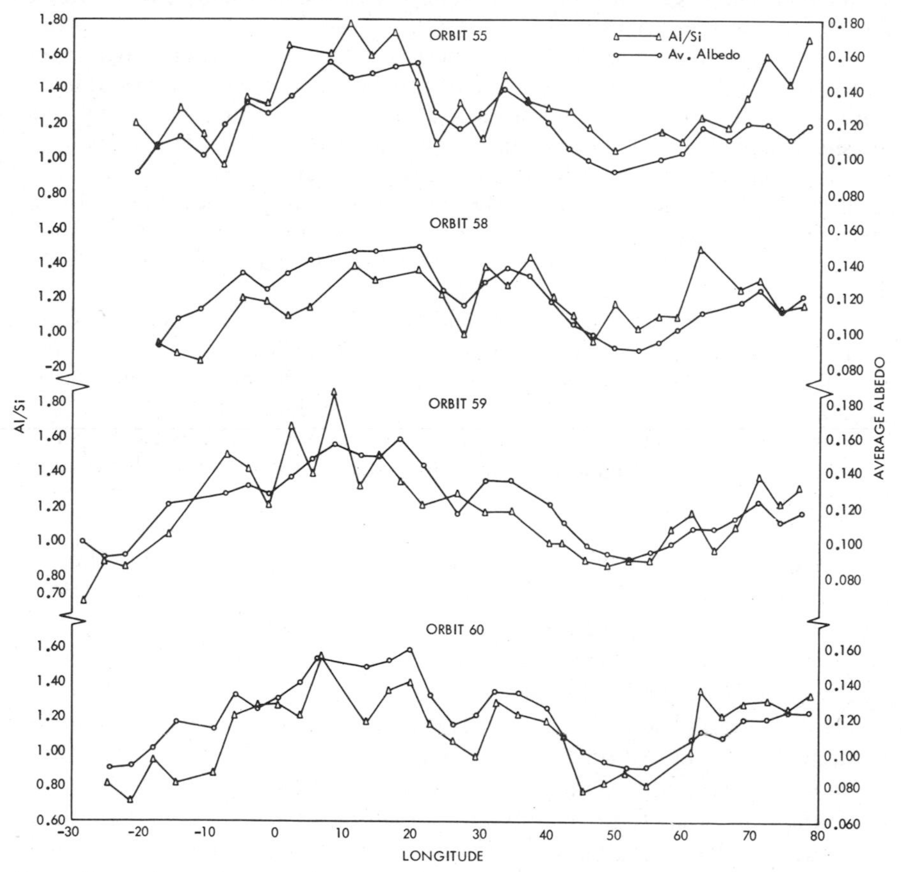

Figure 8.9. Comparison of Al/Si ratios versus optical albedos for several Apollo 16 orbits.

between the albedo and the Al/Si values, although the rate of change is not always similar. In the Apollo 15 data, the main anomalies were observed where small Copernican-type craters (young, rayed, bright albedos) occurred and produced abnormally high albedo values. This anomaly was attributed to highly reflective, finely divided ejecta, rather than to compositional changes. A similar anomaly was noted in the Apollo 16 data around 27° west longitude in a Tranquillitatis embayment north of Theophilus. Four Apollo 16 orbits are plotted. Orbits 58 and 60 show the expected decrease in Al/Si with decreasing albedo. Orbits 55 and 59, on the other hand, show an occasional decrease in Al/Si ratio, although the albedo also decreases. This may record the existence of an old ''weathered'' ray consisting of Al-rich, highland-derived ray material that has lost its high reflectivity.

The X-ray fluorescence experiment was particularly well suited for a study of the general problem of the horizontal transport of lunar material on the surface. In particular, a significant question was whether or not the material in the mare was of highland origin and carried there by electrostatic forces. To appreciate why the X-ray experiment was so useful in casting some light on this problem, consider that the X-ray measurements were shallow, a 100-μm layer of basalt or feldspar would effectively represent infinite thickness.

In a paper published by Gold in 1973 (19), it was reported that horizontal transport by a mechanism like electrostatic charging and levitation had played a large role in the formation of the flat mare basins. However, it appears obvious that if highland material had drifted into the basins to any extent, the difference between the mare and the surrounding highlands seen by the X-ray measurements would have been either totally extinguished or certainly greatly obscured. In actual fact, marked differences were observed.

One outstanding illustration is the crater Tsiolkovsky (see Fig. 8.8). The ratio of the Al in the rim area to that of the basin is about 2 : 1. Further, real differences can be seen in such relatively homogeneous sites as the Serenetatis–Tranquillitatis zones and in the Tranquillitatis basin itself. These findings were substantiated in a paper by Kochariv and Victorov (20) based on the observations obtained with Lunakhod 2 in the crater Le Monnier. Based on the analysis of the returned samples, it was found that the soils of the highland regions are generally like the highland rocks, and in the mare regions the soils are like the maria rocks.

While the X-ray fluorescence experiment provided relatively unambiguous data about large-scale features such as contrasts between the mare and the highlands, the measurements also were employed in looking at small lunar features. Because of the spacecraft's motion, short time (8-

and 16-s) X-ray data were analyzed in order to determine their applicability for mapping these relatively small features. It was understood that fluctuations in solar spectral shape and variations of the look angle with spacecraft altitude made this type of analysis rather difficult; however, some analysis did prove possible and some results of a preliminary nature were obtained (21). Using a technique of trend surface analysis, it was possible to demonstrate that a usable signal could be abstracted from Al/Si ratios compiled for short time periods. The details of the analytical method have been published by Podwysocki in 1974 (21).

As a demonstration, a portion of Mare Serenitatis and Tranquillitatis and their adjacent highlands, observed during the Apollo 15 flight, were chosen. Based on the analysis of 16 second Al/Si data, a fourth-order surface was found to be the highest order that showed a proper level of significant improvement. Podwysocki and associates were able to extract an amazing amount of detailed information about the chemistry of the area, which agrees well with geological observations.

Andre and co-workers (22) used the Al/Si X-ray orbital data to study the chemical character of the Smythii basin. The variations in the surface chemistry in the Smythii region were particularly interesting because: (1) the basins appeared, from photography, to be only partially filled with basalt, and (2) the topographic relief was the greatest measured for any lunar feature.

The purpose of the investigation was to determine the following: How the Al/Si values of the mare soils compared with those for other nearside mare soils; whether the seemingly unflooded western third of the basin consisted of reworked basalt; the explanation of the differences between the Al/Si values for the adjacent terra units to the east and west of Mare Smythii; and finally, how did the ratios compare to those for the other highland areas flown over.

By the time this particular study was undertaken, a number of methods for the processing of the X-ray data had been developed. First, a trend surface was produced using the procedure described by Podwysocki (21). The method was used to depict broad chemical changes in the Smythii region using 409 8-s data points obtained from the Apollo 16 mission. The trend surface showed sharp chemical differences between the extremely high Al/Si values east of Smythii and low values in the northeastern quadrant of the basin where mare basalts had flooded topographic lows. These low values were confirmed by a cluster of low Al/Si ratios from the three Apollo 15 orbits that crossed the area. By contrast, the Al/Si values increased gradually across the western section of the basin into the western terra adjacent to Smythii.

A second approach was to use an Al/Si profile (Fig. 8.10), which em-

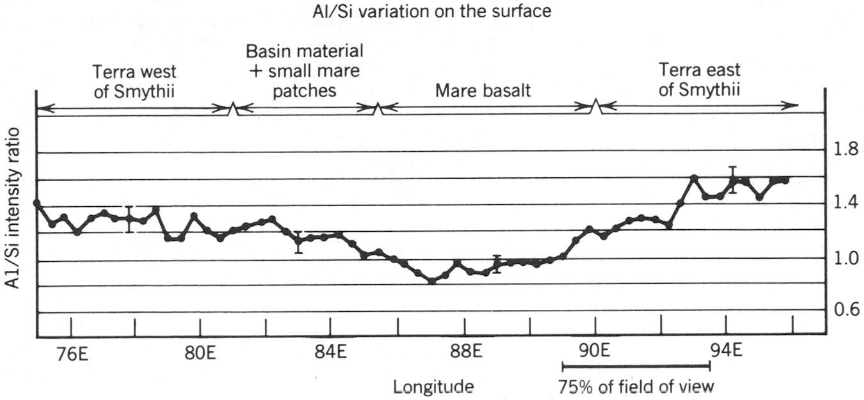

Figure 8.10. Profile of the mean 16 sec Al/Si intensity ratios from the five Apollo 16 orbits versus longitude. Source: Andre et al. (22).

phasized more subtle chemical variations between the flooded and unflooded areas of the basin and between the highlands to the east and west of Smythii. The profile shows that the terra east of Smythii has an Al/Si ratio that is considerably higher than the western terra. The mare regolith was clearly distinguishable from the regolith in the western third of the basin and from the eastern highlands.

Significantly, the differences in Al/Si ratios between the western basin and highlands could not easily be resolved, despite an extremely abrupt change in elevation—an observation consistent with that portion of the basin's actually being uncovered highland material. There were 16-s data points plotted on the profile. These are the average of data points from the same approximate longitude from five orbits where the latitude range was no greater than 2.5°. Such averaging improved the signal-to-noise ratio and resulted in greater precision. Shown are error bars, which are the regional average standard deviation of the mean calculated for each of the average data points on the graph.

The third procedure involved averaging individual data points for each region, excluding points with a composite signal for highland–basin, basin–mare, and mare–highland contacts. The results ae shown in Table 8.8. The difference of 0.21 in Al/Si intensity ratio between the eastern and western highlands adjacent to Smythii represents 15% of the total range of values in the study area from 24 Apollo 16 orbits.

The fourth method consisted of applying imaging techniques to the lunar data sets. These methods have been developed by the U.S. Geological Survey in Flagstaff, Arizona, and are very effective in integrating a

Table 8.8. Comparison of Al/Si Intensity Ratios Averaged for Each of Four Regions[a]

	Average Al/Si Intensity	Approx. Al/si Concentration	Data Points	Std. Dev. of Mean Intensity
Highlands west of Smythii (75E–80E)	1.26	0.55	39	±.03
Basin material (81E–85E)	1.22	0.53	13	±.04
Mare basalt soils (85E–92E)	0.90	0.38	47	±.04
Highlands east of Smythii (93E–98E)	1.47	0.65	47	±.04

[a] These four regions are defined in Figure 8.10.
Source: Andre et al. (22).

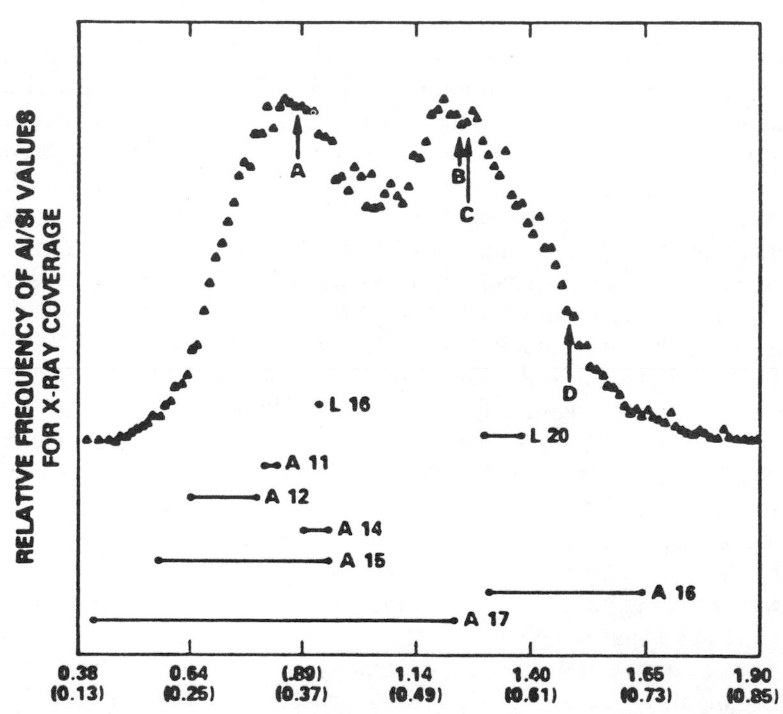

Figure 8.11. Bimodal frequency distribution of Al/Si intensity ratios and (concentrations) after near neighbor smoothing (to achieve the best balance between statistical reliability and spatial resolution). The letters indicate the average values for four regions in the Smythii region. Al/Si ranges from chemical analysis of soils from the Apollo (A) and Luna (L) missions are shown by the horizontal bars. Source: Andre et al. (22).

very large number of data points into a composite image. In this particular study some 9000 data points from Apollo 15 and 16 were used. The nature of the technique has been described by Bielefeld and associates (23). A useful product of this imaging technique was the disclosure of a bimodal frequency distribution of Al/Si intensity ratios as shown in Figure 8.11. The two peaks of this bimodal distribution represent typical Al/Si values for mare regolith and highland regolith in the areas overflown by both missions.

There are numerous other examples of the application of the X-ray fluorescence experiment to the study of small lunar features where the chemical data provided information about processes and subsurface compositions. A study of the small basins and craters was able to provide information on the depth of penetration and the degree of overturning and whether the crust had been breached. In most instances the interpretations required a careful analysis of the data using methods of data reduction like those described above. To compensate for the basically crude precision (compared with analysis in a terrestrial laboratory), it was frequently necessary to "use every trick in the book." However, the reader must bear in mind that we are here dealing with truly remote chemical analysis.

REFERENCES

1. A. L. Turkevitch, E. J. Franzgrote, and J. H. Patterson, *Science,* **165,** 277 (1969).
2. A. P. Vinogradov, Y. U. A. Surkov, G. M. Chernov, F. F. Kornozov, and G. B. Nazarkina, *Cosmic Research* (English transl.), **5,** 741 (1967).
3. J. Carpenter, P. Gorenstein, H. Gursky, B. Harris, J. Jordan, T. Callum, M. L. Ortman, and L. Sodickson, NAS5-11086, 1968.
4. H. W. Kraner, G. L. Schroeder, G. Davidson, and J. W. Carpenter, *Science,* **152,** 1235 (1966).
5. W. H. Tucker and M. Koren, *Astrophys. J.,* **168,** 283 (1971).
6. P. Gorenstein, H. Gursky, I. Adler, and J. I. Trombka, *X-Ray Anal.,* **13,** 330 (1969).
7. E. Eller, private communication (1972).
8. J. R. Arnold, L. E. Peterson, A. E. Metzger, and J. I. Trombka, *Preliminary Science Rept.,* NASA SP-289, 1972, p. 16-1.
9. T. M. Harrington, J. H. Marshall, J. R. Arnold, L. E. Peterson, J. I. Trombka, and A. E. Metzger, *Nucl. Instr. Meth.,* **118,** 401 (1974).
10. R. C. Reedy, J. R. Arnold, and J. I. Trombka, *J. Geophys. Res.,* **26,** 5847 (1973).

11. A. E. Metzger, J. I. Trombka, L. E. Peterson, R. C. Reedy, and J. Arnold, *Science,* **179,** 800 (1973).

12. I. Adler, J. Trombka, J. Gerard, R. Schmadebeck, P. Lowman, H. Blodgett, L. Yin, E. Eller, R. Lamothe, P. Gorenstein, P. Bjorkholm, B. Harris, and H. Gursky, *Preliminary Science Rept.,* NASA SP-289, 1972, p. 17-1.

13. I. Adler, J. Gerard, J. Trombka, R. Schmadebeck, P. Lowman, H. Blodgett, L. Yin, E. Eller, R. Lamothe, P. Gorenstein, P. Bjorkholm, B. Harris, and H. Gursky, *Proc. Lun. Sci. Conf. 3rd,* M.I.T. Press, Cambridge, MA, 1972, p. 2157.

14. I. Adler, J. Trombka, J. Gerard, P. Lowman, R. Schmadebeck, L. Yin, H. Blodgett, E. Eller, R. Lamothe, P. Gorenstein, and P. Bjorkholm, *Science,* **175,** 436 (1972).

15. I. Adler, J. Trombka, P. Lowman, R. Schmadebeck, L. Yin, H. Blodgett, E. Eller, R. Lamothe, R. Oswald, P. Gorenstein, H. Gursky, P. Bjorkholm and B. Harris, *Science,* **177,** 256 (1972).

16. I. Adler, J. Trombka, R. Schmadebeck, P. Lowman, H. Blodgett, L. Yin, E. Eller, M. Podwysocki, J. R. Weidner, A. L. Bickell, R. K. L. Lum, J. Gerard, P. Gorenstein, P. Bjorkholm, and B. Harris, *Proc. Lun. Sci. Conf. (Suppl. 4 Geochim. Cosmochim. Acta),* **3,** 2783 (1973).

17. E. A. Whitaker, in H. H. Hess et al., Eds., *The Surface of the Moon,* The John Hopkins Press, Baltimore, 1965.

18. J. H. Patterson, A. L. Turkevich, E. J. Franzgrote, T. E. Economou, and E. P. Sowinski, *Science,* **168,** 825 (1972).

19. T. Gold, *The Moon,* **7** (3/4) (1973).

20. G. E. Kocharov and S. V. Victarov, *Doklady Akad. Nauk S.S.S.R.* (1977).

21. M. H. Podwysocki, J. R. Weiner, C. G. Andre, A. L. Bickel, and R. S. Lum, *Proc. Lun. Sci. Conf. 5th* (1974).

22. C. G. Andre, R. W. Wolfe, I. Adler, P. E. Clark, J. R. Weidner, and J. A. Philpotts, *Proc. Lun. Sci. Conf. 8th* (1977).

23. M. J. Bielfeld, C. G. Andre, E. M. Eliason, P. E. Clark, I. Adler, and J. I. Trombka, *Proc. Lun. Sci. Conf. 8th* (1977).

24. I. Adler and J. I. Trombka, *Physics and Chemistry of the Earth,* Vol. 10, Pergamon, New York, 1977.

CHAPTER

9

REMOTE IN SITU ANALYSIS OF PLANETARY SURFACES

THE ALPHA-SCATTERING CHEMICAL ANALYSIS EXPERIMENT ON THE SURVEYOR LUNAR MISSIONS

The experiments to be described in this chapter represent the height of scientific ingenuity and imagination and accomplishments that are truly remarkable. Among these was the series of lunar investigations employing alpha backscatter and proton spectroscopy for the performance of in situ chemical analysis. The alpha backscatter experiment was carried for the first time on the Surveyor V spacecraft, and it represented the first successful in situ compositional analysis on the surface of a planetary body other than the Earth. This was then followed by similarly successful efforts deployed by the Sureyor VI and VII spacecraft.

The history of the alpha scattering technique has been summarized by Turkevich and associates in their final report to NASA (1). The large-angle scattering of alpha particles by matter was initially reported by Geiger and Marsden in 1909 (2). Rutherford (3) used this phenomenon as the basis for his nuclear model of the atom.

The use of scattered alpha particles for obtaining chemical analyses of surfaces was originally proposed by Professor S. K. Allison of the Enrico Fermi Institute for Nuclear Studies. The technique was then investigated by A. Turkevich (4) who presented experimental data confirming that the energy spectra of the backscattered alpha particles were characteristic of the elements present in the scattering material. As an outgrowth of these results, a proposal was made that a rugged, compact, analytical instrument could be designed and built for obtaining the chemical composition of the lunar surface under the flight conditions being considered for the Surveyor Lunar Landing missions. The main features of the proposed experiment was summarized by Patterson in 1965 (5). The principles are as follows.

PRINCIPLES

In 1960, Rutherford, Chadwick, and Ellis were able to demonstrate that scattered energetic alpha particles carry off some fraction of their original kinetic energy. The maximum fraction of this remaining initial energy could be calculated from the relationship

$$\frac{T_{\max}}{T_0} = \left(\frac{4_{\cos}\,\theta}{A} + \sqrt{1 - \frac{16}{A^2}\sin^2\theta}\right)^2 \Big/ \left(1 + \frac{4}{A}\right)^2 \qquad (1)$$

where θ is the scattering angle, A the mass number of the scatterer, and $T_{\max}/T_0$ the fraction of kinetic energy remaining in the scattered alpha particle. For scattering angles close to 180°, the backscatter direction, the above expression reduced to:

$$\left(\frac{T_{\max}}{T_0}\right)_{180°} = \left(\frac{A-4}{A+4}\right)^2 \qquad (2)$$

where now $T_{\max}/T_0$ represents the high-energy cutoff or the minimum energy loss associated with a single backscatter collision.

The role of target thickness is shown in Figure 9.1. Where the elemental sample is only a few atoms thick, the energy spectrum of the backscattered alpha particles would consist of a narrow peak whose energy is determined by Eq. (1). For thick samples, the alpha particles scattered at various depths below the surface suffer a loss of energy in the material before and after the scattering event. The distribution is seen to be a

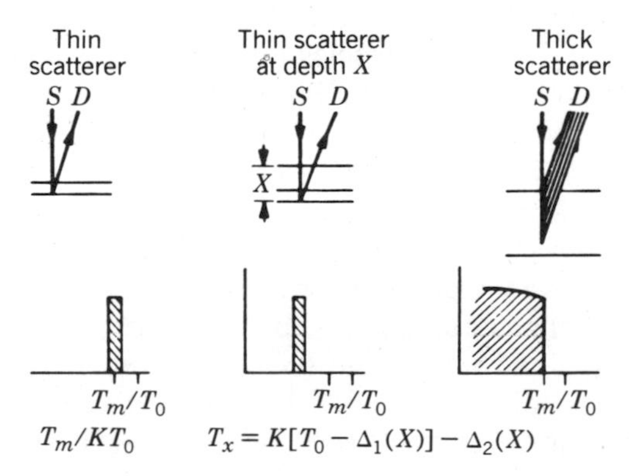

Figure 9.1. Comparison of spectra of alpha particles scattered from thin and thick samples. The scattering geometries are shown in the upper diagram and the lower diagram shows the corresponding spectral shapes. Source: A. L. Turkevich (1).

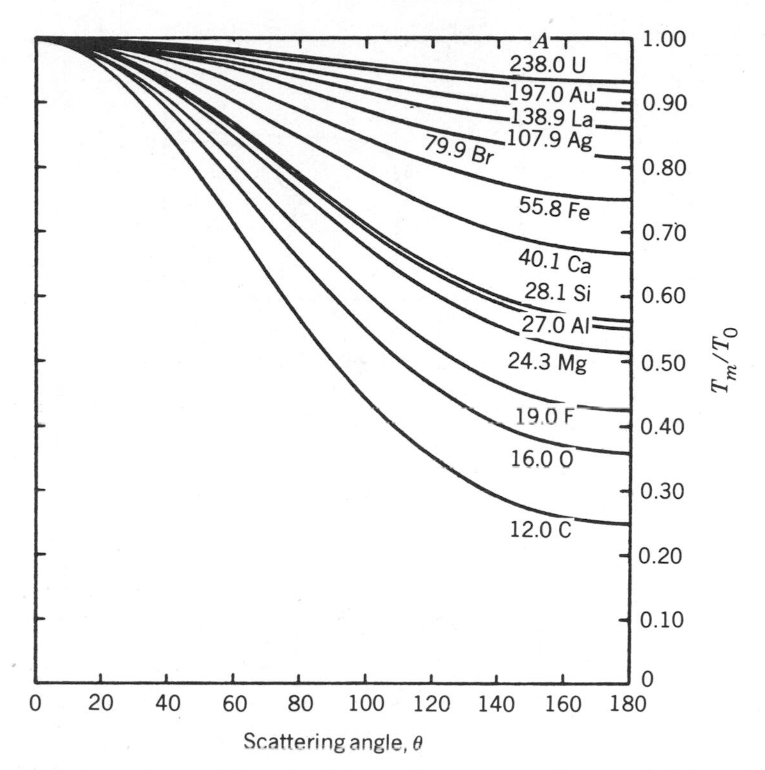

Figure 9.2. Dependence of the ratio of the maximum energy of scattered alpha particles to the initial energy, T_m/T_0; on the laboratory angle of scattering, θ; and on the mass number A. Source: A. L. Turkevich (1).

continuous spectrum terminating sharply at the energy T_m, determined by the mass number of the element.

Figure 9.2 shows a series of plots of T_m/T_0 as a function of scatter range for various mass numbers in accordance with Eq. (1). We see the following:

1. The greatest resolution between mass numbers is at a scattering angle of 180° (although there is little change above 160°).
2. The resolution decreases rapidly with increasing mass number.

Practically, the resolution limit for distinguishing individual elements for the Surveyor instrument package ended at about atomic number 40.

Some examples of alpha spectra (referred to as the library elements) for some light elements and some of the heavier ones are shown in Figure 9.3. The shape, typically, for elements heavier than calcium is a relatively flat plateau terminating in a sharp dropoff to zero at T_m, as required by Eq

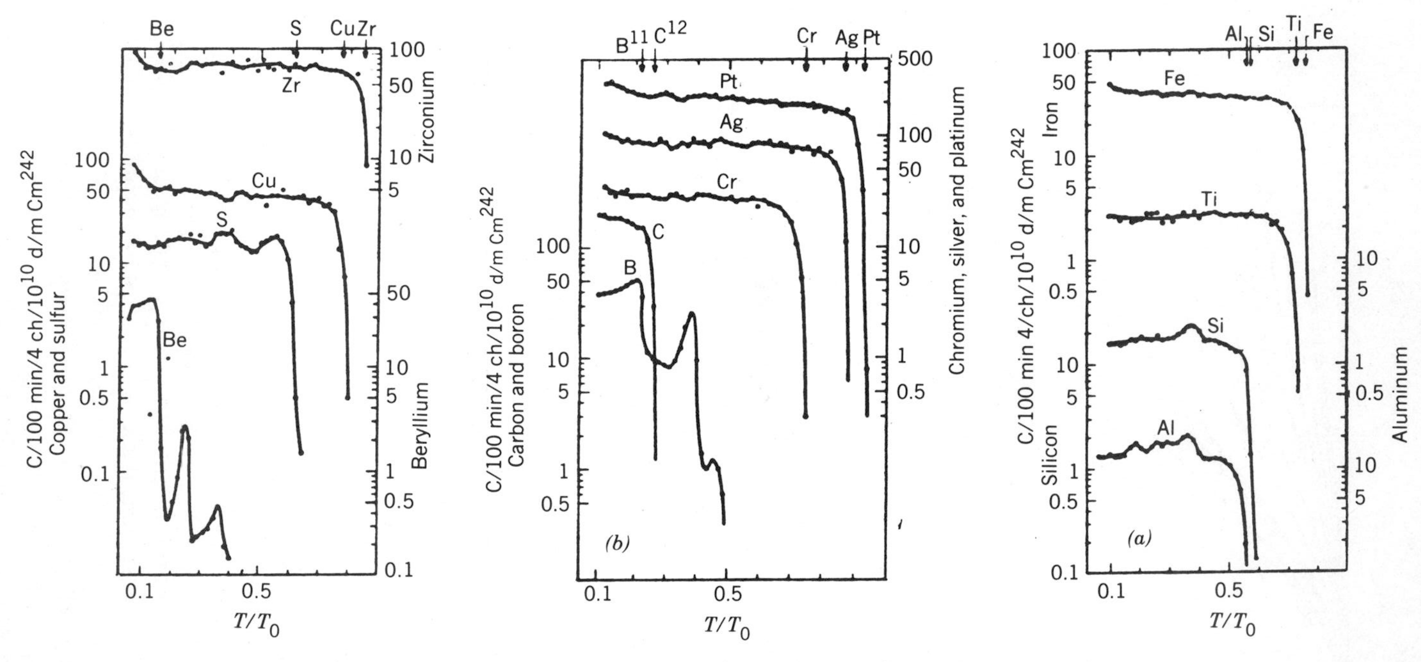

Figure 9.3. Typical alpha-scattering spectra of pure elements. Source: Patterson et al. (6).

(1). The intensity depends on the scattering cross-section for alpha particles of the target nuclei and energy loss of the alpha particles caused by ionization in the sample. In view of the fact that the spectral shape of the heavier elements was nearly constant, it was necessary to include only a few representative spectra in the library. This was, however, not the case for the light elements.

The regularity of the relationship between the endpoint energy fraction and the mass number of the scattering element is shown in Figure 9.4.

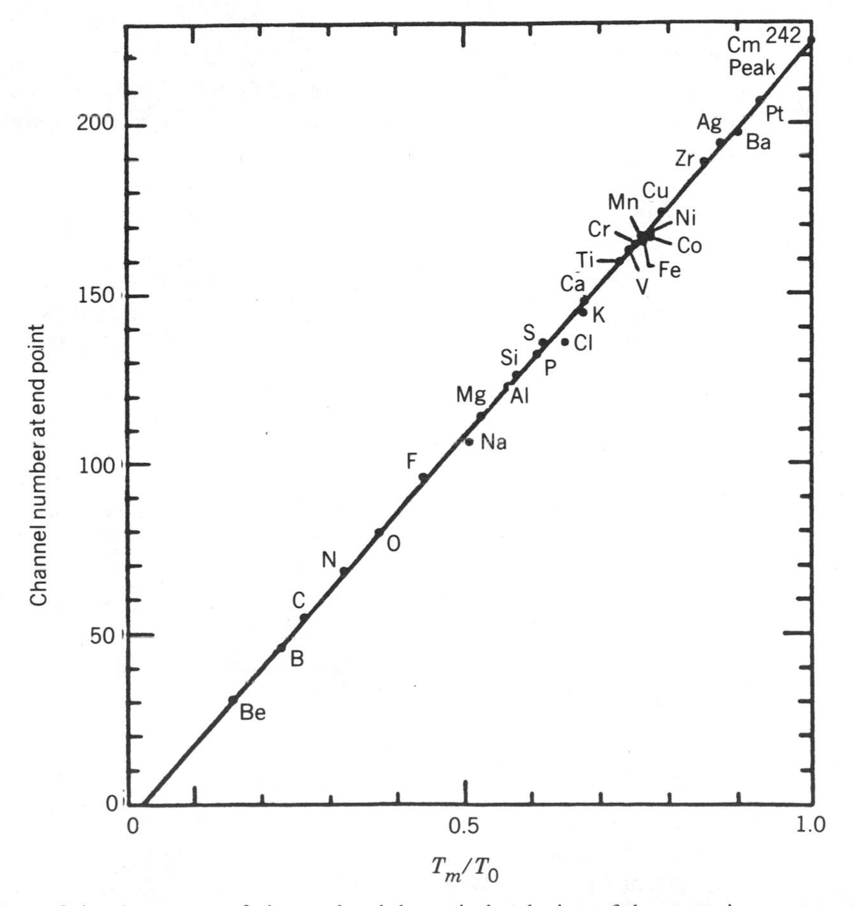

Figure 9.4. Agreement of observed and theoretical endpoints of the scattering spectra of various elements as obtained with research apparatus. The straight line drawn through $T_{max}/T_{0=1}$ at channel 224 (the channel of peak energy of the incident particles) indicates how well the experimental end points agree with the predictions of Eq. 1. The negative intercept at $T_{max}/T_0 = 0$ is consistent with the positive energy threshold of the analyzer. Source: Patterson et al. (6).

The experimentally observed data points are in excellent agreement with the mathematical relationships described above. Significantly, the end-point dispersions are large for the low-atomic-number elements and compressed for the higher atomic number elements. The technique is thus most effective for the low mass ranges. It was found that, with an instrumental resolution of about 3%, it was difficult to resolve elements of atomic number greater than 25. Coupled to this was an additional complication caused by the presence of several isotopes of different elements with the same mass number.

For the high-atomic-number elements, the shape of the spectra and the intensities depend on Rutherford scattering of the incident alpha particles. Given a known geometry, it is, in principle, possible to calculate intensities from the theoretical coulomb scattering cross section and the energy loss characteristics of the scattering substances. However, in practice complications are introduced by an incomplete knowledge of the loss cross sections and multiple scattering phenomena, both effects being most pronounced at low energies.

The geometrical relationships in alpha particle scattering are shown in Figure 9.5. Based on the work of Patterson and co-workers, the following expression has been developed for the scattered at a given angle θ as a function of energy:

$$dI/dT_3 = I_0 n\sigma(T_1)\ dy_1/dT_3 \tag{3}$$

The intensity of the backscattered particles per steradian, at a given energy T per unit emergent energy is a function of I_0, the intensity of the incident particles; n, the number of scattering nuclei per cubic centimeter; $\sigma(T)$, the alpha scattering cross section for the nuclei of interest at some emergent energy T; and dy/dT_3, variation in depth at which the scattering occurs with a change in emergent energy. The scattering cross section in turn is defined as

$$\sigma(T_1) = \frac{5.184 \times 10^{-27} Z^2}{T_1^2(\text{MeV})} \frac{1}{\sin^4\theta/2} \tag{4}$$

and thus depends on the atomic number Z of the scatterer, the energy T of the alpha particle at the point of recoil, and the scattering angle θ.

This expression tells us that it is essential to know the nature of the energy loss by the alpha particle due to ionization and to define the trajectories of the alpha particle in the target. Patterson and associates (5) have pointed out, in regard to the spectral shapes, that the observed character of the spectra for high Z elements are "consistent with Rutherford scat-

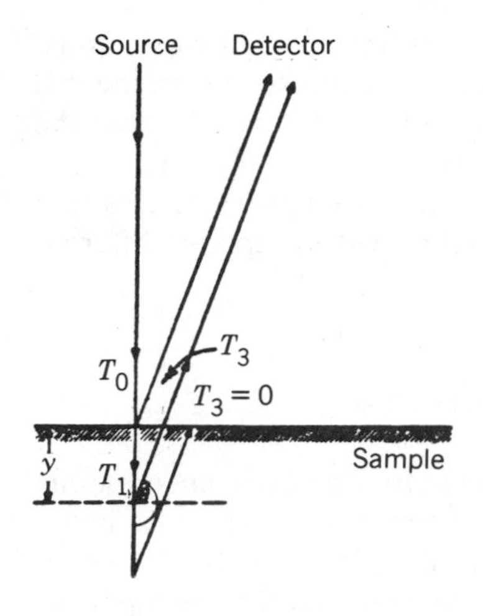

Figure 9.5. Geometrical relationships in alpha particle scattering from a thick target. The distances from the source and detector are assumed to be very large (centimeters) in comparison with the penetration of the alpha particles into the sample (micrometers). Source: Patterson et al. (6).

tering cross section and the stopping powers of the target to the extent that they are known. Comparisons of the intensities of Rutherford scattering is best done near the high-energy endpoints of the respective scatterers.'' It has been shown that in this region of the spectrum, the intensities of the alpha particles entering the detector (in the backscatter direction) has a $Z^{3/2}$ dependence. For element lighter than Na, however, the regularity disappears and the scattering intensity actually becomes higher than predicted. Furthermore, the scattering varies irregularly from element to element.

The structure in the spectra of the low-Z elements is apparently due to the departure from the $Z^{3/2}$ dependence. This has been attributed to nuclear interactions with the target nuclei.

The alpha scattering also offers an additional mode for obtaining analytical information. It is known that the energetic alpha particles will produce characteristic proton spectra as a consequence of (alpha, proton) reactions with a number of elements. B, N, F, Na, Mg, Al, and Si yield protons with a useful range of energies for measurement consistent with known nuclear masses and energy levels. The energy spectra and proton yields depend on the cross sections for these reactions. The method is limited. For incident alpha energies up to 6 meV, such nuclides as ^{10}Be, ^{12}C, ^{16}O, ^{17}O, and ^{18}O do not yield protons. In addition, for elements with nuclear charges greater than 20, the coulomb barrier becomes great enough to greatly lower the (alpha, proton) cross-sections.

The alpha–proton method is a useful supplemental approach to the alpha scatter mode. In general, the proton intensities are frequently much lower than that of the scattered alpha particles. The sensitivity can be improved, however, by using a detector system sensitive only to protons. It has been observed in the analysis of rocks, for example, where there is a relatively low abundance of Al and Na relative to Si and Mg, that the scattered alpha spectra are insensitive to variations in Al and Na concentrations. On the other hand, because they extend to higher energies, the proton spectra have characteristic shapes and provide a superior differentiation for these two elements.

INSTRUMENTATION

Having examined the principles of the alpha scattering experiment, let us now look in some detail at how the actual implementation was done in the series of flight experiments flown on the Surveyor V, VI, and VII. These experiments have been described by Turkevich (1). The equipment, shown in Figure 9.6, included a sensor head deployed to the lunar surface, a digital electronics package kept in a thermally controlled compartment on the Surveyor spacecraft, a deployment mechanism, and a standard sample assembly. The total weight of the assembly, including the mechanical and electrical interface substructures and cabling, was 13 kg. Power dissipation was remarkably low, about 2 W, increasing to 17 W during heater use.

Figure 9.6 shows internal construction of the alpha scattering head in cutaway view. The important details include the alpha detectors, the proton detectors, and the radioactive source of alpha particles. The apparatus was a box approximately 17 × 16.5 × 13 cm in size. The bottom of the box was a plate 30.5 cm in diameter, a device intended to keep the box

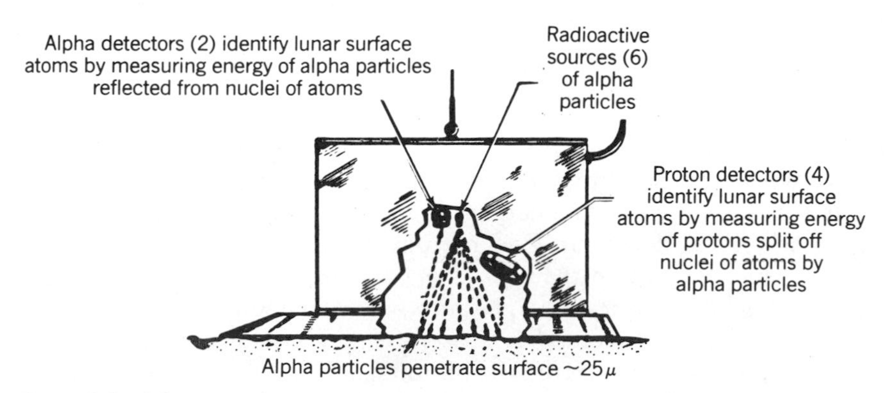

Figure 9.6. Diagrammatic view of alpha-scattering sensor head internal configuration. Source: Turkevich et al. (1).

from sinking into the lunar soil should it prove to be soft. The bottom of the sensor head had a sample port about 11 cm in diameter. Six 1-cm alpha sources were recessed 7 cm above the opening. These alpha sources were collimated so that the alpha particles were directed to the surface material through the opening in the sensor head. In order to prevent the alpha particles, recoiling from the sources themselves, from reaching the surface of the sample area, each source was covered by a thin aluminum oxide film. The two alpha particle detectors were arranged to detect the backscattered alpha particles at an angle of about 174° from the original direction. These alpha particle detectors were silicon-surface barrier types, 0.2 cm, with evaporated gold front surfaces. In addition, thin films were mounted on the collimation masks to prevent the detectors from excessive light and alpha contamination.

Four proton detectors were also included. These were lithium drifted silicon detectors, 1 cm^2 each. The detectors were protected from the alpha particles by 11-μm-thick gold foil. In order to measure only those events associated with the sample, the proton detectors were backed by guard detectors in anticoincidence mode. Solar events registering in the guard and primary detectors were rejected by the electronics.

Two techniques were used to provide calibration of the electronics. One was a pulser supplying electronic pulses of two known magnitudes at the input stage of the alpha and proton detectors on command from the earth; the other was a small amount of the element einsteinium, an alpha emitter, located on the gold foil facing each proton detector and on the thin films located in front of the alpha detectors.

The device for deployment is shown in Figure 9.7. The deployment mechanism, in addition to its use for translating the sensor head to the surface, was also used to stow the sensor during flight and to provide both background and calibration measurements prior to the surface measurements. As the figure shows, there was a standard assembly covering the circular opening in the bottom of the sensor head during flight and landing. In addition to serving as a dust and light barrier, their assembly also helped to evaluate the instrument performance immediately after landing. Before deployment to the surface, the sensor head was held suspended about 50 cm above the lunar surface in order to obtain background data. A typical operational sequence consisted of:

1. Measurements were made in the stowed position, the data being compared to prelaunch numbers.
2. The supporting platform and standard were moved aside and the sensor head suspended about 50 cm above the lunar surface to obtain backgrounds (cosmic ray, solar protons, and possible surface radioactivity).

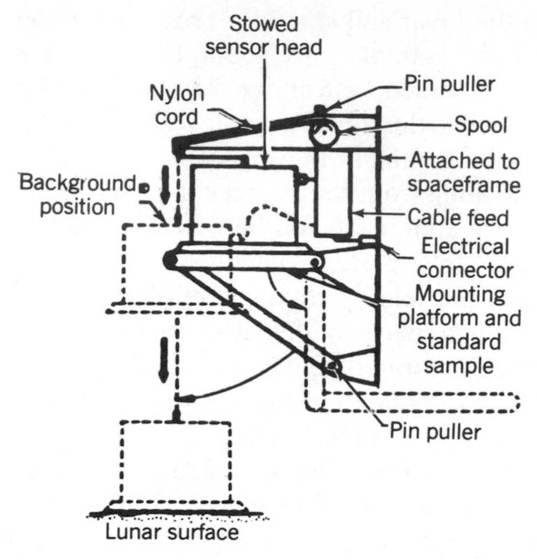

Figure 9.7. Deployment mechanism for the alpha backscatter experiment. Source: Turkevich et al. (1).

3. Finally, the sensor head was lowered to the lunar surface for the accumulation of surface data.

RESULTS

The first Surveyor spacecraft (Surveyor V) to carry the alpha experiment landed in the southwest portion of Mare Tranquillitatis on September 11, 1967. The spacecraft landed on a slope of about 19.5° on a crater wall. In an account of the mission, Turkevich and co-workers (6) summarized the data accumulation times. This is shown in Table 9.1.

Table 9.1. Science Data Accumulation

Operational Configuration	Accumulation Time (min)
Transit	20
Stowed (standard sample)	75
Background	170
Lunar surface sample 1	1056
Lunar surface sample 2	4005
Calibration	281
Total	5607(93.5 h)

Two sets of measurements were made of the standard sample (an analyzed glass) referred to above. The first measurements were made during lunar transit and the second after touchdown on the lunar surface. The data shown in Figure 9.8 were taken on the Moon over a 60-min measurement period. The error bars displayed are for a 1σ statistical error. The calculated results are shown in Table 9.2. A comparison with the standard chemical analysis shows a remarkable agreement.

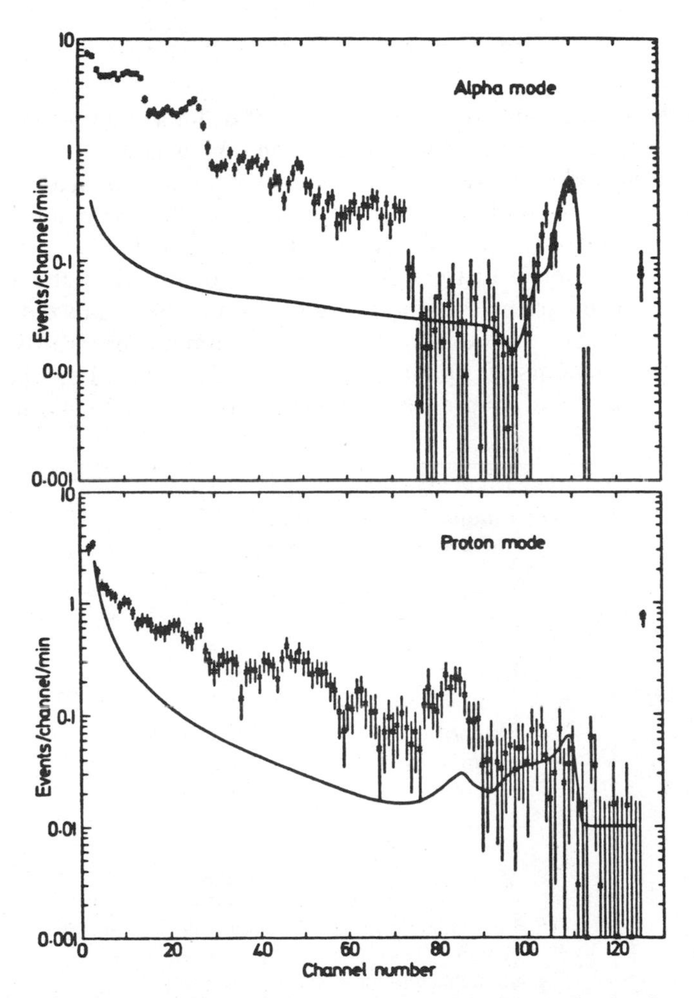

Figure 9.8. Measurement of a standard glass sample on the Moon after landing. The time interval was 60 min. The error bars on the observed points are for a 1σ statistical error. The peaks on the right are due to ^{254}E. Source: Turkevich (1).

Table 9.2. Analysis of the Standard Glass Sample on the Moon in Atomic %

Element	Surveyor V	Standard
O	56.4	58.6
Na	7.3	7.7
Mg	7.6	8.5
Al	2.0	1.5
Si	20.2	17.2
Fe	8.5	6.5

The first lunar sample data taken in both the alpha and proton modes is shown in Figure 9.9. The total accumulation time was 900 min (15 h). The curves are plotted with the number of events per channel on a logarithmic scale as a function of channel number (energy). The prominent peaks in both modes at approximately channel 110 are due to the einsteinium. There are a number of distinct energy breaks identified by the arrows at the top of the diagram. Figure 9.10 shows computer-analyzed data for both modes. The observed lunar sample spectrum has been broken down into eight components: C, O, Na, Mg, Al, Si, "Ca," and "Fe." Turkevich has stated that "Ca" represents elements $28 < A \leq \sim 45$ and "Fe"

Table 9.3. Chemical Composition of the Lunar Surface of Surveyor V Site

Element	Atomic %[a]
Carbon	< 3
Oxygen	58.0 ± 5
Sodium	< 2
Magnesium	3.0 ± 3
Aluminum	6.5 ± 2
Silicon	18.5 ± 3
$28 < A < 65$[b]	13.0 ± 3
(Fe, Co, Ni)	> 3
$65 < A$	> 0.5

[a] Excluding hydrogen, lithium, and helium. These numbers have been normalized to approximately 100%.
[b] This group includes, for example, S, K, Ca, and Fe.
Source: Turkevich et al. (1).

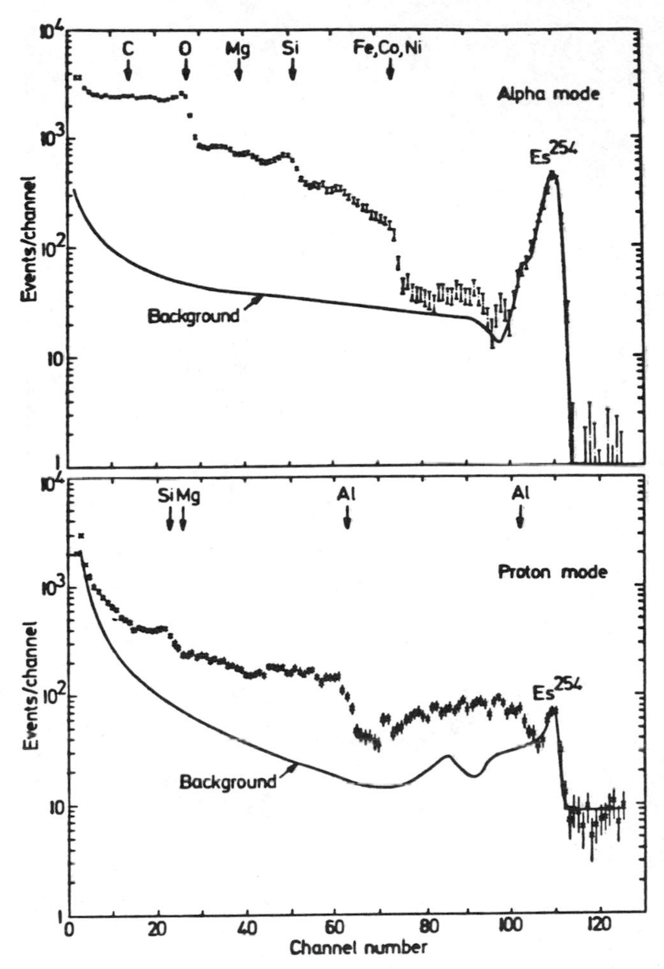

Figure 9.9. First lunar sample data taken on the Moon for a 900-min measurement time. The experimental points are crosses and the error bars a 1σ statistical error. The peaks at about channel number 110 are due to ^{254}E. Source: Turkevich et al. (1).

represents elements $45 \leq A < 65$. The computer fit was performed by first subtracting the background and possible heavy element contribution. The resolution of the spectra into only eight elements provides an excellent fit to the observations with very few systematic deviations. The one region of questionable fit is between channels 63 and 74 in the alpha mode. For this reason, the elements were reported as a composite. Table 9.3 shows the estimates of the chemical composition at the lunar surface of the Surveyor V site.

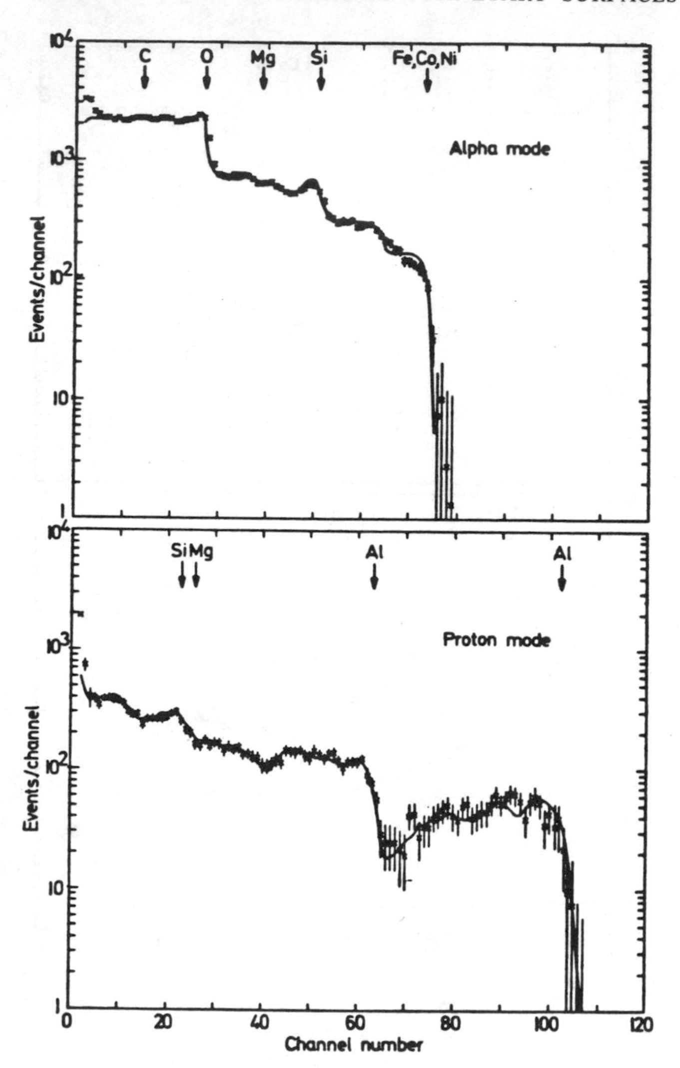

Figure 9.10. Computer analysis of the first lunar samples. The smooth curves are calculated spectra based on an eight-element library. Source: Turkevich et al. (1).

The alpha-scattering experiment flown on Surveyor VI was a duplicate of the one flown on the Surveyor V. A report of the results was made by Turkevich and associates (7). Once again, a flat area along the equatorial zone was chosen as a site. In this instance, the landing site was Sinus Medii. The equipment, except for minor problems, functioned as well, and it was possible to obtain a comparison of the chemical compositions

Table 9.4. Comparison of Chemical Composition at the Surveyor V and VI Sites in atomic T[a]

Element	Surveyor V[b]	Surveyor VI
Carbon	< 3	< 2
Oxygen	58 ± 5	57 ± 5
Sodium	< 2	< 2
Magnesium	3 ± 3	3 ± 3
Aluminum	6.5 ± 2	6.5 ± 2
Silicon	18.5 ± 3	22 ± 4
"Calcium"[c]		6 ± 2
"Iron"[d]	13 ± 3	5 ± 2

[a] Excluding elements lighter than Be.
[b] Surveyor V results were for the total of atoms heavier than Si; a lower limit of 3% was set for "Fe."
[c] "Ca" denotes elements with mass numbers between approximately 30 and 47 and includes, for example, P, S, K, and Ca.
[d] "Fe" denotes elements with mass numbers between approximately 47 and 65 and includes, for example, Fe, Ni, and Co.

at the Surveyor V and VI sites. These results are shown in Table 9.4. The constancy of results was an important observation.

The Surveyor VII mission was distinct from the Surveyor V and VI in that the landing site was in a highland region near the rim of the Crater Tycho rather than in the equatorial mare region. A specific intent was to sample material considered to be part of the Tycho ejecta blanket. The mission was also somewhat different in that the surface sampler (8) provided a means of moving the alpha-scattering instrument from one position to another, thus making it possible to obtain data from three types of samples: undisturbed soil, small rock, and finally a disturbed soil area.

As an interesting sidelight, the surface sampler played a key role when the alpha experiment deployment device failed to function properly. It was used to force the sensor head to the lunar surface and to provide the critically needed shade when the sensor head showed signs of exceeding the specified values of operational temperatures.

While the alpha-scatter experiment was essentially the same as the two previous ones, there were small differences. The preparation of the alpha sources was modified. The plates containing the curium were coated with carbon by vacuum evaporation in order to prevent aggregate recoil. The technique proved to be partially successful. The new alpha sources were also more intense than those flown previously, thus making it possible to use shorter accumulation times.

The Surveyor VII touchdown occurred in January 1968 in an area less than 1 diameter north of the rim of Tycho. Ultimately, three different samples were measured. The first sample was undisturbed soil near the spacecraft. The second sample was an exposed rock described as 5 × 7 cm in size, visible in the TV display prior to the beginning of the surface-sampler operation. The third sample was in a trenched area previously prepared by manipulating the surface sampler.

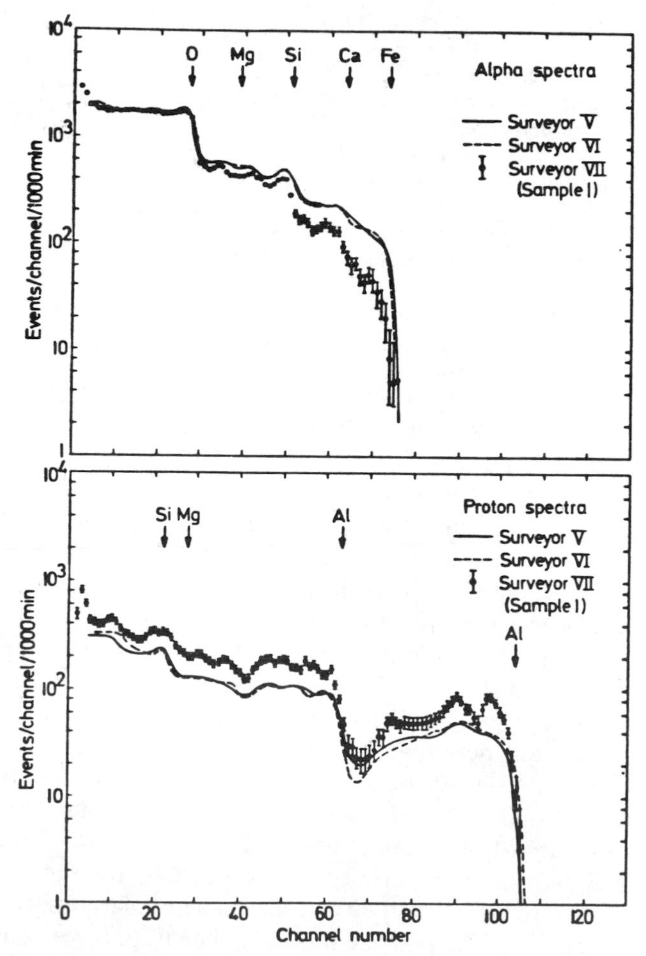

Figure 9.11. Comparison of Surveyor V, VI, and VII lunar sample data. Source: Turkevich et al. (1).

An interesting observation reported by Frangrote and co-workers (9) is that the overall counting rate for the alpha mode for the rock sample was about double that for sample one. This fact and the TV return demonstrated that the rock sample was well centered in the sample area of the sensor head and, in fact, probably intruded slightly into the sensor head. By contrast, the intensity from sample 3, the trenched area was lower than nominal, leading to the inference that the sample under examinations was partially surface material and partially within the trench.

While there were some problems such as delay in deployment and some instrumental drift because of the higher operating temperatures, the program of chemical analysis was much broader and more productive than the previous missions due to the greater sample variety.

An interesting comparison of the spectra for the three missions is shown in Figure 9.11. This diagram includes only sample one from the Surveyor VII effort. In addition to the background corrections, the curves have been normalized by multiplication factors that make the curves match in the oxygen region of the alpha mode. The only significant difference between the highland samples and the mare samples is the lower content of the iron group of elements (see Chapter 8 for a discussion of the orbital experiments). Table 9.5 shows a comparison of the chemical composition at the various landing sites.

Some interesting and valuable inferences were drawn from the Surveyor VII analyses, which were subsequently reenforced by the orbital measurements and the returned samples. The investigators speculated

Table 9.5. Chemical Composition of the Lunar Surface at the Surveyor Landing Sites in Atomic %

	Mare Sites		Highland Sites
Element[a]	Surveyor V	Surveyor VI	Surveyor VII
Carbon	< 3	< 2	< 2
Oxygen	58 ± 5	57 ± 5	58 ± 5
Sodium	< 2	< 2	< 3
Magnesium	3 ± 3	3 ± 3	4 ± 3
Aluminum	6.5 ± 2	6.5 ± 2	8 ± 3
Silicon	18.5 ± 3	22 ± 4	18 ± 4
"Ca" }		6 ± 2	6 ± 2
}	13 ± 3		
"Fe" }		5 ± 2	2 ± 1

[a] "Ca" includes mass numbers between 30 and 47. "Fe" includes mass numbers between 47 and 65.

that the lower content of the "Fe" group elements were a contributing factor to the higher albedos of the highland regions; the lower "Fe" group concentrations, if truly characteristic of the highlands (we now know this to be true), could mean a bulk density of subsurface, highland rocks smaller than those of comparable maria material. As a terrestrial analogue, the continental highlands are less dense that the basaltic ocean bottoms.

In summary, these alpha-scatter experiments were a truly impressive example of the application of known physical principles to the solution of a particularly difficult and challenging problem. They yielded excellent analytical results, in situ from inaccesible regions of the Moon. These experiments have demonstrated the potential for applications to future Solar System exploration.

LUNOKHOD 1—INVESTIGATION OF THE LUNAR SURFACE

Among the various remarkable accomplishments making up space age exploration was the landing of an automatic roving vehicle by the Soviet Union to investigate the lunar surface (10). The rover called Lunokhod 1 successfully performed a variety of studies for a period of 7 lunar days (equivalent to 7 terrestrial months). The automated station, active during the lunar day, was quiescent during the lunar night because of its use of solar power.

The vehicle had an Earth weight of 756 kg and consisted of a self-propelled chassis and a hermetically sealed instrument. The instrument package contained TV cameras, as well as a number of scientific devices to perform a variety of functions. The instrument complement consisted of a means for the determination of the physical properties of the lunar soil: an X-ray fluorescence instrument, RIFMA (the Soviet acronym for an X-ray fluorescence device using radioactive sources) for determining the chemical composition of the surface layer; a corner reflector for optical location of the vehicle; a patrol dosimeter to determine the radiation environment in the vicinity of the Lunokhod; an X-ray telescope for astronomical observations; and an instrument for low-energy gamma-ray measurements. Of these devices, emphasis will be placed here on the X-ray fluorescence measurements.

The Luna 17 carrying the Lunokhod landed on the lunar surface in November 1970 in Mare Imbrium, and shortly thereafter it began to rove the surface, transmitting data. Figure 9.12 shows the construction of the Lunokhod 1. Shown are the hermetically sealed equipment bay and the position of the various instruments. The Lunokhod was designed so that it

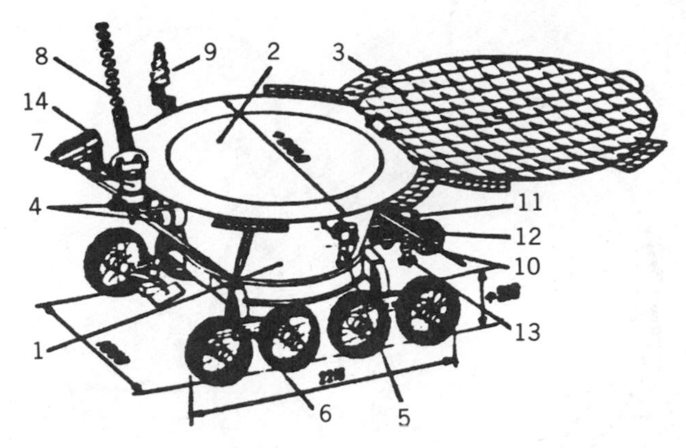

Figure 9.12. A schematic diagram of the Lunokhod 1: 1, the hermetically sealed instrument bay; 2, the cooling radiator; 3, the solar batteries; 4, television porthole; 5, TV cameras; 6, the wheel units; 7, the driving mechanism for the high gain antenna; 8, the high gain antenna; 9, the low gain antenna; 10, the spike antenna; 11, radioisotopic heat source; 12, the ninth wheel; 13, equipment for the physical-mechanical soil properties; 14, the corner reflector. Source: Petrov (8).

could be moved forward and backward at two different speeds, and turn either on the spot or while in motion. The performance of the vehicle was monitored by a set of sensors that continuously measured the roll and pitch of the Lunakhod, the currents to the driving motors, and number of revolutions and temperature of the wheels. The length of the path traversed was determined by the number of revolutions of the driving wheels. To quote G. I. Petrov (10),

The geometrical parameters of the working parts of Lunokhod, its specific pressure on the soil, its driving characteristics, parameters of elastic suspension and the shape of the supporting surface of the wheels, all these factors allowed the Lunokhod to move confidently over a surface with quickly crumbling soil, to overcome steep upgrade slopes, to override craters and obstacles such as single rocks or ridges of rocks comparable with the size of its working parts.

Figure 9.13 is a sketch of the Lunokhod's route on the fourth, fifth, and sixth lunar days (about a 3-month period). The distance covered was over 3 km. The points at which chemical analyses were performed are shown by the letters P. The RIFMA instrument was used to perform a rapid analysis of the soil at these sites. By using the acquired data, it was possible to determine the amounts of Al, Ca, Si, Fe, Mg, Ti, and other elements. The RIFMA instrument is described below in greater detail. This description is based on that given by Kocharov and co-workers (11). The details can be seen in Figure 9.14*a* and *b*. The sensor head of the

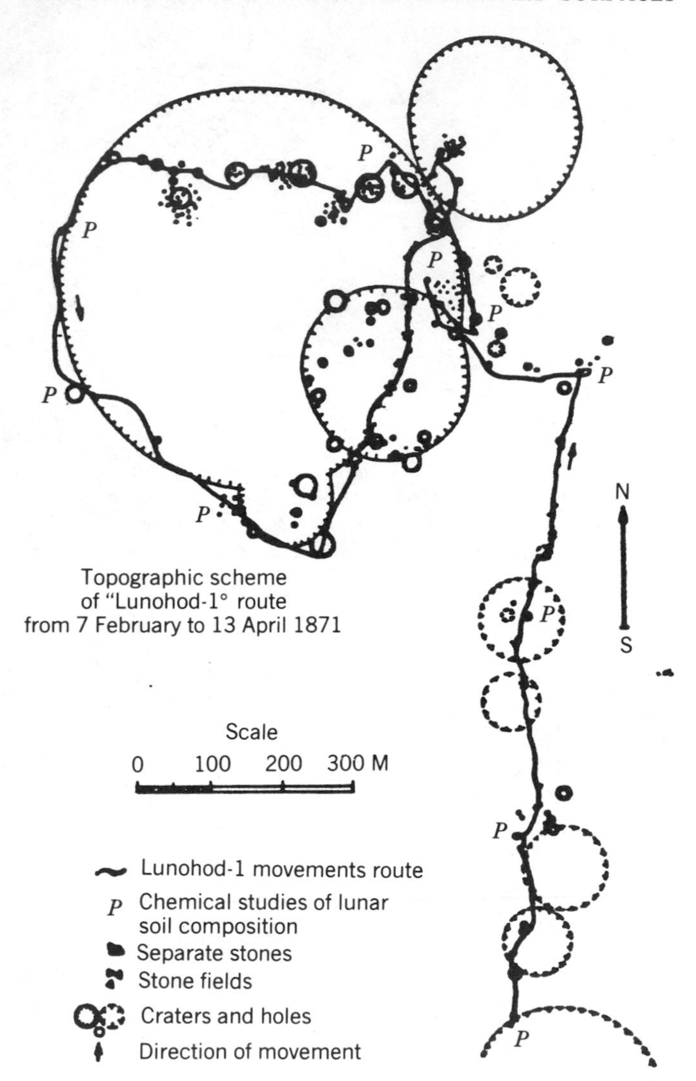

Figure 9.13. The route of the Lunokhod 1. Shown is the nature of the terrain and the points at which analysis was performed. Source: Petrov (8).

RIFMA spectrometer is shown in Figure 9.14*a*, with the relationship between the radioactive sources and the sensor head containing the X-ray detectors. The radioactive source was tritium with a high specific activity. Because of the range of energy emission they were able to excite fluorescence X rays with an energy of less than 10 keV, a spectral region containing the K spectral lines of the main rock-forming elements (see Table 9.6).

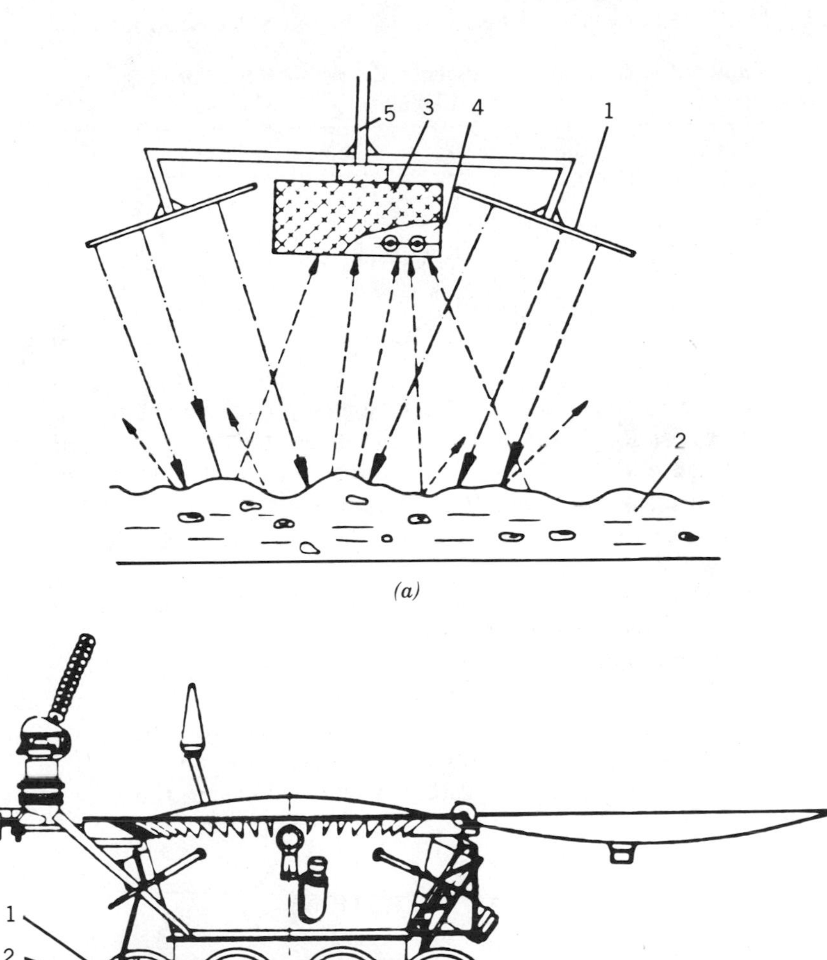

Figure 9.14. (*a*) Schematic view of the RIFMA spectrometer sensor head. 1, The radioisotopic source; 2, the lunar surface; 3, the sensor head; 4, the X-ray detector; 5, the mounting. Source: Petrov (8). (*b*) Lunakhod 1 (a generalized view). 1, the sensor head of the RIFMA; 2, the radioactive source. Source: Petrov (8).

Table 9.6. K X-ray Fluorescence Energy of Rock-Forming Elements

Element	Energy (keV)	Element	Energy (keV)
Mg	1.254	Ca	3.691
Al	1.487	Ti	4.510
Si	1.740	Fe	6.403
S	2.308	Ni	7.477
K	3.313		

The detectors consisted of gas-filled, sealed proportional counters with thin windows for soft X-ray transmission (the details were not supplied). Several of these detectors were contained in the RIFMA and were pointed at the surface (see Fig. 9.14). Because proportional counters cannot resolve the X rays of adjacent elements, selective filters were used in the nondispersive mode.

The RIFMA electronics consisted of a number of components normally used in conventional spectrometry, such as a charge-sensitive preamplifier, an amplifier, a multichannel pulse-height analyzer, and a stable high-voltage supply for the proportional detectors. The multichannel analyzer was a 64-channel instrument. While the analog to digital processor was on the Lunokhod, the memory and display was on Earth, thus increasing the reliability of the system, while decreasing the weight and volume.

LUNAR OPERATIONS

The Lunokhod roved about the lunar surface crossing or passing craters of various ages. Along the way, details of large craters such as floors, slopes, and rims were studied. The Lunokhod crossed crater ejecta, rock fields, and so on, and the RIFMA was turned on. In the process, features like nondisturbed regolith, the subsurface layer, and separate rocks were analyzed. As a technological study, RIFMA was used with the Lunokhod in motion to demonstrate that analysis could be performed while the Lunokhod was moving.

DATA

Some hundreds of spectra were accumulated. Because of the peculiarities of the spectrometer and the ability to record and play back the spectra in

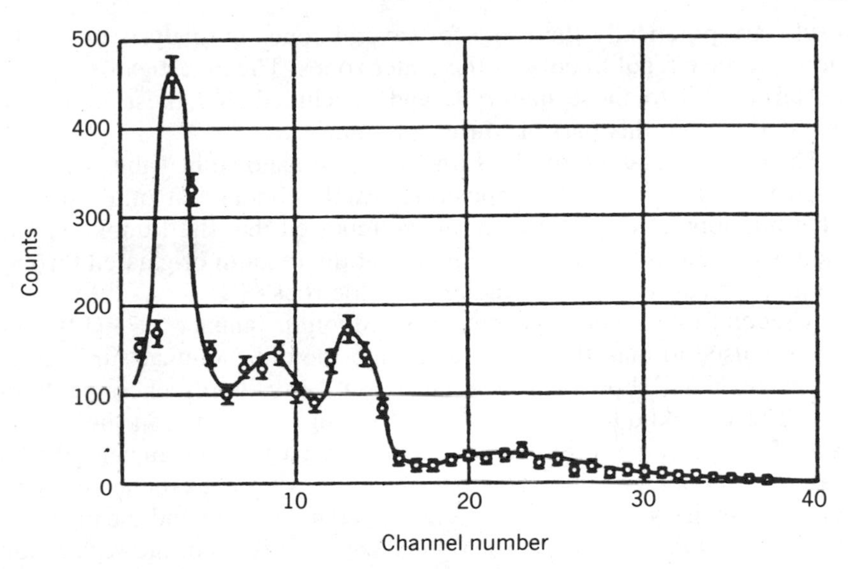

Figure 9.15. Spectrum (without filter): peaks numbered in order from left: 1st, group of light elements (Mg, Al, Si); 2nd, a line seen in spectra of all counters, including those shielded from the surface; 3rd group of Ca and K; 4th (around channel No. 23), Fe range. Source: Kocharov et al. (9).

real time, it was possible to control and, in some cases, to increase the quality of the spectra. Figure 9.15 is an example of a typical spectrum. A number of features are shown, which have been identified as follows: Numbering the peaks from left to right, the first peak is an envelope for the light elements (Mg, Al, and Si). The second peak was seen in the spectra of all the counters, including those shielded from the surface. This second peak was considered as coming from dE/dx of the cosmic rays. The third peak contains the elements Ca and K, and the fourth peak, around channel 23, is from the Fe range of elements.

The proper evaluation of the spectra obtained required consideration of a number of parameters, such as the thermal regime of the sensor head, the orientation of the Lunokhod relative to the lunar surface and the Sun, and the nature of the lunar surface being analyzed.

RESULTS

The Lunokhod continued to function for several months, accumulating a great many experimental data. A set of results was obtained for the chemical composition of some relatively flat areas of the lunar plains. These results are practically the same for several zones of analysis of nondisturbed surface regolith outside the crater zones. The investigators applied the term fines to these materials and concluded that these data were characteristic of this part of Mare Imbrium.

The results obtained by the Lunokhod 1 are shown in Table 9.7, compared with the chemical composition results obtained at other sites by other missions. The investigators have reported that the Lunokhod data and results affirm the general opinion that the regolith originated through the mechanical reduction of mainly basaltic rocks.

A second Lunokhod was sent to the Moon in January 1973 (12). This one was used to map the chemical composition of the lunar surface in a "mare–highland" boundary, again using an X-ray fluorescence spectrometer. The Lunokhod was put down into a region situated in the eastern part of Mare Serenitatis near the southern part of an ancient, 55-km crater, Le Monnier. The principal objective was to examine the lunar surface chemistry in the contact zone between the mare and the highland. In addition to the chemistry, the optical characteristics of the surface and the magnetic fields were also explored. With regard to the chemical composition, an attempt was made to establish a correlation between the morphological peculiarities and the nonhomogeneity of the chemical composition.

The X-ray fluorescence spectrometer on this second mission was designated as RIFMA-M. It was, according to the investigators, modified and improved. Special attention was given to the determination of iron, because they felt that it was the clearest indicator of differences between the highlands and the mare. The concentration of Fe in the highlands was about 3–5%, while the mare values were about 12–15%. The ratios of values found for various elements, when the highlands were compared to the mare, were: Mg = 1.1, Si = 1.4, Ca = 1.2, and Fe = 2.4; thus the selection of iron as the best indicator of the transition from mare to highlands.

The improvements in the RIFMA spectrometer for Lunokhod 2 had as a special objective the optimization of the determination of Fe. This involved a radioactive source to more efficiently excite the Fe. A new type of radioactive source was used, consisting of a combination of tritium, ziroconium, and tungsten. In addition, a detector was mounted in the sensor head, which efficiently measured X rays in the 5–30-keV region.

Figure 9.16 shows the track of the Lunokhod 2. The circled numbers are the points at which analyses were done. Table 9.8 shows the chemical composition along the route. In brief, the Lunokhod effort demonstrated that there were considerable variations in the abundance of some chemical elements in the upper layer of the regolith. The iron content in the

Table 9.7. Chemical Composition of the Lunar Surface

Element	Mare Tranquillitatis			Sinus Medii, Surveyor 6,	Oceanus Procellarum, Apollo 12		Tycho Rim, Surveyor 7		Mare Foecunditatis, Luna 16		Mare Imbrium, Lunokhod 1,
	Surveyor 5,	Apollo 11									
	Fines	Fines	Rocks	Fines	Fines	Rocks	Fines	Rocks	Fines	Rocks	Fines
Si	21	20	20	23	20	19	21	21	20	20	20
Fe	9	12	14	10	13	17	4	3	13	15	12
Ca	10	8	7	9	7	8	13	13	9	7	8
Al	8	6	6	8	7	6	11	14	8	7	7
Mg	3	5	5	4	7	7	4	< 2	5	4	7
Ti	4	5	6	2	2	2	< 0.4	< 0.7	2	3	< 4
K	—	0.1	0.2	—	0.3	0.05	—	—	0.08	0.12	< 1
Na	0.5	0.4	0.4	0.6	0.3	0.3	0.5	0.3	0.3	0.2	—

bottom of part of the crater LeMonnier was about 6 wt%, but it fell to about 4.0 ± 0.4 wt% in the highland region. The vehicle did encounter a 20-km tectonic break where the underlying rock on the slopes appeared to be exposed, and the Fe value was about 8 wt%.

The surface points 1, 2, 3, 5, and 6 listed in the table appear to be intermediate between highland and mare surfaces, evidence for mixing of highland and maria materials.

As the investigators point out, the Lunokhod 2 activity was the first and, as yet, only detailed examination of a mare–highland contact zone.

Table 9.8. Lunar Surface Chemical Composition (wt.%) along the Route of Lunokhod 2[a]

Point No.	Place of Analysis	Elements				
		Aluminium	Silicon	Potassium	Calcium	Iron
1	Rim of crater 40 m diam. on plain near landing site	8.8 ± 1.0	24 ± 4	< 1	8.0 ± 1.2	6.1 ± 0.7
2	Plain; near crater 13 m in diam.			< 1	7.8 ± 0.8	6.2 ± 0.6
3	4.5 km to S of landing site			< 1	8.3 ± 0.9	4.9 ± 0.4
4	Place of the deepest penetration into hilly highland region	11.6 ± 0.9	22 ± 3	< 1	9.1 ± 1.2	4.0 ± 0.4
5	Plain; bottom of crater Le Monnier			< 1	8.5 ± 0.9	6.1 ± 0.6
6	Plain; bottom of crater Le Monnier; 5.5 km to E of point 5			< 1	8.1 ± 1.1	6.5 ± 0.7
7	Western slope of tectonic break			<1	7.8 ± 1.0	7.6 ± 0.9
8	Eastern slope of tectonic break			< 1	7.8 ± 0.9	8.2 ± 0.9

[a] Measurements were carried out in various modes, so that in a number of places the light element contents were not determined.

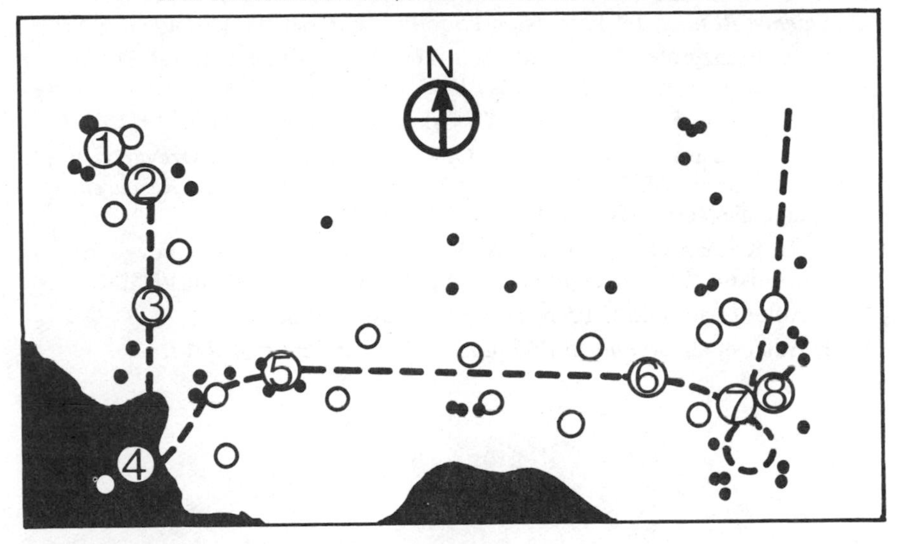

Figure 9.16. The Lunokhod 2 route. The circled numbers are the points where analyses were performed. The dashed line is the highland region. Source: Kocharov et al. (10).

They have compared their results to those reported by the Apollo 15 and 16 investigators (13, 14) and found that if one makes allowances for the difference in spatial resolution, the two investigations yielded consistent results.

REFERENCES

1. A. L. Turkevich, W. A. Anderson, T. E. Economou, E. S. Frangrote, H. E. Griffin, S. L. Grotch, J. H. Patterson, and K. P. Sowinski, *Jet Propulsion Laboratory Tech. Rept. 32-1265,* 1968, p. 119.
2. H. Geiger and E. Marsden, *Proc. Roy. Soc.,* **82A,** 495 (1909).
3. E. Rutherford, J. Chadwick, and C. D. Ellis, *Radiation from Radioactive Substances,* Cambridge University Press, Cambridge, 1930.
4. A. Turkevich, *Science,* **134,** 672 (1961).
5. J. H. Patterson, A. L. Turkevich, and E. Franzgrote, *J. Geophys. Res.,* **70,** 1311 (1965).
6. A. L. Turkevich, E. J. Franzgrote, and J. H. Patterson, *Science,* **158,** 635 (1967).
7. A. L. Turkevich, J. H. Patterson, and E. J. Franzgrote, *Science,* **158,** 1108 (1967).

8. R. F. Scott and F. I. Roberson, *Surveyor VII Mission Report. Part 11: Science Results,* Jet Propulsion Laboratory, Pasadena, CA, 1968, p. 9.
9. E. J. Franzgrote, J. H. Patterson, and A. L. Turkevich, *Jet Propulsion Laboratory Tech. Rept. 32-1264,* 1968, p. 241.
10. G. I. Petrov, *Space Research XII,* Akademie-Verlag, Berlin, 1972.
11. G. E. Kocharov, S. V. Victorov, O. M. Voropayev, A. Yu Dzevanovskaya, G. V. Kirian, V. V. Petrov, and V. A. Sakulsky, *Space Research XII,* Akademie-Verlag, Berlin, 1972.
12. G. E. Kocharov, S. V. Victorov, V. P. Kovalev, G. A. Matveev, and V. I. Chesnokov, *Space Research XV,* Akademie-Verlag, Berlin, 1975.
13. I. Adler et al., GSFC Preprint X-641-72-246 (1972).
14. I. Adler et al., *Apollo 16 Preliminary Science Report* (1973).

CHAPTER

10

MARS ATMOSPHERIC AND SURFACE MEASUREMENTS

The extraordinary character of the Martian exploration program, even prior to the Viking landers, was summarized by K. F. Weaver in his article "Journey to Mars" (*National Geographic,* Feb. 1973).

> No human traveler has yet gone to Mars. But thanks to a little blue-winged satellite named Mariner 9, I can write about the red planet's cold and tortured face almost as confidently as if its landscape lived in my memory. The details are all to be found in a flood of information gathered by the first spacecraft to orbit Mars. Circling twice each day since Nov. 13, 1971, Mariner 9 has photomapped the planet's entire surface, probed its atmosphere, taken its temperature and assayed its chemistry.

The accomplishments became even more remarkable with the Viking spacecraft, which provided a closeup view of the Martian planet, examined the surface for evidence of life, and performed XRF analysis of the Martian soils.

Some of the earliest observational data came via the earth-based reflectance spectroscopy already discussed in Chapter 7. As we have seen, the results led to a deduction that the regions observed were combinations of ferric oxide and mafic silicate rocks.

During the Mariner 9 mission, thermal emission spectra of Mars were measured by the infrared interferometer–spectrometer (IRIS), which supplied data for the 5–50-μm region. The spectra recorded during the first 100 orbits showed strong absorptions, which the investigator team reported as characteristic of a SiO_2 content of 60 ± 10 wt.%. Based on these observations, Hunt and co-workers (1) and Logan and associates (2) suggested that the clay mineral montmorillonite could be a major component of the Martian dust clouds (montmorillonite is a common weathering product of mafic minerals and would be consistent with basalts on Mars). They also reported that hematites and carbonates were excluded as major components of the Martian dust clouds but that salts, such as sulfates, could be present in substantial amounts. These interpretations were borne out by the subsequent Viking XRF results, which will be discussed later.

ATMOSPHERIC STUDIES

An important study performed during the Viking mission involved the observations of the behavior of water vapor in the Martian atmosphere (obviously one of the more significant substances). These were done from the orbiting component of the Viking mission. Prior to the Viking mission, the only knowledge of the behavior of atmospheric water vapor came from Earth-based observations. This information was limited with regard to the spatial and seasonal dependence. It was known that the atmospheric vapor varied seasonally, apparently reaching a maximum in each hemisphere at some time after the summer solstice. There also appeared to be a diurnal variation of the water-vapor column abundance at midlatitudes during at least a part of the season of maximum vapor content.

The basic questions involving the present day behavior of Mars condensates and the extent to which the planet has retained its primitive fraction of water have been too difficult to answer from terrestrial observations because of a number of factors. These have been the inaccessibility of the extreme polar regions from the Earth, the poor resolution of the measurements, and limitations in sensitivity.

Both Viking spacecraft carried instruments for mapping the water vapor. These provided a capability for measurements with sufficient spatial and temporal resolution and coverage to give the investigators insight into answers to the questions listed above. The instruments and results have been detailed by C. B. Farmer and associates (3).

Figure 10.1 shows the flight instrument and the optical configuration. It was designed to measure the absorption of solar radiation using the strong lines at the center of the 1.4-μm combination vibration–rotation bands of water vapor. The instrument was a grating spectrometer of five channels, which in the normal (locked) mode of operation was centered on three

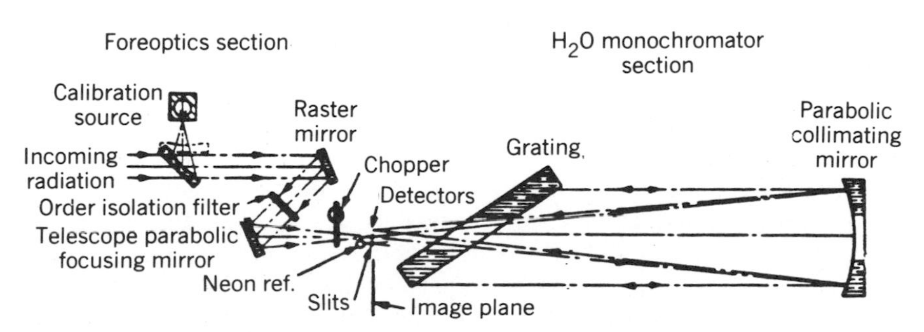

Figure 10.1. Optical configuration of the instrument used for the Mars atmospheric water detection (MAWD). Source: Farmer et al. (3).

absorption lines and two nonabsorbing or continuum regions between the lines. The exit apertures were defined by the PbS detectors, which were operated at low temperatures (200 K). The cooling of the detectors was accomplished by means of a flexible copper strap connected to a plate radiating out into space. The grating (1200 lines/mm) was operated in the first order for the five water vapor channels, the second-order diffraction being rejected by a silicon-cut filter. In order to maintain the precise frequency location of the water vapor channels against possible shifts due to the thermal distortion of the instrument optical alignment, a neon reference system was employed. This involved operating the grating in the second order, using a set of silicon photodiodes as detectors. Motion of the neon-lamp exit-plane image relative to the silicon reference detectors was sensed and used to drive a cam, which repositioned the grating to preserve the correct wavelength alignment.

While the instrument was used as a five-channel monochromator in its normal mode, it also had the capability, by the use of a wavelength servo-system, to scan the grating through a small angular range. The second mode was used occasionally to record the continuous spectrum of the atmosphere. The spectral range was 8 cm^{-1} on either side of the nominal position of each of the five channels, thus yielding a composite spectrum of about 40 cm^{-1}.

The purposes of the second mode were:

1. To verify that the Martian atmospheric spectrum in the region of water vapor absorption was not contaminated by absorptions due to other constituents, for example, weak transitions of CO_2.
2. To validate the theoretical spectrum generated for Martian conditions, based on extensive laboratory measurements of the 1.4-μm region water vapor transition strengths, frequencies, and pressure-broadened half-widths, on which the quantitative interpretation of the received data was based.
3. To determine to what extent, if any, absorption of lines of solar origin might affect the distributed continuum radiation at the nominal line and continuum frequencies. The solar spectrum was obtained by orienting the scan platform to view the Sun's radiation falling on a small, diffusely reflecting plate mounted on a structural member of the spacecraft.

Farmer and co-workers (3) in their description make clear that in viewing the planet, the spectrometer obtains a measurement of radiance, in each of the five channels, of reflected solar radiation that has passed twice through the Martian atmosphere. To determine the absolute radiance values, the channel responses were corrected by means of periodic calibrations done by inserting a mirror into the input beam so that the detectors viewed an illuminated cavity. The ratios of the radiances in the line channels to the continuum radiance yielded three values of absorbance from which the line-of-sight water vapor content was derived. It was found that the contribution of surface emission to the received radiation was negligible at the chosen wavelengths of operation.

DATA ANALYSIS

A characteristic of many of these remote observations so far described is that, unlike many laboratory measurements, there are a number of parameters that cannot be controlled but, rather, must be corrected for, frequently on the basis of laboratory studies. In the Mars atmospheric water detection study (Mawd), the scattering by atmospheric particulates represented a difficulty to be dealt with. In the absence of scattering by these particulates, the vertical-column abundance of water vapor W was obtained from the measured line-of-sight abundance along the optical path W' and acknowledge the Sun–Planet–spacecraft geometry. The relationship used was $W = W'/\eta$, where the airmass factor of the observations was $\eta = \sec i + \sec e$, and i and e were the incidence and emission angles, respectively. It was observed that particulates in the Mars atmosphere did affect the formation of the absorption spectrum of the selected wavelengths, the general result being an underestimate of the water content. However, for most of the regions of the planet, the effects of scattering could be ignored for observations made at $\eta < 4$ (3).

The line-of-sight water vapor content of the Martian atmosphere could be determined directly from the observed channel absorptions by employing an empirically derived calibration, made with the flight instrument, prior to launch. The operating assumption was that the absorption was dependent only on the abundance of water vapor along the optical path. This assumption was beset with some difficulty because, as indicated by Farmer, the absorption is dependent on pressure (and to some extent temperature). Laboratory simulation of Martian water vapor quantities under representative conditions of partial pressure and temperature would require prohibitively long absorption paths.

As a consequence of the above, the method chosen was to compare the measured channel absorption values with the theoretical values generated for a wide range of conditions of pressure and temperature. The set of theoretical curves developed was stored in the form of a three-dimensional table (A_η versus W, P, and T) as part of the routine Mawd data processing. The table was generated by using standard algorithms for the spectral extinction coefficients of lines mixed with collision and Doppler profiles and the known (measured) instrument spectral response function. The molecular parameters, such as transition strengths, collision-broadened widths, ground-state energies, and foreign and self-broadening coefficients, were determined from laboratory measurements. A total of 81 transitions arising from the five vibration–rotation bands that occur in the 1.38-μm water vapor spectrum were included in the final computation. The resultant theoretical spectra and the corresponding channel absorption values were verified in the laboratory over the limited range of pressures and temperatures that could be achieved. Figure 10.2 shows the calculated spectra. It was found that there was a good fit between observed and computed spectra. The investigators state that this provides strong evidence, although not proof, of the validity of the application of the spectral calculations to the physical conditions of the Mars atmosphere.

To report all the results in detail are beyond the scope of this text. Some of the observations, however, will be reported because they are so interesting from an analytical point of view. Figure 10.3 is one such example. A number of scans were taken of the solar diffusion plate during a series of revolutions of the Voyager 1. The principal purpose for these

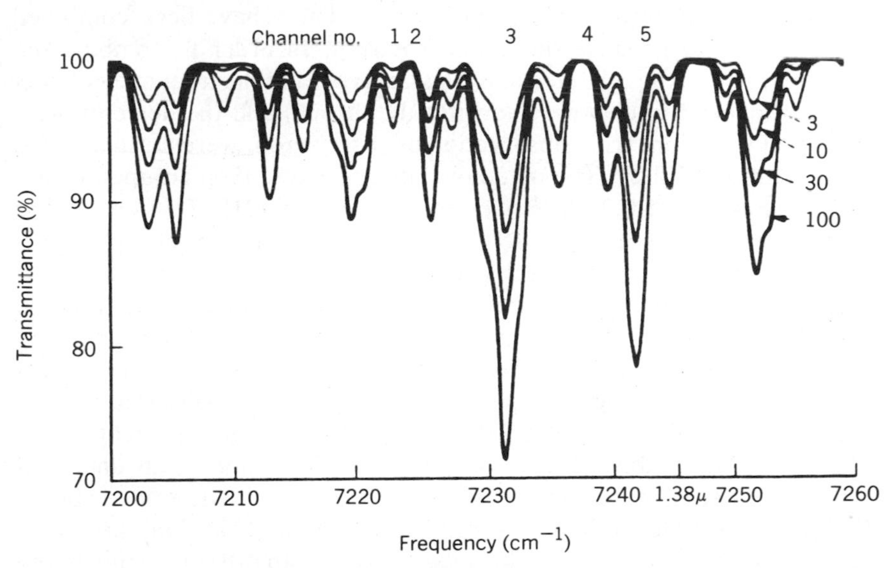

Figure 10.2. Theoretical water vapor spectra in the frequency range 7200–726 cm^{-1}. The numbers 1–5 show the spectral positions of the five detectors. The curves are for vapor of 3, 10, 30, and 100 precitable μm (pr) at a total pressure of 6.9 mbar and a temperature of 225 K. Source: Farmer et al. (3).

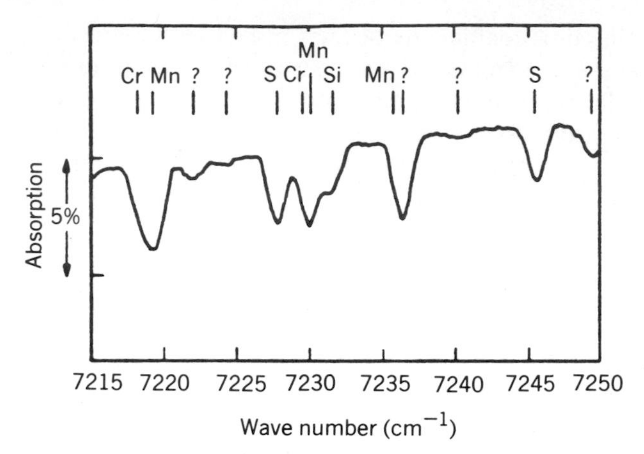

Figure 10.3. Solar spectrum obtained from wavelength scans of the diffuser plate on VO-1 revolutions 55, 107, 124, and 210. Source: Farmer et al. (3).

scans was to monitor the spectral distribution of the continuum radiation covered by the water vapor measurements. Farmer and co-workers report that they revealed a number of solar absorption which had not previously been observed. Each of the observations on revolutions 55, 107, 124, and 210 represent four repeated scans. These have been combined and averaged to produce the solar spectrum. In order to combine the individual scans, the relative channel gain levels and the frequency offsets were normalized to channel 3 (arbitrarily chosen) and the detector radiance level variations from observation to observation, were normalized to the revolution 55 data. The detector radiances were then grouped in frequency intervals of 0.2 cm^{-1} over the total range 7215–7250 cm^{-1}. The frequencies, intensities, and identification of solar absorptions appearing in the spectra are shown in Table 10.1.

A second example is shown in Figure 10.4. The spectra shown come from a number of wavelength scans of the planet taken with the Voyager 1 instrument during the first period of synchronous orbit. A typical spectrum is shown in the figure (second from the top). The observation was made from Voyager 1 (revolution 45) viewing an area of Mars centered at 10° W, 82° N, at about 2 h from periapsis. The planet scan data was analyzed in a similar manner to that for the solar spectra described above. Because the wavelength scan was only 12 min in total duration, the resulting spectrum was noisier than the solar spectrum. In order to estimate the water content, it was necessary to correct for the pressure, temperature, and solar spectrum. The corrected atmospheric spectrum is shown in the same figure. A spectrum fitting program was used to determine the values

Table 10.1. Identification, Frequency, and Intensity of Solar Absorptions on Mars

Line	Identification	Frequency (cm^{-1})	Intensity (cm^{-1}) ($\times 10^3$)
1	Cr	7218.38	30
2	Mn	7219.46	53
3		7222.08	14
4		7224.57	5
5	S	7227.88	45
6	Cr	7230.05	51
7	Mn		
8	Si	7231.58	23
9	Mn	7236.42	47
10			
11		7240.35	4
12	S	7245.66	34
13		7249.57	30

Source: Farmer et al. (3).

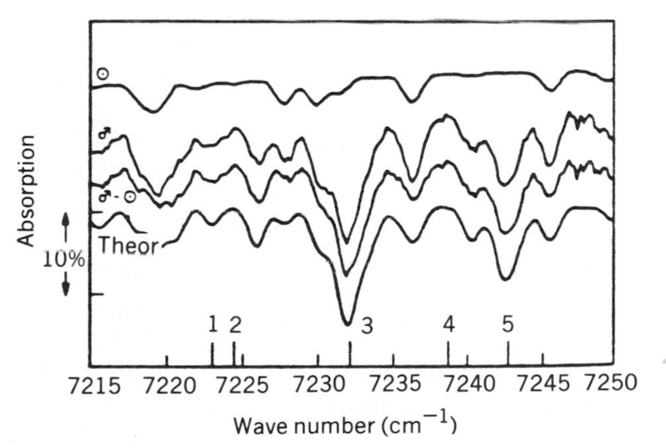

Figure 10.4. Spectra of the Sun (top), Mars (at 10°N, 83°W) (second curve), and Mars corrected for the solar features (third curve). The bottom curve is the theoretical spectrum for the conditions giving the best fit to the atmospheric spectrum. The curves have been offset for clarity. Source: Farmer et al. (3).

of *W*, *P*, and *T*, which gave the best fit to the corrected Mars spectrum. The theoretical spectrum calculated for these conditions is shown as the bottom curve in Figure 10.4.

The differences between the theoretical and corrected observed spectra were 0.5% or less. No absorption lines other than the expected water vapor lines appeared in the Martian atmospheric spectrum. As was previously pointed out, the close agreement between the observed and computed spectra, under Martian atmospheric conditions, was taken as an important verification of the algorithms and molecular parameters used to generate the transmittance tables on which the analysis of the Mawd locked grating mode data was based.

The water measurements thus far described led to interesting and important results about the Martian atmosphere. The summary offered by Farmer (3) includes the following.

During the seasonal period from the northern summer solstice to the following equinox, the water vapor undergoes a gradual redistribution. The latitude of maximum column abundance moved from the northern polar area to the equatorial latitudes. The total global vapor content remained approximately constant at the equivalent of about 1.3 km^3 of ice.

The peak vapor abundances observed were found to occur over the dark circumpolar region, an area inaccessible to Earth-based observers. A maximum vapor column abundance of about 100 pr-μm were measured. These results and corresponding measurements of the local surface temperature showed that the residual polar caps are composed of water ice and at the season of maximum polar vapor content, the atmosphere above the ice is at or close to saturation.

High-resolution observations suggest that local variations in vertical column abundance of vapor occurs only where there are known topographic features characterized by abrupt elevation changes.

There are still outstanding questions about the mechanisms that control the seasonal redistributions of water vapor and its possible hemispheric migration, and the presence of a major subsurface reservoir of water ice at other than polar latitudes. At the time of writing their report, the investigators felt that additional measurements during the extended flight would provide additional insights.

ENTRY SCIENCE—THE MARTIAN ATMOSPHERE

Among the scientific payloads of the Viking 1 and 2 landers were mass spectrometers that were designed to measure the composition of the martian atmosphere at high altitudes. The experimenters, A. O. Nier and

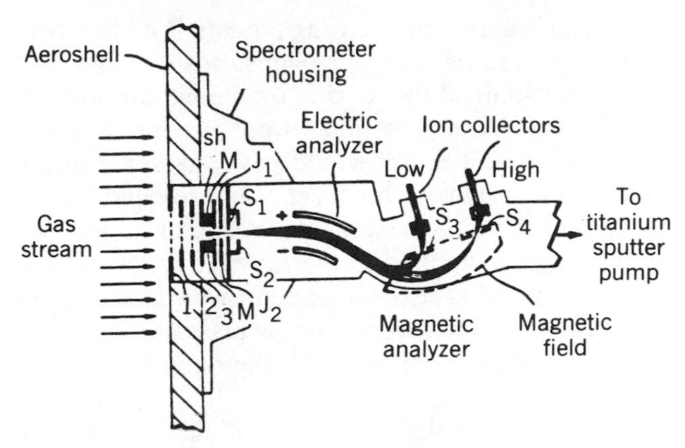

Figure 10.5. Schematic diagram of the Viking upper atmosphere mass spectrometer mounted on the aeroshell of the lander. Until descent to the planet, the instrument was sealed by a cap covering the ion source and pumped continuously. Source: Nier and McElroy (4).

M. B. McElroy, have described these investigations in detail (4). The mass spectrometer was mounted at the apex of the Viking aeroshell; as the vehicle descended toward the Martian surface, the spectrometer began to make measurements at altitudes over 100 km in the range of 1–50 atomic mass units (amu).

The compositional measurements employed a double focusing magnetic deflection type spectrometer, which has been described in detail elsewhere (5). The Mattuch–Herzog geometry (6) employed had special advantages for space flight because it permitted efficient use of the available space, weight, and power, and yet offered good performance characteristics. The electron bombardment ion source used in the Viking instrument was similar in design to sources previously used with success in the study of the Earth's atmosphere by rockets and satellites. A schematic drawing of the instrument is shown in Figure 10.5. In principle, the ambient gas entered the ionizing region of the spectrometer through three high-transmission grids. The ions were formed by a well-collimated electron beam (M in the figure). Once formed, the ions in the source region were directed to the right by an electric field, due in part to the repelling voltage on grid 3, and in part to field penetration through the slit to the right of the electron beam. Grid 1 was maintained at the same potential as the instrument housing, grid 2 was held at negative potential, selected to cancel fields produced outside by the positive potentials of the ion box sh and the ion repeller grid. The potentials on grids 2 and 3 were varied in a linear way in response to a change in the potential of the ion box sh. A similar linear relationship applied to the focusing potentials on the focusing plates J_1 and J_2 and for the electric analyzer plates, positive and negative.

Mars scans were obtained by varying the potentials applied to the ion box exponentially with time. The limits were chosen to ensure that ions reaching the high-mass collectors would have masses in the range 7–49 amu. The field of the magnetic analyzer was provided by a permanent magnet. The mass of the ions reaching the collector varied inversely as a function of the voltage V applied to the ion box. In order to restrict the range of voltages required, a second collector was used to capture ions in the 1–7-amu-mass range. Thus high and low mass numbers were sampled simultaneously. A complete spectrum for the range 1–49 amu was accumulated in a 5-s interval.

The ions were detected using conventional electrometer type amplifiers—the input stages using field-effect transitors. The use of electrometers was in response to the weight restrictions imposed by the Viking project (electron multipliers would have given significantly enhanced sensitivities). The actual sensitivity was such that the nitrogen (amu = 28) gave a signal better than 2×10^{-5} A/Torr of pressure in the ion source. Background fluctuations of the amplifiers were of the order of $1–5 \times 10^{-14}$ A.

The calibration procedures were built on the experience on the earlier Atmospheric Explorer. The instrument, with an exposed ion source, was pumped by a liquid helium pump separated from the spectrometer by an orifice of accurately known dimensions. Gas was admitted to the chamber through a molecular leak of known conductance from a reservoir of high but known pressure. Thus the equilibrium pressure of gas in the chamber surrounding the mass spectrometer ion source could be computed accurately by using the known rate of flow into the chamber and the known rate of removal through the orifice to the liquid helium-pumped chamber. Calibration runs were made for CO_2, N_2, Ar, CO, O_2, He, and H_2. The CO_2 runs covered the pressure range of 2×10^{-7} to 3×10^{-4} Torr. Reduced pressure ranges were employed for the other less abundant gases.

Once calibrated, a cap was installed over the ion source and the mass spectrometer evacuated through a small tube attached to the cap. The tube was pinched off after the instrument was evacuated and baked. A small Ti sputter ion pump, permanently attached to the analyzer portion of the instrument vacuum housing, was operated almost continuously for the remaining life of the instrument. The procedure was designed to ensure an instrument pressure of less than 10^{-8} Torr for the almost 2-y period between calibrations and entry into the Martian atmosphere.

The cap was cut off flush with the surface of the aeroshell, exposing the ion source (Fig. 10.5) immediately following the separation of the lander from the orbiter. The cutting device has been described by Thorness and Nier (7).

The instrument was additionally pumped by the essentially zero-pressure environment during its $2\frac{1}{2}$ h descent to a measurable atmosphere. The mass spectrometer was turned on 60 min prior to the expected encounter with the measurable atmosphere to enable it to warm up. The useful measurements were made over an approximate height range of 100–200 km. The lander was moving at a speed of 4.5 km s and the normal to the instrument ion source was within 10° of the velocity vector. Thus, as Figure 10.5 shows, the instrument practically was seeing a high-speed gas beam from the normal direction.

Nier and McElroy (4) write that there are problems and complications associated with the measurement of ambient gas densities by using instrumentation on high-speed spacecraft. It has been observed that a stagnation effect, resulting in a substantial increase in particle density, is known to occur in the ion source of forward-looking instruments. The stagnation ratio can be calculated in terms of the ambient and instrument temperature and the spacecraft velocity for instruments with ideal pinhole openings. A correction factor must be employed for the more open source used on the Viking. This factor depends on the geometry of the ion source. Because of the complexity of the electrode and grid structure of the Viking instrument, the correction factor was determined experimentally, using a molecular beam simulation. In the Viking instrument a correction factor of 1.17 was determined and applied to the data.

RESULTS

As we shall see, the instrumental design was a sound one. After about a year in transit to Mars, the mass spectrometer provided useful data. The range of height over which data was collected was 120–200 km for the Viking 1 and down to 115 km for Viking 2. Twelve spectra were analyzed for each entry. Results were derived for CO_2, N_2, Ar, CO, and O_2. Carbon dioxide was observed to be the major constituent for all altitudes below 180 km. The variation with altitude of the density of CO was used to obtain relatively direct information on the temperature of the upper atmosphere over the height range studied.

A typical mass spectrum at an altitude of about 140 km taken by the Viking 1 lander is shown in Figure 10.6. The spectrum includes all the

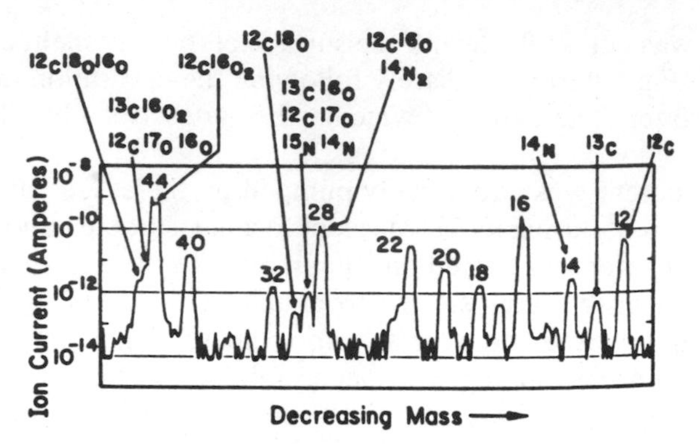

Figure 10.6. Mass spectrum observed at the 140-km altitude during the descent of Viking 1. Source: Nier and McElroy (4).

important fragments that came from the ionization and dissociation of CO_2 (note the mass peaks at 44, 28, 22, 16, and 12 amu). These were associated with $CO^+{}_2$, CO^+, $CO_2{}^+$, O^+, and C^+. The spectrum also included peaks, associated with Ar^+ and Ar^{2+}, at 40 and 20 amu. The 14-amu peak was attributed to N^+ formed by the dissociative ionization of N_2. Molecular nitrogen was also found to make a significant contribution to the peak at 28 amu. The peak at 32 amu was due to O_2. The mass numbers 17 and 18 amu were probably due to terrestrial water released from the surface of the ion source by interaction with the incident gas stream, a phenomenon that is known and was invariably seen on the mass spectrometer carried on sounding rockets and earth orbital satellites.

The mass spectrometric data was used to arrive at a number of important conclusions with regard to the isotopic compositions of carbon, oxygen, and nitrogen in the Martian atmosphere. There were some 13 useful spectra, 8 from Viking 1 (taken between 120 and 180 km) and 5 from Viking 2 (taken between 125 and 150 km). Some of these isotopic features are shown in Figure 10.6. The information on the isotopic composition was derived on the basis of the data at mass numbers 46, 45, and 44 amu. The results are shown in Figure 10.7.

In summary, the CO_2 was found to be the major constituent of the Martian atmosphere at all heights below 180 km. The thermal structure was found to be complex and variable, with the average temperature below 200°K for both the Viking 1 and 2 flights. The atmosphere was observed to be mixed, to heights over 120 km. The isotopic composition of carbon and oxygen in the Martian atmosphere is similar to that in the terrestrial atmosphere. Nitrogen-15 was enriched by a factor of 1.62 ± 0.16 relative to the terrestrial atmosphere.

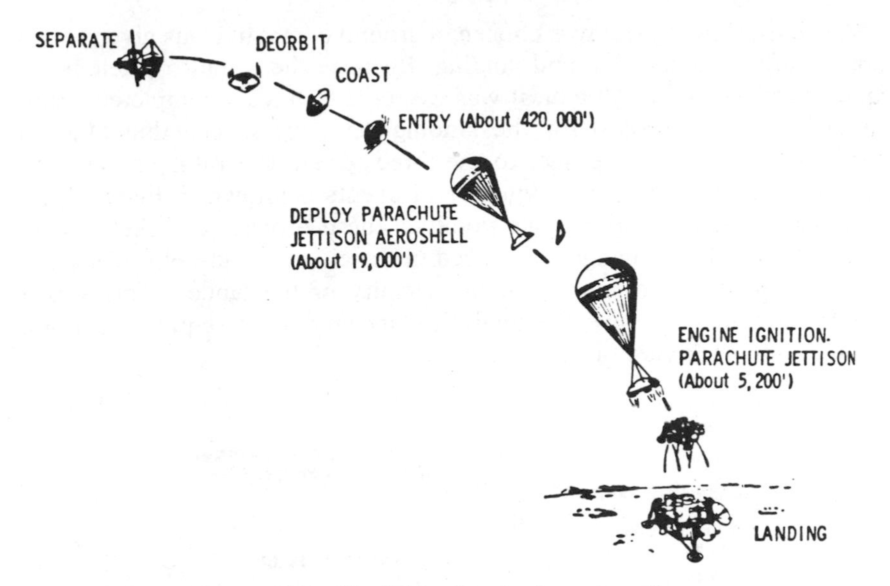

Figure 10.7. The Viking Lander descent profile.

SURFACE SCIENCE

In order to establish an appropriate framework for an understanding of the Viking surface science, it is essential that we examine the Viking spacecraft and something about the Martian mission. An excellent summary has been provided by G. A. Soffen (8), which will be briefly summarized here. With regard to the spacecraft, each Viking consisted of an orbiter and a heat-sterilized lander within its capsule. The function of the orbiter was to transport the lander to Mars; to carry reconnaissance instruments for aid in the selection of suitable landing sites; to perform as a relay station for the lander data and to perform a number of scientific investigations. The periapsis (defined as the orbital point nearest the center of attraction) was located over candidate landing sites to permit maximum viewing resolution and relay of lander data. Three reconnaissance instruments of the orbiter were mounted on a scanning platform. These consisted of a pair of high-resolution TV cameras, an infrared spectrometer for water vapor measurements, and an infrared radiometer for thermal measurements of the surface and atmosphere. All of these instruments were sighted along a common axis to permit mapping of a common area of the planet. They were employed during the first 4 wk for landing site selection. Subsequently, they were used to investigate the Martian atmosphere and surface over a large part of the planet and its satellites.

Once the landing site was chosen, a ground command was given to the spacecraft for separation and landing. Because the round trip telemetry signal required about 40 min, it was essential to have a completely automated system on board for the landing maneuvers. The thin Martian atmosphere made it necessary to use three sequential braking systems to assure a soft landing. The sequence of events is shown in Figure 10.8. Because the final braking stage involved the use of retro-rockets, there was concern about changes produced in the chemical and physical character of the Martian surface in the vicinity of the lander. This was a special problem because the analytical measurement required samples from the upper layers of the surface.

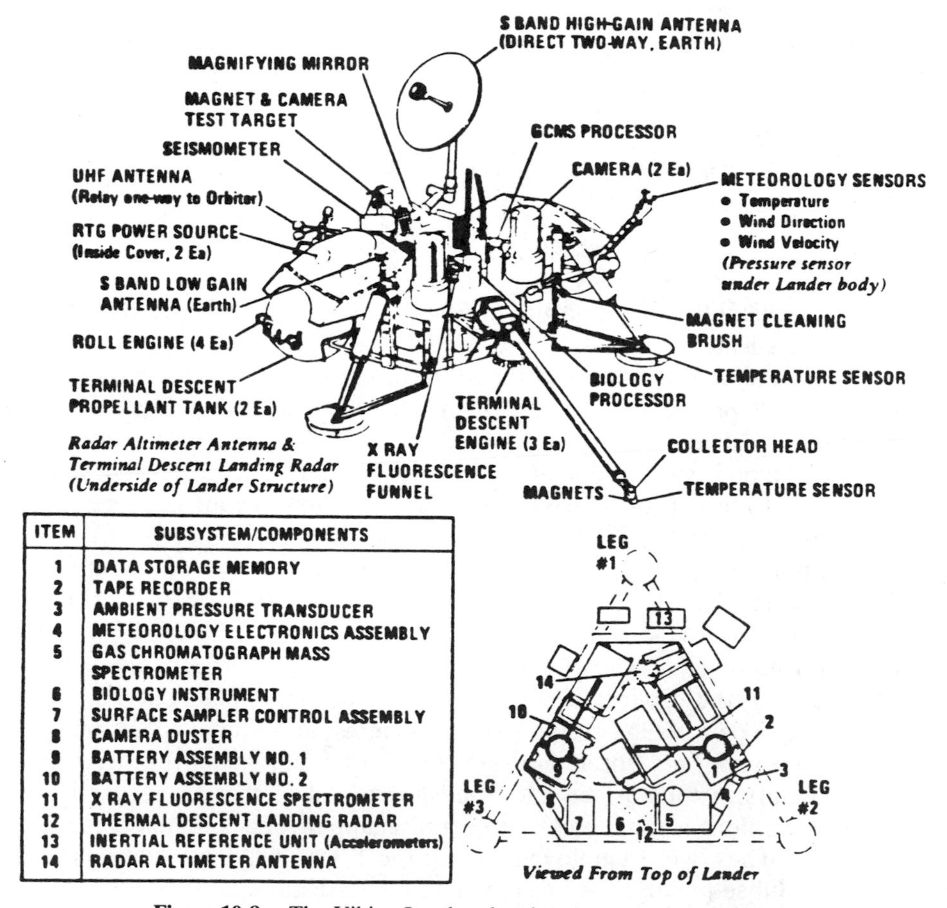

ITEM	SUBSYSTEM/COMPONENTS
1	DATA STORAGE MEMORY
2	TAPE RECORDER
3	AMBIENT PRESSURE TRANSDUCER
4	METEOROLOGY ELECTRONICS ASSEMBLY
5	GAS CHROMATOGRAPH MASS SPECTROMETER
6	BIOLOGY INSTRUMENT
7	SURFACE SAMPLER CONTROL ASSEMBLY
8	CAMERA DUSTER
9	BATTERY ASSEMBLY NO. 1
10	BATTERY ASSEMBLY NO. 2
11	X RAY FLUORESCENCE SPECTROMETER
12	THERMAL DESCENT LANDING RADAR
13	INERTIAL REFERENCE UNIT (Accelerometers)
14	RADAR ALTIMETER ANTENNA

Figure 10.8. The Viking Lander showing the surface subsystems.

Thus a great deal of effort went into the development of a retro-rocket engine that would have a minimum of chemical and mechanical effects on the surface. The retro-rocket selected used purified hydrazine as a propellent. This type of fuel obtains its energy from the exergonic breakdown of NH_3 into hydrogen and nitrogen, thus avoiding the contamination of the surface by uncombusted hydrocarbons. In addition, to avoid overheating and excessive erosion of the surface, the rockets used an 18-nozzle design to spread the impacting gas over a broad area.

Numerous tests were conducted in the Mars simulation chamber to examine the chemical, thermal, and mechanical influences on various models of a Marslike surface. It was ascertained that this design would cause the surface to be heated no more than 1°C at the hottest place, and that no more than 1 mm of surface would be stripped away by the exhaust gases.

One important concern was the injection of nitrogen (or NH_3) into the Martian soil. This was related to the intention to analyze the atmosphere soon after arrival at the surface. Several test of the residence time of gases suggested that the analysis of the atmospheric constituents should be delayed for the first 3 days to allow soil outgassing to come to equilibrium.

As an interesting sidelight, the first Viking landed within 1° of the horizontal. The second lander apparently landed with one leg on a rock and was tipped about 8° to the horizontal.

The lander shown in Figure 10.8 was an extraordinarily complex and sophisticated vehicle. It was a horizontal platform structure 0.5 m thick and 1.5 m across. It weighed about 600 kg and housed the scientific instruments and their attendant computers, tape recorder, data system, power system, transmitter, and receiver. The three analytical instruments; biology, X-ray fluorescence spectrometer, and gas chromatograph-mass spectrometer (GCMS), were mounted within the structure. Two identical facsimile cameras were mounted on top, along with a three-axis seismometer and a meteorological boom with weather sensors. The electric power came from two radioisotope thermoelectric generators using ^{238}Pu that could provide 70 W of continuous power. Peak power loads were supplied by rechargeable batteries. A sampling arm and scoop were mounted on the front side for delivering samples to the analytical instruments.

The transmission and receiving antennas for communication to the earth were also mounted on top of the lander. The lander was able to communicate directly with the Earth through the large movable parabolic S band antenna or through the orbiter. The linkage time changed with the geometry of the Earth and Mars, but characteristically was several hours each day. The distance between the Lander and the Orbiter determined the data rates, which were either 4000 or 16,000 bits/s, and the linkage time varied from 15 min to a little less than 1 h. Thus, on a single pass the total data that could be returned from the lander was between 10 and 50 million bits. This was a serious constraint on the photography, which required 10^6–10^7 bits for every image. Other investigations such as seismology and meteorology were also limited, but the analytical experiments were usually accommodated.

The lander was constructed so that it could operate with complete autonomy. At the time of the landing, the on-board computers were programmed to perform many days of operation if the lander could not be commanded from the Earth. It would have carried out a complete mission of picture taking, obtained samples from a preselected site, analyzed them, recorded the weather and seismic activity, and relayed the data to the Earth. However, both landers operated without flaw, and so the programmed mission was overwritten beginning with the first command link.

One of the very important requirements was the ability to respond to data. Essentially nothing was known in advance about the local surface topology, chemistry, or biology. Thus it was necessary to be able to modify the experiments as they progressed. The ability to make changes based on received data required about 2 weeks because of the need to prepare software, to check for errors that might have disastrous consequences, and for sending and verifying commands. As Soffen (8) points out, the automated laboratory was over 400,000,000 km away, and it was possible to command it once a day at best.

With this rather brief description of the Viking Lander we shall now examine some of the analytical studies.

MOLECULAR ANALYSIS

The composition of the atmosphere in the vicinity of the Viking lander was determined by means of a combined gas chromatograph–mass spectrometer combination. A summary of the effort has been written by T. Owen and co-workers (9). A description of the original instrument, subsequently somewhat modified, was published previously by D. M. Anderson and associates (10). Figure 10.9 is a schematic block diagram of the entire instrument package. The gas-flow path is shown by the oblique parallel lines. The solid lines represent the power, control, and data circuits. GC denotes the gas chromatographic components, while MS represents the components of the mass spectrometer.

The mass spectrometer was the key component of the system, and for a number of reasons the Nier–Johnson double focusing was selected. Its

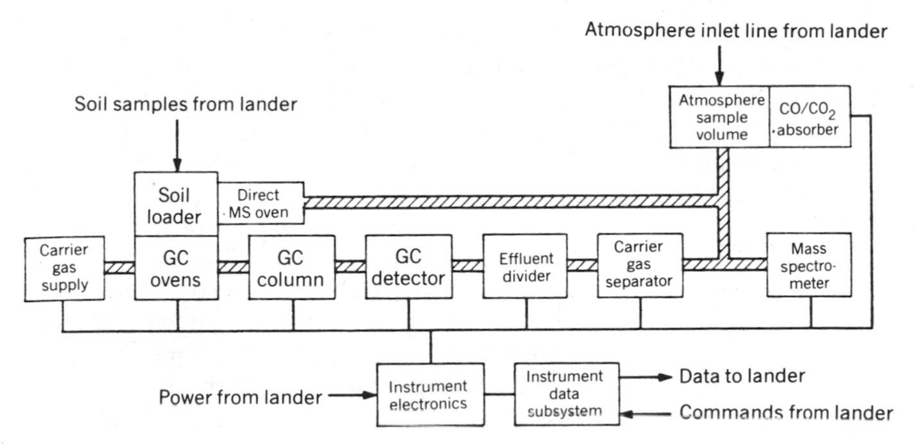

Figure 10.9. Functional block diagram of the gas-chromatograph–mass-spectrometer system. The gas-flow path is shown by the solid lines. Sources: Owen et al. (9), Anderson et al. (10).

intrinsically high resolving power was traded off for compactness. Small radii of deflection (3.8 cm for the magnetic sector and 4.7 cm for the electric sector) were selected, which led to a compact and relatively light instrument. By scanning the accelerating voltage rather than the magnetic field, the designers were able to use a constant magnetic field, furnished by a permanent magnet. This, in turn, led to simpler electronics. The use of a variable accelerating field facilitated the mass identification from the values of the accelerating voltages, which could be readily monitored and transmitted to the Earth.

The choice of a magnetic deflection spectrometer over other instruments such as time-of-flight and quadrupole analyzers followed from the fact that is best understood in terms of design and performance. Its resolution and sensitivity is not so dependent on complex electronic circuitry, some of which requires relatively high power levels. There were, however, limitations due to the relatively limited ratio of high to low mass covered in a single sweep with a fixed magnetic field. A range of e/m of 12–200 was selected. At the low end, it was necessary to sacrifice the detectability of hydrogen and helium. However, it was desirable to cover the range of the majority of the simple organic molecules, including those formed in the thermal degradation of the more complex ones.

The resolution of the spectrometer was such that it could differentiate between m/e 199 and 200 when their relative intensities differed by less than a factor of 10.

Where soil samples were to be studied by heating the soil sample into the carrier gas stream, there was a set of eight pairs of cylindrical ovens of

0.015- and 0.150-mL capacity arranged on a circular motor-driven holder. Any one of the ovens could be switched into a working position, where it then formed part of the gas line between the hydrogen supply and the gas chromatographic column. A pressure regulator and valves assured that gas flowed through the oven when it was energized for about 30 s prior to gas chromatographic separation.

The chromatograph consisted of two columns in series, a packed (152×0.127 mm i.d., 10% SF 96 on Chromosorb W), followed by an open tubular column (61 m $\times$ 0.5 mm i.d. stainless steel capillary coated with 10 : 1 SF 96 : Igapal CO 880). The combination was selected to assure that the permanent gases and water would be separated as far as possible from the organic compounds of interest—difficult to detect in the presence of a large amount of another constituent. Such a large component would trigger an overload protective device.

The overload protective device was actuated by one of two redundant systems, the total ion monitor and the ion pump current. The redundancy was necessary because a failure would permit a large amount of gas, other than hydrogen, such as CO_2 or H_2O, to pass through the palladium separator and enter the mass spectrometer at a rate far beyond the pumping speed of the ion pump. The protecting device consisted of a stream splitter, which vented a large fraction ($> 99.9\%$) of the gas chromatographic effluent to the martian atmosphere. Finally, the detector selected was of a type that responded to a wide variety of gas species. Thus, although its sensitivity was not as high as the total ion monitor of the mass spectrometer, it still provided a reasonable gas chromatogram to back up the mass-spectrometer results.

There were three options for analyzing the Martian atmosphere:

1. the admission of the atmosphere directly to the mass spectrometer,
2. the use of chemical scrubbers to reduce CO and CO_2, and
3. repeated scrubbing procedures, progressively admitting fresh samples to the gas reservoir in order to build up the partial pressure of the trace gases.

By the fourth and fifth days after the landing of the Viking 1 Lander, a total of six atmospheric analyses were performed at approximately 6-h intervals. In the first four analyses, CO and CO_2 were removed, while in the last two, samples of unaltered atmospheres were used. Analysis of these spectra confirmed the presence of nitrogen and argon in the Martian atmosphere. This had been reported by Nier and led to the discovery of ^{36}Ar (11). The ratio of $^{36}Ar/^{40}Ar$ was found to be 3.34×10^{-4} (one tenth the terrestrial value of 3.43×10^{-3}). The total abundance of ^{36}Ar and ^{40}Ar in the Martian atmosphere relative to the planet's mass were 0.0075 and 0.08, respectively, of comparable terrestrial values.

As a summary of the observations, the investigators have reported that they confirmed the discovery of N_2 and ^{40}Ar by the Entry Science Team (previously described). They also detected N_2, Kr, Xe, and the primordial isotopes of Ar. The noble gases were found to exhibit an abundance pattern similar to that found in the terrestrial atmosphere and the primordial component of meteoritic gases. The isotopic ratios of $^{14}N/^{15}N$, $^{40}Ar/^{36}Ar$, and $^{129}Xe/^{132}Xe$ were distinctly different for terrestrial values, thus implying different evolutionary histories for the volatile elements on the two planets. Based on the noble gas abundances, it is felt that at least 10 times the present atmospheric amount of N_2 and 20 times the CO_2 abundance were released by the planet over geologic time. Similarly, the outgassing of large amounts of H_2O must also have occurred. This was used as an explanation for the higher surface pressure and the abundance of water required in some earlier period to have produced the dendritic channels observed on the Martian surface.

THE SEARCH FOR ORGANIC SUBSTANCES

Another and highly significant application of the GCMS was to look for the presence of organic compounds in the Martian surface material. If such compounds were found, the objective then was to determine their structure and abundances. This question was ultimately tied to the general question of life on the planet. The structure of organic molecules, if they did in fact occur, might provide clues to the nature of the biotic or abiotic synthesis occurring on Mars (Biemann and associates, 12).

To achieve the goal, a technique that was both sensitive and of high specificity and broad applicability was required. Accordingly, the gas chromatograph coupled to a mass spectrometer was chosen (see previous section). The usual terrestrial approach would have been a digestion of the surface material to be analyzed by wet chemical processes, followed by solvent extraction and, possibly, chemical separations. For technical reasons, such an approach was beyond the capability of the Viking lander. In this framework of the Martian mission, where nothing was known prior to the landing, a general or group identification could be considered a major advance were there an encounter with a complex mixture.

For these reasons, thermal volatilization (with or without thermal degradation) of the possible organic compounds was chosen as the most reliable approach. The expectation was that by identifying the degradation products one could then identify the parent compounds in the sample.

The original instrument plans, previously described, included a sample oven directly monitored by the mass spectrometer so that it was possible to detect more complex and less volatile substances that would not pass through the gas chromatograph. However, this was eliminated during the design and testing, to simplify the final instrument package. In addition, the number of ovens connected to the GC was reduced from 8 to 3. This did reduce the flexibility, since each oven could only hold one soil sample. It was also discovered during transit to Mars that one oven in each instrument was not operable. However, because of the apparently uniform surface composition, this did not prove to be a problem.

The final version of the GCMS as described by Biemann (12) is shown in Figure 10.10. The sample ovens were in a circular holder that could be positioned in preset sequence on command from the instrument's internal logic system—modifiable within limits by ground command where the experimental results suggested a change.

There were two positions for the ovens, the load position and the analysis position. In the load position, the oven was directly under the sampling system, which was programmed to deliver 1–2 cm^3 of surface material that had been previously ground and passed through a 0.3-mm sieve. A mechanical poker pushed the material through a funnel into the

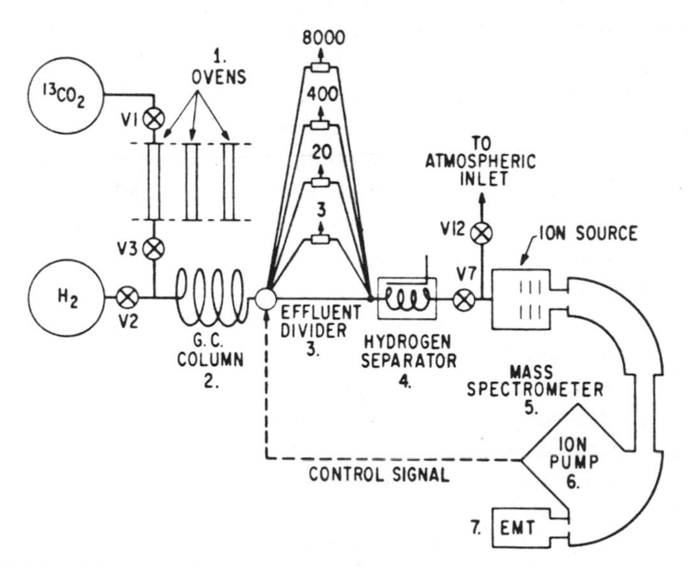

Figure 10.10. Schematic diagram of the Viking Lander GC-MS. Source: Biemann et al. (12).

oven. Terrestrial tests with soils such as finely ground basalts showed that the oven would fill. In the Mars case, there was no monitor of this process, but the assumption was made that the ovens were filled. Once "filled," the oven was moved to the analysis position, where the lines leading to the valves V1 and V3 were clamped to both ends of the oven. A gas-tight seal was achieved by pressing circular knife edges, into gold rings, into which the gas lines terminated, at each end of the oven.

Each oven could be heated to 50°, 200°, 350°, or 500° in 1–8 s and held there for an elapsed time of 30 s. The valves V1 and V3 were opened for 30 s prior to the heating of the ovens and then closed immediately after the heating period. Any volatile material evolved from the sample was swept into the GC column with 2–3 mL of CO_2 labeled with ^{13}C (99+% isotopically pure). CO_2 was used rather than H_2 as the carrier gas for the GC in order to avoid catalytic or thermally induced reduction of the possible organic material in the sample.

The performance of the instrument was demonstrated prior to the mission by the use of a laboratory version, which corresponded closely to the flight instrument. A sample of Antarctic soil was heated to 200° and 500° and analyzed. The chromatogram (a total ion current plot) is shown in Figure 10.11. A large number of compounds were identified. The code in the figure and the corresponding compounds are listed in the accompanying tabulation. AR represents the aromatic hydrocarbons, O the oxygen-containing compounds, N the nitrogen-containing compounds, and HC the alliphatic hydrocarbons. With respect to the sensitivity of the instrument, the quantities of benzonitrile and benzofuran (based on the peaks observed) were estimated to correspond to 150 and 43 ppb, respectively.

Code	Compound	Code	Compound
AR-1	Benzene	O-8	Phenol
AR-2	Toluene	O-10	Dibenzofuran
AR-3	Xylene (benzene-C)	N-1	Acetonitrile
AR-5	Benzene-C	N-2	Propionitrile
AR-9	Naphthalene	N-3	Pyridine
AR-10	Methyl naphthalene	N-4	Benzonitrile
AR-11	Biphenyl	N-5	Toluonitrile
O-1	Furan	S-1	Thiophene
O-2	Acetone	S-2	Methylthiophene
O-4	Methyl vinyl ketone	S-3	Thiophene-C
O-5	Methyl furan	S-4	Benzothiophene
O-6	Furane-C	S-5	Sulfur dioxide
O-7	Benzofuran	HC-1	Cyclohexene

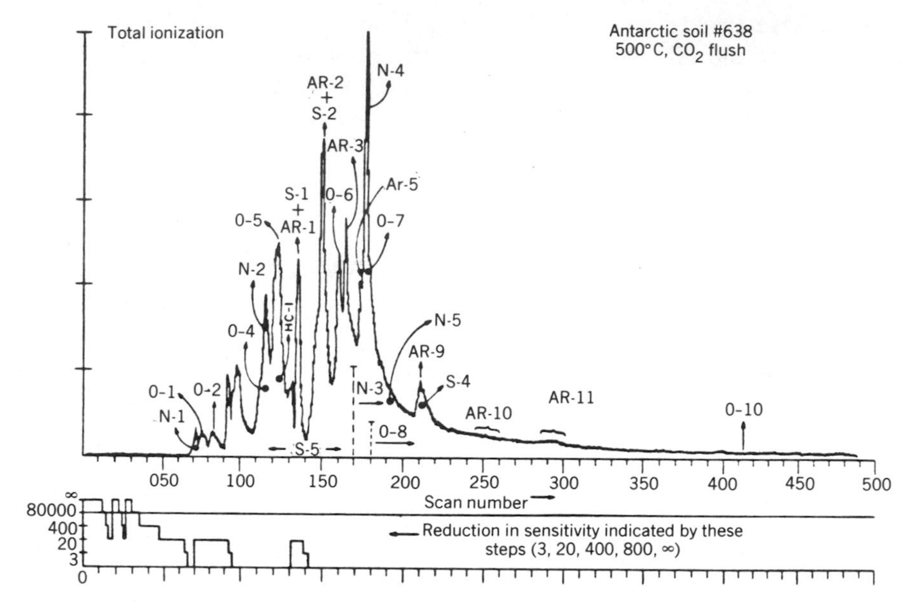

Figure 10.11. Gas chromatogram obtained with a sample of Antarctic soil. Source: Biemann et al. (12).

RESULTS

Four samples of the Martian surface and subsurface were analyzed for both organic and inorganic volatiles. Two samples were obtained at each landing site. The Viking Lander 1 sampled Chryse Planitia and Viking Lander 2 was landed at Utopia Planitia. The results of the analysis at the site of Viking Lander 1 is shown in Figure 10.12, where all the gas chromatograms are plotted to the same scale.

The chromatograms are shown as the summed intensity of all ions above m/e 47 (to eliminate the effects of water and carbon dioxide) in each scan versus scan number along the abscissa. The chromatogram at the top represents a blank or background achieved by measurements during the cruise to Mars. The sharp peak observed in the first chromatogram from the Martian site was identified as methyl chloride. The identification was based on the mass spectrum recorded at that point and the fact that in the mass chromatograms all the ions characteristic of CH_3Cl showed the same sharp maxima.

Calculations by the investigators (12) showed the concentration to be about 15 ppb with respect to the sample. This proved to be the only clearly identifiable peak in the chromatogram. Ultimately, this peak was

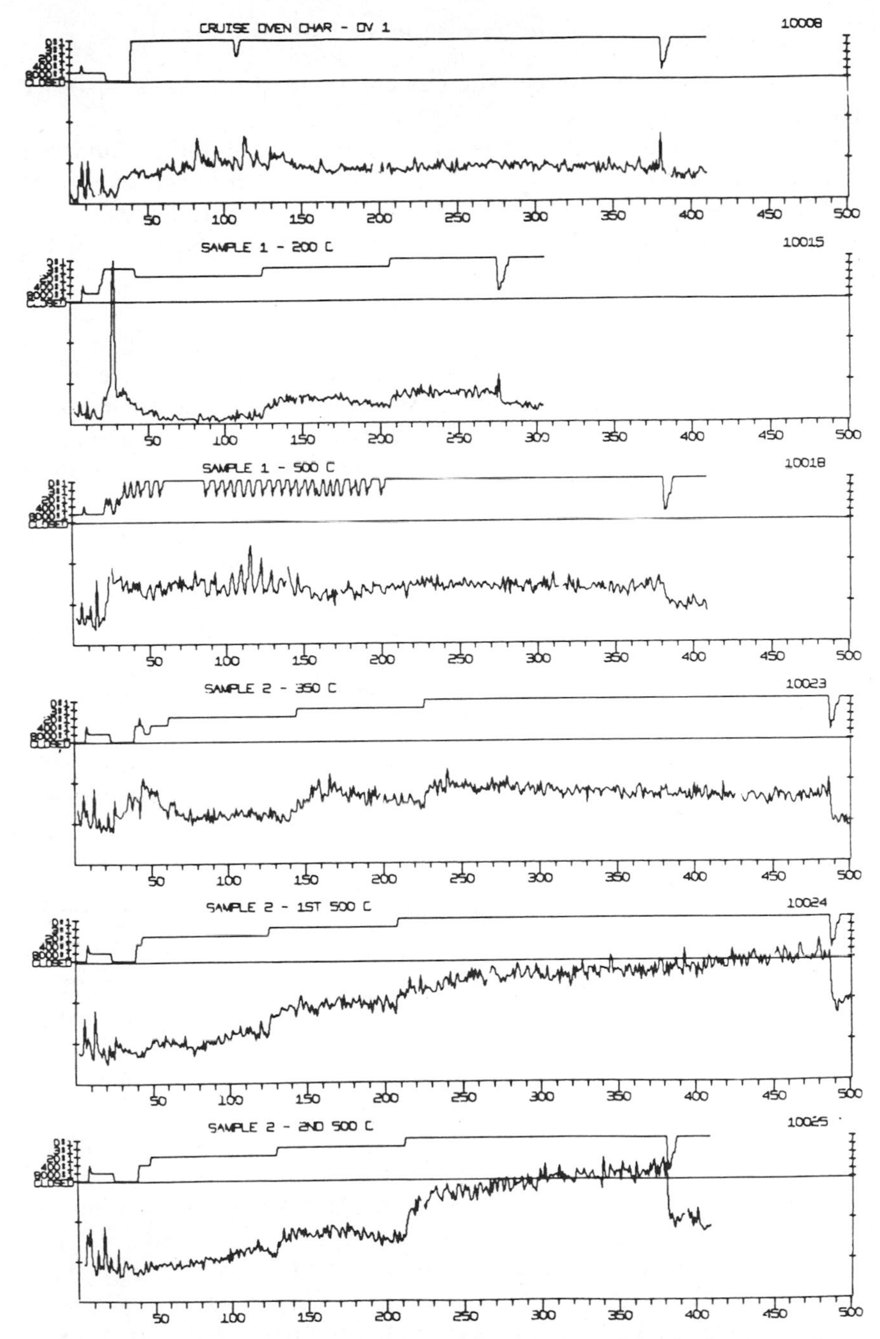

Figure 10.12. Gas chromatogram of cruise blank and five Viking lander 1 experiments, plotted as the sum of all ions above *m*/*e* 47. Source: Biemann et al. (12).

attributed to terrestrial sources such as chlorinated solvent or adsorbed traces of methanol and HCl.

In summary, a total of four Martian samples, one surface and one subsurface at each of the two Viking Lander sites, were analyzed by the GCMS. In none of the samples could any organic matter be detected—the detection limits being generally of the order of parts per billion for a few substances closer to parts per million. The evolution of water and carbon dioxide was observed, but not other organic gases. The absence of organic molecules was a very significant occurrence on silicate and other mineral surfaces irradiated with UV shorter than 320 nm. The use of the xenon lamp during the experiment was optional and under command from the Earth. Also commandable from the Earth was the injection of approximately 80-μg of water vapor (or some multiple) into the test chamber at the start of the experiment.

The atmosphere in the test chamber was labeled by the injection of $^{14}CO_2$ and ^{14}CO (92 : 8 by volume) with a total radioactivity of 22 μCi at the start of the experiment. The resulting pressure increase was 2.2 mbar over the initial ambient of 7.6 mbar at both landing sites. Since the Martian atmosphere contained about 95% CO_2 and 0.1% CO, the injection of radioactive gases increased the partial pressure of CO_2 and CO by 28% and 23-fold, respectively.

The sample was incubated for 120 h at temperatures that ranged from 8° to 26°C in the experiments. The lamp was then turned off and the chamber brought to 120°C while the radioactive atmosphere was vented. The next step involved the heating of the chamber to 635°C to pyrolyze the organic matter in the sample. The volatile products with a large amount of $^{14}CO_2$ and ^{14}Co desorbed from the soil grains and walls of the chamber were swept by a stream of He into a column packed with a mixture of 25% cupric oxide and 75% Chromosorb-P (a form of diatomaceous earth). This column, operated at 120°C, retained organic matter larger than CH_4, but allowed all but a small fraction of the CO_2 and CO to pass into the radiation counter where their radioactivity was measured. The investigators refer to this as peak 1.

The column temperature was then raised to 640°C, this higher temperature causing the release of organic compounds and their oxidation to CO_2 by the CuO in the column. The radioactivity of this gas was identified as peak 2. It represents the organic matter, if any, synthesized from ^{14}CO or $^{14}CO_2$ during the incubation. Peak 2 also contained a small fraction of CO and CO_2, which failed to elute with peak 1. The radioactivity of this fraction, designated as peak 2(0), was subtracted from peak 2 in order to estimate the amount of C fixed in the organic matter. The size of the peak 2(0) was known from prior laboratory tests carried out on heat-sterilized soils (or with soils absent) in test chambers by using flight-configured columns.

A total of 10 experiments were performed on Mars, six at the Chryse site (numbered C1 to C6) and four at the Utopia site (U1 to U4). Of these experiments, U4 was not usable because of apparent valve failure. A summary and statistical analysis of all the usable results can be seen in Figure 10.13 and Table 10.2. Experiment C5 was performed on a surface sample acquired before solar conjunction and stored in the hopper of the soil distribution assembly for 69 sols before the C5 measurements were begun (1 sol = 1 Martian day = 24.65 h). Since three test chambers of the Viking Lander 1 had already been used, the experiment was run in what has been described by the investigators as a soil-on-soil mode. The C5 experiment was run to determine the effect of water vapor injection, followed by the evaporation of the water at an elevated temperature, on the soil activity. Because the test chamber contained two soil aliquots, two injections of water vapor were made. After 4 h the chamber was vented and the temperature brought to 120°C for 1–2 min, and then it was dropped to about 90°C for about 112 min. Beyond this, the temperature was lowered to 17°C + 1° for the remainder of the experiment.

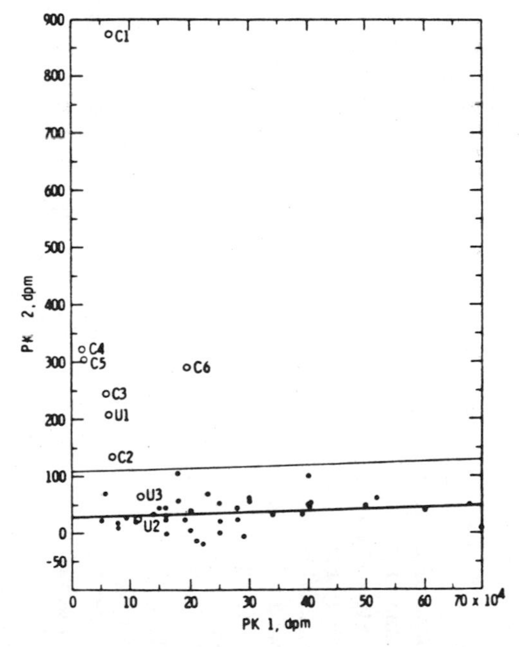

Figure 10.13. ^{14}C fixation results from Mars (open circles) compared with laboratory data obtained with heat-sterilized soils or no soils (solid circles). Source: Horowitz et al. (14).

Table 10.2. Carbon Fixation Statistics[a]

Experiment	Sample	Conditions	Incubation Temperature (°C)	Disintegrations per Minute ± Standard Error: Peak 1	Peak 2	Peak 2(0)	Peak 2 − Peak 2(0)	*P*
C1	Sandy Flats 1, fresh	Light, dry	17 ± 1	67,464 ± 536	873 ± 10	30.7 ± 27	842 ± 29	~0
C2	Sandy Flats 1, stored 19 sols	Light, dry, 175°C heat treatment	15 ± 1	69,536 ± 545	136 ± 12	30.8 ± 27	105 ± 29.5	2×10^{-4}
C3	Sandy Flats 2, fresh	Light, dry	13–26	61,027 ± 527	245 ± 8.9	30.5 ± 27	214.5 ± 28	10^{--14}
C4	Sandy Flats 3, fresh	Light, dry	16 ± 2	18,545 ± 381	318 ± 15	29.3 ± 27	289 ± 31	6×10^{-21}
C5	Sandy Flats 3, stored 69 sols	Light, H_2O, 90°C heat treatment	17 ± 1	20,295 ± 395	304 ± 11	29.4 ± 27	275 ± 29	10^{-21}
C6	Sandy Flats 3, stored 139 sols	Light, H_2O	15 ± 2.5	193,803 ± 864	289 ± 15	34.3 ± 27	255 ± 31	10^{-16}
U1	Beta 1, fresh	Dark, dry	15 ± 3	64,845 ± 527	209 ± 15	30.6 ± 27	178 ± 31	5×10^{-9}
U2	Beta 2, fresh	Light, H_2O	18 ± 1.5	113,845 ± 690	25 ± 8.4	32.0 ± 27	−7 ± 28	0.5
U3	Under Notch Rock, fresh	Dark, dry	10 ± 2	118,309 ± 400	68 ± 23	32.2 ± 27	36 ± 35	0.15

[a] Samples are identified by sampling location, acquisition number, and age at the start of the experiment. Conditions indicate whether the lamp was on or off, whether water vapor was injected, and whether the sample was heated before starting incubation, *P* is the conventional one-tailed probability that a positive deviation from peak 2(0) as large as or larger than that found would be obtained by chance.
Source: Horowitz et al. (14).

It was also necessary to compensate for the reduced pressures in the lander gas bottle, which after 160 sols of use and normal leakage dropped to about 10% of the initial value. Thus three 20-μl injections were used. As the table shows, the results of C5 were not very different than those for C4. The low first peaks were attributed to the depletion of the radioactive gas. The second peaks of C4 and C5 were statistically identical. The investigators (13) felt this to be an important result, considering the different thermal histories.

Another type of experiment is shown by C6. This one tested the effect of water vapor injection without heating or evaporation. It repeated U2, which appeared to suggest that water vapor inhibits the fixation reaction (14). The procedure differed from U2 and C5 in that the order of addition of water vapor and radioactive gases was reversed.

Based on the data described above, Horowitz and associates (13) have concluded that a fixation of atmospheric carbon occurs in the Martian surface materials under conditions that approximate the Martian ones. The highest activity (see Table 10.2) was seen in experiment C1, where about 10 pmol of CO (or 30 pmol of CO_2) was fixed. The activity, although small be terrestrial standards, was significant. As Horowitz et al. state, (13) "indeed, when it is recognized that the only identified non-biological organic synthesis, the surface catalyzed photoreaction referred to earlier, has been eliminated in designing the experiment, the results are startling."

However, a biological interpretation of the results was considered unlikely in view of the thermostability of the reaction. The second peaks of the experiments C4, C5, and C6 were not significantly different from each other, even though the thermal histories were very different. The three samples were aliquots from the same scoopful of surface material. In C4 the material was examined fresh. In C5 it had been stored for 69 sols at temperatures between 10 and 24°C before testing and, furthermore, it had heated for nearly 2 h at 90°C before the incubation began. In C6 the sample had been stored for 139 sols at temperatures from 5°C before the experiment began. Thus with these histories and the observation that the ground (the habitat of possible organisms), temperatures do not reach 0°C at any time of the year at either site, it was not reasonable to assume that the constancy of response could have a biological origin.

As an additional piece of evidence found in C2, a sample from the same spot that had supplied the C1 sample was heated to 175°C for 3 h before the start of incubation. The activity was considerably reduced, but not to the level of sterile soils. Thus while it appears that the agent reacting is somewhat heat labile, it is not as labile as one would expect a living organism to be.

THE LABELED RELEASE EXPERIMENT

The Viking Labeled Release Experiment (G. Levin and P. A. Straat, 15) proved to be one of the more fascinating investigations performed on the Martian surface. The effort has been described in summary by Soffen (8) as being by far the most complex of all the investigations. The results of the experiment showed no unambiguous observation of any recognizable life and, as we will see, some of the results appear to show an absence of biology in the samples tested. However, the experiments did yield significant results about the chemical nature of the Martian surface. Of the various measurements, one at least might have a biological explanation. One of the experiments showed that the Martian soil had an agent capable of rapidly decomposing the organic chemicals used in the medium. One thing was certain–the surface of Mars was obviously highly reactive, containing at least one and probably several highly oxidizing substances. While inorganic reactions might be involved to explain the data seen, the possibility of biological processes is still a subject for debate.

The labeled release (LR) life detection experiment was developed to detect heterotrophic metabolism. It involved monitoring radioactive gas evolution following the addition of a radioactive nutrient containing seven ^{14}C-labeled organic substrates to the surface material. A detailed description has been furnished by H. P. Klein (16). Two modes of activity were tested, involving two different concepts of Martian biology. In one instance, the fundamental assumption made was that the only limiting factor to the growth of a Martian organism was water. It was assumed that the nutrients, possibly as simple organic compounds, formed photochemically were already present in the Martian surface and that any organism present would be dormant under the Martian environment. It was felt that once sufficient moisture was supplied, the dormant organisms would be stimulated into metabolic activity. Such activity was to be measured by an analysis of the atmosphere above the incubator system by means of gas chromatography.

For the second mode, the basic assumption was that a significant fraction of the Martian biota was composed of heterotrophic organisms and, therefore, that the addition of organic compounds was considered necessary to initiate a metabolic response. It was also felt that this response would be found only in an aqueous environment. The presence of a large number of different organic and inorganic compounds was assumed not to inhibit this metabolism.

The first assumption (the nonnutrient mode) was tested in the following manner: The martian samples were incubated in the presence of the Martian atmosphere to which additional carbon dioxide, krypton, and helium were added to bring the total pressure to approximately 200 mbar to facilitate subsequent gas handling. Approximately 0.5 cc of nutrient solution was introduced into the incubation cell in such a manner that the nutrient did not come into contact with the samples. The atmosphere became saturated with water at the incubation temperature of 8–15°C. After an incubation period of about 7 days for this phase of the experiment, it was terminated. The experiment was performed twice, once at each landing site, and the observations were the same for each site. There was physical desorption of some gases and chemical generation of oxygen observed, but nothing in the data suggested metabolic activity. Thus the results were deemed negative with regard to biological activity.

Klein (16) in a review has outlined some of the constraints that exist in the interpretation of the data. If the initial assumptions were valid, one could only conclude that the assayed samples contained no metabolizing organisms. However, if the assumption was incomplete because it did not consider that some source of energy was necessary to stimulate metabolic activity, then under the conditions of the experiment a negative result would not preclude the presence of a living organism. There were other questions about the experimental conditions and whether they prevented the development of biological signals. These questions involved incubation temperatures and whether the incubation period was long enough.

The second series (wet nutrient methods) of experiments were done as follows: The experiment was performed three times for periods of 200 (Viking 1), 31 (Viking 2), and 116 (Viking 2) sols. The Martian atmosphere was present for only a portion of these incubation periods: 13, 19, and 78 sols, respectively. The rest of the time the atmosphere was approximately 200 mbar and the incubation temperatures were in the 8–15°C range. Again, while some gas changes were observed in the three trials, none of these fit the criteria for biological activity. As Klein reported, on the basis of the original assumptions for the experiment and provided the conditions of testing were adequate, it is possible to conclude that no viable organisms were present in the samples. However, a negative finding did not rule out the presence of chemosynthetic organisms. It was possible that even if the original assumptions were correct, some of the experimental conditions such as temperature, pressure, and artificial atmosphere may have precluded positive biological findings in the experiments.

RESULTS

Levin and Straat (17) published results of the LR experiment that had been performed prior to conjunction. These experiments were reported as showing a rapid increase in radioactive counts on the addition of the radioactive nutrient (^{14}C-labeled organic substrate) to a fresh surface sample. This response was similar at both landing sites, and the magnitude of the evolved counts was consistent with utilization of one of the labeled carbon atom positions available in the nutrient. While the radioactivity evolved continuously, even over a incubation period of over 60 sols, there was no evidence of an exponential growth. In one cycle, involving a surface sample obtained from under a rock, the initial labeled response was essentially identical in kinetics and magnitude with those of other active samples. Levin has stated that if a chemical reaction is responsible for the LR results, its activity did not depend on recent ultraviolet activation. Further, the active agent(s) in the Mars sample appeared to be stable to 18°C but was completely inactivated by 3 h of heating at 160°C, and its activity was also substantially reduced following similar treatment at 50°C. According to Levin, this behavior is consistent with biological response although, as we have seen, others have proposed that a limited number of nonbiological reactions could produce such results.

An example of the observations is shown in Figure 10.14. This is a plot of the labeled release data from third sample analysis on VL-2. This particular sample was collected from the Martian surface after the sampler arm had been used to push aside a rock. Figure 10.15 shows the LR data following a second injection to the sample.

The extraordinary release of radioactive gas has proved both exciting and puzzling. There are conflicting interpretations, leaving open the vital

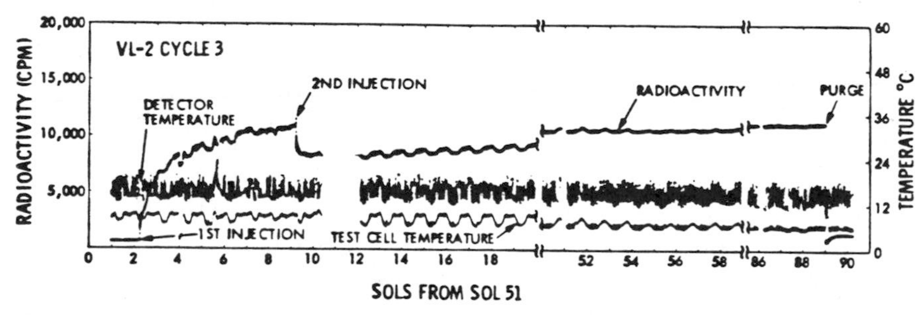

Figure 10.14. Plot of LR data from third sample analysis on VL-2. An active sequence was used on a fresh sample that was acquired from surface material exposed by pushing aside a rock. Radioactivity was measured at 16-min intervals throughout the cycle except for the first 2 h after each injection, when readings were taken every 4 min. Radioactivity data include a background count of 659 cpm prior to the onset of the cycle. Data obtained in the single-channel counting mode between sols 53 and 60 have been corrected to the dual-channel mode of operation for comparison with the remainder of the cycle and with data from previous cycles. Data from sols 61 and 62 were lost. Detector and test cell temperatures were measured every 16 min. Source: Levin and Straat (15).

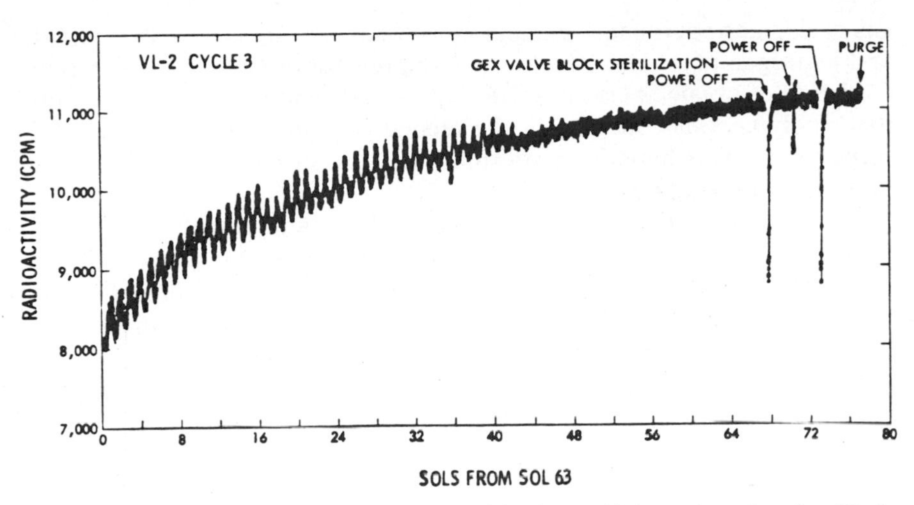

Figure 10.15. Plot of LR data following second injection to third sample analyzed on VL-2. Data after second injection on sol 60 were expanded from data presented in Figure 1. The pronounced drops in radioactivity occurring between sols 66 and 74 correspond to sharp drops in test cell temperature resulting from brief periods in which the power was turned off or during which the valve block in the Gas Exchange (GEX) experiment was sterilized. Throughout the indicated period, radioactivity was measured at 16-min intervals. Source: Levin and Straat (15).

question regarding the existence of life on Mars (16). There have been a number of subsequent studies in terrestrial laboratories in which an attempt has been made to produce models and mechanisms to imitate the behavior observed on Mars.

Banin and Rishpon have examined the question (18) and stated that the intrinsic complexity of the biology package of the Viking landers, originally considered as necessary to recognize the complex phenomena associated with viable life forms, may in actuality make it very difficult to reach definite conclusions. The prevailing opinion, they state, is that, almost certainly, biological activity of the type recognized on the earth was not the cause, and the behavior observed was due to chemical reactions. As for the mechanisms, they are still far from being established. It was possible, for example, that the Martian minerals might play a role, but the nature of the Lander experiments provided no direct clues to the minerals in the Martian surface. A number of hypotheses have been proposed, of which Banin and Rishpon discuss the following three:

1. The soil contains a variety of oxidized compounds, including superoxides, which release oxygen upon contact with water (based on the

gas exchange experiment), gamma iron oxide (or a similar metallic oxide), and hydrogen peroxide, which cause the release of CO_2 by the catalytic oxidation of organic acids (the LR experiment) and possibly carbon suboxide polymers, accounting for the second peak in the Pyrolytic Release Experiment. This hypothesis presumes that the surface of Mars is essentially anhydrous and self-sterilizing (19).

2. The unweathered fraction of the soil contains OH^- and $CO^{2-}{}_3$ formed by some photochemical weathering process (20). Once water is added to the soil, the adsorbed OH is released to the solution, forming H_2O_2 that then decomposes releasing O_2. The CO_2 release in the Gas Release experiment is then apparently due to formate oxidation by the H_2O_2 in the solution. This is presumably also the case for the LR experiment. The source of the H_2O in the soil is water frost acting on a freshly weathered surface of primary mineral.

3. The soil contains a substance whose grains are permeated with micropores that were formed, perhaps by migration and accumulation of gases released from the crystal itself by radiation. The gases, oxygen and CO_2, are occluded in the pores and can only be released after the samples are wetted (21). On release, these gases can oxidize the organics in the LR medium and thus account for the gas release measured in the Gas Exchange experiment.

Common to the first two assumptions, results are explained by the presence of small amounts of active chemical ingredients in the Martian soil, but no account is taken of the major mineral components in the soil. In addition, all three hypotheses involve a completely waterless, self-sterilizing Martian soil.

In 1978, Banin and Rishpon (22) presented experimental evidence to support their own view that the Viking biology results might be explained by the presence of clay minerals of the montmorillonite family as a major component of the Martian soil. As a consequence of their extensive and active surface area, they were capable of adsorbing and releasing gases and catalyzing the decomposition and/or synthesis of organics, as the Viking measurements disclosed. In their 1979 paper (18), Banin and Rishpon attempt to establish the presence of smectite clays (nontronite and montmorillonite) from the results of the elemental chemical analysis determined by the X-ray fluorescence spectrometry (XRFS), which will be discussed below. Although direct mineralogical identification was not possible on the landers, a computerized search of the terrestrial mineral mixtures showed that a best fit could be obtained for a mixture of weathered ferrosilicate minerals, with the clays nontronite and montmorillonite as major components, mixed with soluble sulfate salts. A representative mineral assemblage, which appeared likely, contained about 50% nontronite, 17–20% montmorillonite, 13% kieserite, 7% calcite, and 1% rutile. Banin and Rishpon emphasized that only a direct mineralogical characterization of the soil itself could establish the correctness of this suggestion.

There were additional observations supporting the presence of the smectites. Light absorption and light scattering were done in the IR, visible, and UV, measured during the 1971 and 1973 Martian dust storms. Following these were analyses of soil reflectance and atmospheric absorption in the visible and UV by the Viking Landers. In 1973, Hunt (23) showed that montmorillonite provided the best match to the spectra obtained from the Martian dust obtained by the Mariner orbiters in the 6–12-μm region. Pollack and co-workers (24) determined from the UV and IR measurements of the Mariner 9 orbiter data, obtained during the 1971 global dust storm, that the solid particles suspended in the atmosphere were mainly silicates. Banin and Chen (unpublished data) found that the general shape of the absorption index curves of clays is similar to Martian dust. The shoulder in the Mars dust absorption curve at 240–260 nm coincides with a typical absorption peak of montmorillonite at 245 nm (25), and nontronite at 260 nm (26), and some indication of the absorption band at 200–210 nm appears as well for the two clays and their mixtures. Bragg (27) analyzed the visible and near-IR spectral reflectance of the soil around the Viking landers and compared it with reflectance characteristics of various minerals. Based on these comparisons and additional data, he identified nontronite as a major constituent of the soil.

Banin and Rishpon (18) have also examined the geological scenario. They point out that smectite clays are the product of alteration of volcanic materials and basic igneous rocks. They conclude that, while such weathering would not occur under present Martian conditions, observations do appear to indicate that there was considerable water in the past. This leads to the inference of early formation in the Martian history (the period of intensive fluvial activity). The uniformity of composition at the two widely separated Viking sites is due to rock abrasion and then transport by the large-scale atmospheric wind storms.

The consequences of the presence of the martian smectite clay minerals is due to the fact that they have platy, thin particles with a high specific surface area of 800 m^2/g. As a consequence of isomorphous substitution during their formation, they carry excess negative charge on their surface and act as a cation exchanger with capabilities in the range of 80 to 120 mEq/100 g. Oxygen atoms arranged hexagonally are exposed to the mineral's surface, which intensively absorbs water. Water is bound also as solvation shells on exchangeable ions. Intensive adsorption phenomena of ions and neutral molecules are characteristic of these clays. The mineral surface is also active in various catalytic reactions such as oxidation, dehydrogenation, and polymerization. On the Earth, smectite minerals are extremely active, and thus their presence in the Martian soil could lead to a host of reactions with the addition of water or solutes to the soil.

Banin and Rishpon undertook to simulate the Viking Biology Experiment results using smectite clays. To do this experiment, they used the smectite clays nontronite and montmorillonite, which contained iron and hydrogen as adsorbed ions. They found that radioactive gas was released from the medium solution used in the Viking LR experiment when it was reacted with the clays at rates and quantities similar to those measured by the Viking landers on Mars. When the active clay mixed with soluble salts was heated to 160°C in a CO_2 atmosphere, the decomposition activity was reduced considerably, in accord with the Martian observations. They have postulated that the decomposition reaction in the LR experiment is an iron-catalyzed formate decomposition on the clay surface.

The main features of the Viking Pyrolytic Release (PR) experiment were also simulated, with the iron-clays giving a relatively low "1st peak" and a significant "2nd peak." They have concluded, therefore, based on various Martian soil properties and the simulation experiments, that the smectite clays are a major and active component of the Martian soil. It appears that many of the results of the Viking Biology Experiment can be explained on the basis of the surface activity in catalysis and adsorption.

THE VIKING GAS-EXCHANGE EXPERIMENT (GEX)

The GEX was one of three experiments making up the Viking Lander biology instrument complement (Oyama and Berdahl, 19). Its function was to periodically sample the head space gases above a Martian surface sample incubating under dry, humid, or wet conditions, and to analyze the gases using the gas chromatograph. The GEX was engineered to distinguish between gas changes due to microbial metabolism and those produced by purely chemical reactions or physical phenomena such as sorption and desorption, by a recycling of the soil sample. The premise was that chemical or physical reactions would be reduced or even eliminated in subsequent cycles. A biological system, on the other hand, would either continue or increase. The recycling was accomplished by flushing the test chamber of gas and nutrient and then adding fresh nutrient and test gas and continuing the incubation.

Five surface samples were tested by the GEX experiment, two from the VL-1 Chryse site (Sand Flats area) and three from the VL-2 Utopia site. The first sample from Chryse was tested as received, the second sample was heated at 145°C for 3.5 h before testing. In the Utopia samples, the first and second were tested as received, while the third was heated at 145°C for 3.5 h before testing.

A 1-cc soil sample was placed in the 8.7-cc test cell and the cell was sealed. Following the addition of test gas (91.65% He, 5.51% Kr, and 2.84% CO_2), which brought the cell pressure to about 200 mbar. Three options were exercised: (1) no moisture was added, (2) enough nutrient was added to humidify the soil (about 0.5 cc), and (3) sufficient nutrient was added to wet the soil (about 2 cc). The nutrient was a complex mixture of organic compounds and inorganic salts.

The recharge system was performed as follows: At the end of the test cycle, the test cell was flushed of test gas and nutrient by passing helium through the cell and out of the drain valve (approximately 0.5 cc of nutrient remained in the cell sump and 0.35 cc remained in each cubic centimeter of soil.) Following this procedure, the test gas and fresh nutrient were then added.

The experiment then called for a termination sequence prior to receiving another soil sample (the GEX had only one test cell). The cell was flushed of test gas and nutrient (as in the recharge sequence described above), except that the drain valve was left open and with no helium flow, for 3 or more sols in order to dry the soil. With the drain valve open, the inside of the test cell was open to the Martian environment.

After termination, a fresh surface sample was added to the test cell. The cell was sealed and heated to 145°C for 3.5 h, with 1 cc of helium flowing through the test cell and out of the drain valve. The test cell was allowed to cool, the drain valve closed and the helium flow stopped.

The GC system had the capability of measuring H_2, Ne, N_2, Ar or CO, NO, CH_4, Kr, CO_2, N_2O, and H_2S. Ar could not be separated from CO by the GC column employed. It was assumed, however, that the Ar/CO peaks measured were Ar, because the Ar abundance had been estimated to be at least an order of magnitude greater than the CO in the martian atmosphere, and the CO contribution considering the Ar concentration, was not measurable by the instrument.

RESULTS

Figure 10.16 is an example showing the gas changes that occurred in the cell head space throughout the four cycles of incubation of the first Sandy Flats' sample. Trapped Martian atmospheric gases and the test gases were present in the cell before the humidification performed in cycle 1.

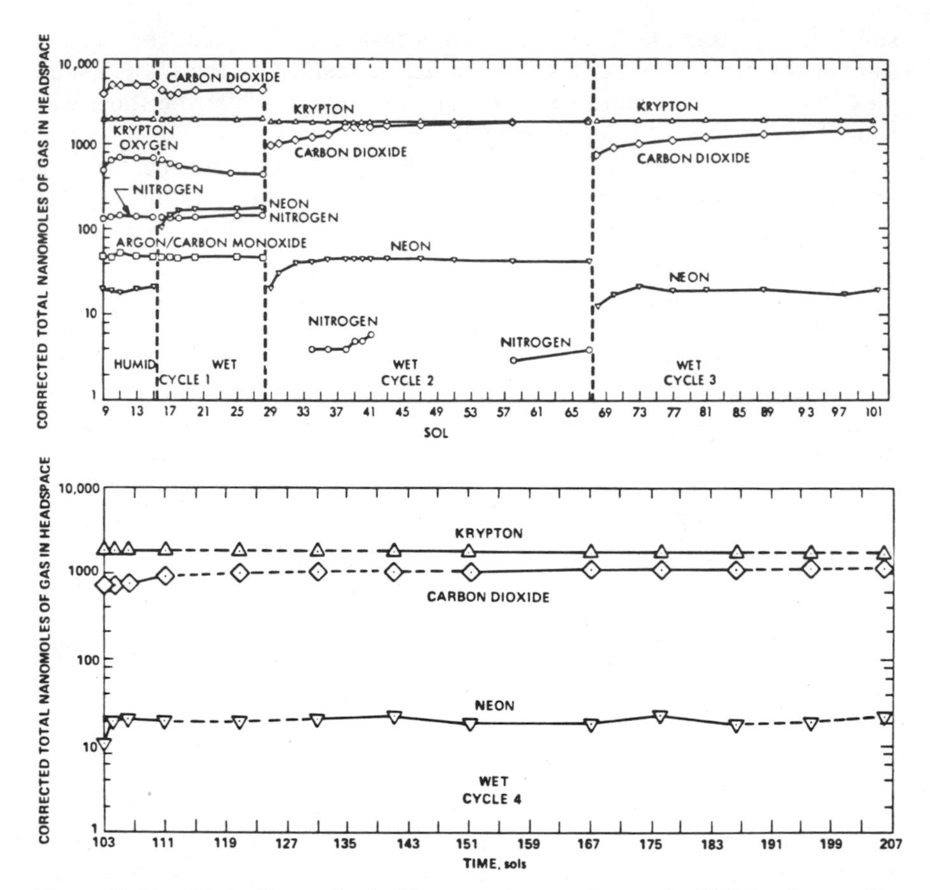

Figure 10.16. VL-1, Chryse Sandy Flats sample, gas changes in GEX headspace. The corrected total nanomoles were calculated as follows. The GC detector data were sampled at 1-s intervals, digitized, and fitted to a skewed Gaussian distribution from which peak heights were obtained. The gas in the headspace was obtained from the radio of the sample loop volume to the total headspace volume. The gas composition was corrected for cumulative sampling losses by referencing absolute changes in the krypton values for successive samples. Corrections were made for pressure sensitivity in this flight instrument caused by a partial restriction in the gas sampling system, which prevented total evacuation of the sample loop to ambient pressure before filling (three times) from the test cell. Source: Oyama and Berdahl (19).

The quantities of these gases were calculated to be 61 nmol of N_2, 4 nmol of O_2, 40 nmol of Ar, and 3600 nmol of CO_2. Oyama and Berdahl compared these values to those obtained from the sol 9 analysis. Without exception, the concentration of all gases proved to be greater than expected upon humidification of the soil sample. The greatest change in concentration was in the O_2, from an expected 4 nmol, it actually went to 520 nmol. Virtually all gas production ceased by sol 11.

Based on laboratory studies, Oyama and Berdahl (19) concluded that the early, rapid gas changes produced by the humidification were associated only with physical and chemical phenomena; as a consequence, they were unreliable indices of biological activity. To add to this, the investigators had never seen native terrestrial soils evolve O_2 in the dark. The suggestion made was that the extremely fast production of gases indicated either desorption processes or unusually rapid chemical reactions. One of the questions raised was whether the physical and chemical processes observed were caused by bringing the cold Martian soil to a warmer environment. However, as Oyama has pointed out, the soil after aquisition had been exposed to a temperature of 9°C for at least 0.28 h, so that most of the thermal degassing had already occurred prior to the soil being dumped into the test cell. Thus it seemed more likely that humidification rather than desorption was the driving mechanism.

From Figure 10.16, it can be seen that the O_2 production ceased between sol 11 and 15, and after the injection of additional nutrient on sol 16, the concentration of O_2 began to decrease. This was attributed to the uptake of O_2 by the ascorbate ion in the nutrient. It is apparently well known that ascorbate ion has the capacity to react with O_2. In contrast to the O_2 changes, the N_2 and Ne continued to slowly increase. N_2 had been shown to be desorbed very slowly from sterile terrestrial soils in the same way it appeared to be desorbed from Martian soils.

The wet mode data showed that only CO_2 evolution persisted beyond the first cycle. Generally, all of the CO_2 curves were similar except for the first wet cycle. In that cycle, a decrease lasting for 1 sol occurred after the addition of more nutrient—thought to be caused by a reabsorption of CO_2 into an alkaline Martian sample. Following this, there was an accelerated increase, which lasted for a few sols, followed by a slower log linear increase persisting to the end of the cycle. The first few gas analyses following a recharge were made before the bulk of the carbonate reserve in the wet soil had equilibrated with the atmosphere in the head space. The relatively large swing observed for CO_2 equilibrating in the beginning of the cycle was in contrast to the slow log linear increase in CO_2 in the head space for the rest of the cycle. To the investigators, the persistence of the CO_2 log linear increase over 4 wet cycles suggested a slow oxidation of the organics of the nutrient by an oxidant in the Martian soil. It is apparent that the oxidant was not very soluble, because it had not washed out by three recharges of aqueous nutrient. From a kinetic standpoint, the log linear CO_2 production would be expected from a diffusion-limited system in which the movement of organics in the liquid phase was the rate-limiting step.

Figure 10.17 is a demonstration of the remarkable achievement of the Viking program, The astonishing ability to vary a scientific experiment over the enormous distances involved and in response to prior observations. It shows the gas changes in the head space over two incubation cycles on a sample collected at the Utopia site. In this instance, the nominal sequence was changed to test the thermal hypothesis of the immediate gas release when the Martian sample was placed in the test cell.

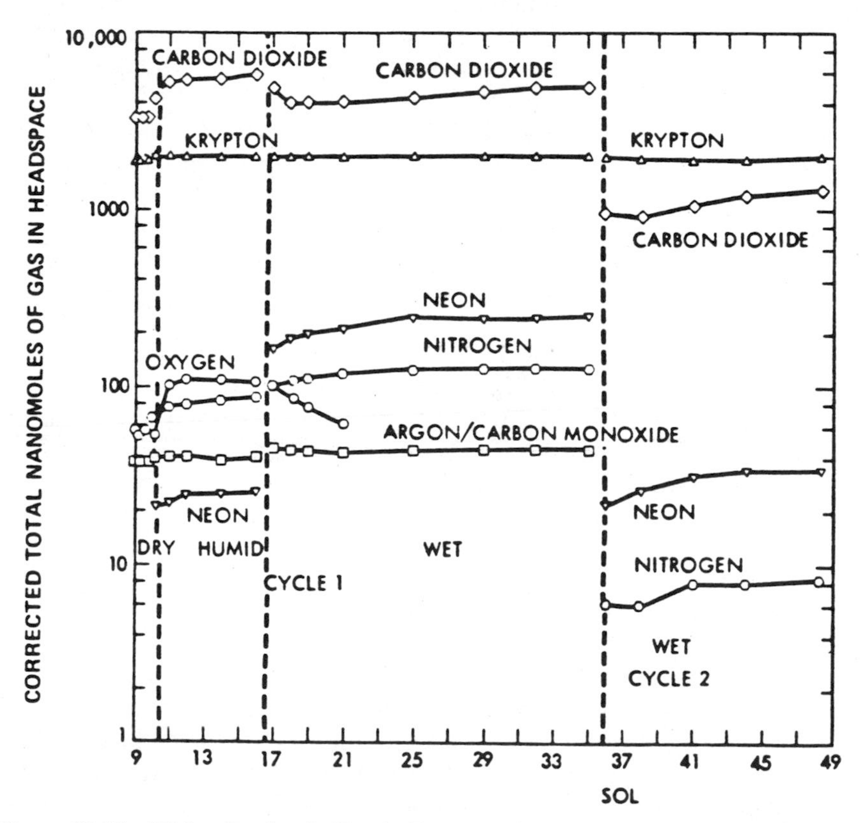

Figure 10.17. Viking Lander 2, Utopia Beta sample, gas changes in the head space. Source: Oyama and Berdahl (19).

The experiment protocol was modified so that the option in the initialization sequence involved the analyses of the dry soil before humidification. No significant changes in the head space gases were observed. Within 2.8 h after humidification, however, CO_2, Ar, N_2, and O_2 were already being released.

In summary, Oyama and Berdahl state that

> the response of the Martian surface samples of water vapor resulting in O_2 output is ascribed to superoxides in the Martian surface material. The desorption of most of the CO_2, Ar, and N_2, from the Martian surface samples is inversely related to previous water exposure of the soil surface. The reactions of superoxide and water vapor at low temperature in the Martian environment could account for presence of superoxides that form H_2O when exposed to liquid water. The residual superoxide suggests a very active atmosphere, likely to give rise to a whole variety of C–O combinations. The slow log linear production of CO_2 is attributed to the direct oxidation of organics by gamma Fe_2O_3 which has been shown to produce the catalytic oxidation of organics by H_2O_2.

Thus all the gas changes observed can be most easily explained by chemical reactions that require no biological processes.

REFERENCES

1. G. R. Hunt, L. M. Logan, and J. W. Salisbury, *Icarus,* **18,** 459 (1973).
2. L. M. Logan, G. R. Hunt, and J. W. Salisbury, in C. Kaer, Jr., Ed., *Infrared and Raman Spectra of Lunar and Terrestrial Minerals,* Academic, New York, 1975.
3. C. B. Farmer, D. W. Davies, A. L. Holland, D. D. Laporte, and P. E. Doms, *J. Geophys. Res.,* **82,** 4225 (1977).
4. A. O. Nier and M. B. McElroy, *J. Geophys. Res.,* **82,** 4341 (1977).
5. A. O. Nier and J. L. Hayden, *Int. J. Mass Spectrom. Ion Phys.,* **6,** 339 (1971).
6. J. Mattuch and R. Herzog, *Z. Phys.,* **89,** 786 (1934).
7. R. B. Thorness and A. O. Nier, *Rev. Sci. Instrum.,* **33,** 1005 (1962).
8. J. Soffen, *J. Geophys. Res.,* **82,** 3959 (1977).
9. T. Owen, K. Biemann, D. R. Rushneck, J. E. Biller, D. W. Howarth, and L. A. LaFleur, *J. Geophys. Res.,* **82,** 4635 (1977).
10. D. Anderson, K. Biemann, L. E. Orgel, J. Oro, T. Owen, G. Shulman, P. Toulmin III, and H. C. Urey, *Icarus,* **16,** 111 (1972).
11. T. Owen and K. Biemann, *Science,* **193,** 801 (1976).
12. K. Biemann et al., *J. Geophys. Res.,* **82,** 4635 (1977).

13. N. H. Horowitz, G. L. Hobby, and J. S. Hubbard, *J. Geophys. Res.*, **82,** 4659 (1977).
14. N. H. Horowitz, G. L. Hobby, and J. S. Hubbard, *Science,* **194,** 1321 (1972).
15. G. V. Levin and P. A. Straat, *J. Geophys. Res.,* **82,** 4663 (1977).
16. H. P. Klein, *J. Geophys. Res.,* **82,** 4677 (1977).
17. G. V. Levin and P. A. Straat, *Science,* **194,** 218 (1976).
18. A. Banin and J. Rishpon, *J. Mol. Evol.,* **14,** 133 (1976).
19. V. I. Oyama and B. J. Berdahl, *J. Geophys. Res.,* **82,** 4669 (1977).
20. R. I. Hugenin, *Icarus,* **28,** 203 (1976).
21. M. D. Nussinov, Y. B. Chernyak, J. L. Ettinger, *Nature,* **274,** 859 (1978).
22. A. Banin and J. Rishpon, *Proc XXI Cospar Plenary Meeting, Insbruck, Austria,* 1978, p. 9.
23. G. R. Hunt, L. M. Logan, and J. W. Salisbury, *Icarus,* **18,** 459 (1973).
24. J. B. Pollack, D. Colburn, R. Kahn, J. Hunter, W. V. Camp, C. E. Carlston, and M. R. Wolf, *J. Geophys. Rev.,* **82,** 4479 (1979).
25. A. Banin and N. Lahav, *Nature,* **217,** 1146 (1968).
26. Y. Chen, D. Shaked, and A. Banin, *Clay Miner.,* **14,** 93 (1979).
27. S. L. Bragg, M.Sc. Thesis, Washington University, St. Louis, MO (1977).

CHAPTER

11

THE VIKING X-RAY FLUORESCENCE EXPERIMENT

Clark and co-workers in their 1977 paper (1) have pointed out that the original science payload was heavily oriented to the exobiological aspects of the Martian mission. No XRF instrument was included among the instruments in the early stages of planning. In 1969, Baird and Clark (2) proposed the incorporation of an XRF geochemical analyzer, which led to a feasibility study supported by NASA headquarters. A successful demonstration test and the support of the planetary science community, inside and outside of NASA, led to a broadening of the science base of the Viking mission, and the XRF experiment was added to the payload in the spring of 1972. A science team was selected, designated as the inorganic chemical analysis team, and given the responsibility for the development of the experiment, including the instrument, the data analysis methods, and mode of operation.

The team had to face some special problems because the inorganic experiment was being added to the mission some 2 y later than the other experiments, and thus had a considerably shorter time for developing instruments. The spacecraft design was already firm, which imposed serious constraints on the instrument size and configuration. There was, as Clark tells us, no opportunity to influence the design of the surface sampling system or schemes for sample processing or instrument deployment. Multiple sample handling had to be dealt with by retaining analyzed material within the volume alloted to the instrument, and at that time it was not possible to include a multichannel analyzer because of weight limitations.

However, as we shall see, the experiment developed by the team met the scientific objectives of detecting and permitting quantitative analysis of about a dozen major, minor, and trace elements in the Martian soil. In addition, the investigators have analyzed what appears to be duricrust fragments, determined an upper limit for Ar in the Martian atmosphere, measured the soil-bulk density, established a new upper limit for thickness of iron oxide coatings on mineral grains, and demonstrated the highly adhesive nature of the Martian soil.

INSTRUMENT DESCRIPTION

The Viking Lander XRF experiment involved the measurement of the fluorescent X rays from a sample irradiated by two radioactive sources, the analysis being performed in the energy dispersive mode using four miniature proportional detectors. The instrument is shown in the following three figures. Figure 11.1 is a photograph of a spare flight unit. Some of the salient features to be seen are the delivery funnel, in practice capped with wind spoiler baffles (to deal with high winds during sample loading), the sample analysis cavity, the soil dump cavity, and the assorted electronics. Figure 11.2 shows a cross section of the analysis section showing the radioactive sources, the proportional counters, detectors, and the analysis section with the flag calibrator in the activated position. Figure 11.3 shows the details of the sample cavity and calibration targets. The cavity is shown filled with lithic and crustal fragments in the 2–12-mm size range. Also shown are the plaque target for the ^{109}Cd source and the flag target for the ^{55}Fe source.

The characteristics of the proportional counters (PC) are given in Table 11.1. These miniature PC's were custom designed for the instrument. The gas compositions listed in Table 11.1 were selected to minimize escape peak interferences and to optimize spectral response for the typical element abundances for geologic samples. The PC windows were unsupported beryllium foil, with the exception of PC1, which had a 5-μm pure Al foil on an aluminum supporting grid to transmit Mg and Al fluorescent X rays and to filter out the interfering Si fluorescence signal. The only reliable method for sealing these thin windows to the counter body was epoxy binding, but this produced some problems. All other counter parts were joined and sealed either by welding, brazing, or cold welding.

The proportional counters proved to be a serious problem in the design of the equipment. It was found early in the developmental testing that the prototype detectors showed an unfortunate change in gain in response to temperature variations. It appeared that the stability required by the mis-

Table 11.1. Characteristics of Proportional Counters

PC	Gas Composition	Window Material
1	20% Ne, 75% He, 5% CO_2	5-μm Al
2	10% Xe, 73% Ne, 10% He, 7% CO_2	25-μm Be
3	40% Xe, 47% Ne, 10% He, 3% CO_2	50-μm Be
4	40% Xe, 47% Ne, 10% He, 3% CO_2	25-μm Be

Source: Clark et al. (1).

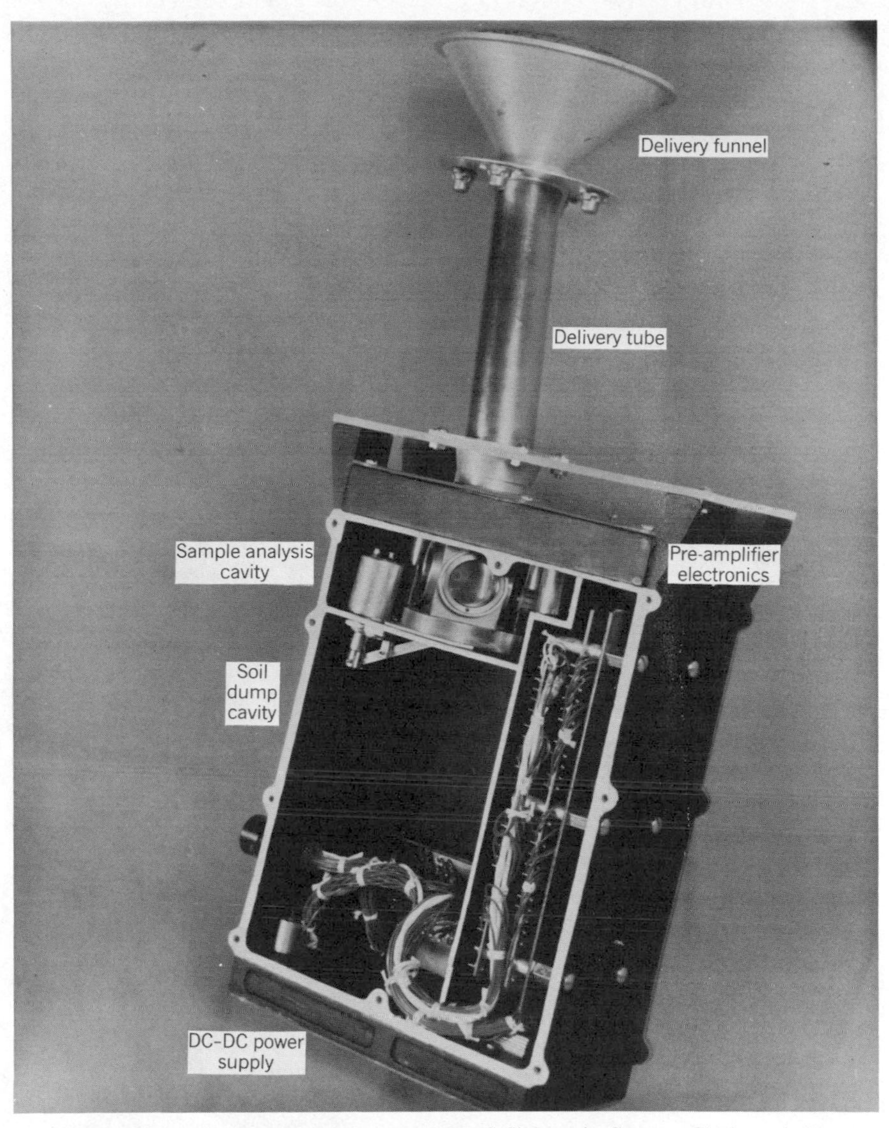

Figure 11.1. Photograph of the spare XRFS flight unit. Source: Clark et al. (1).

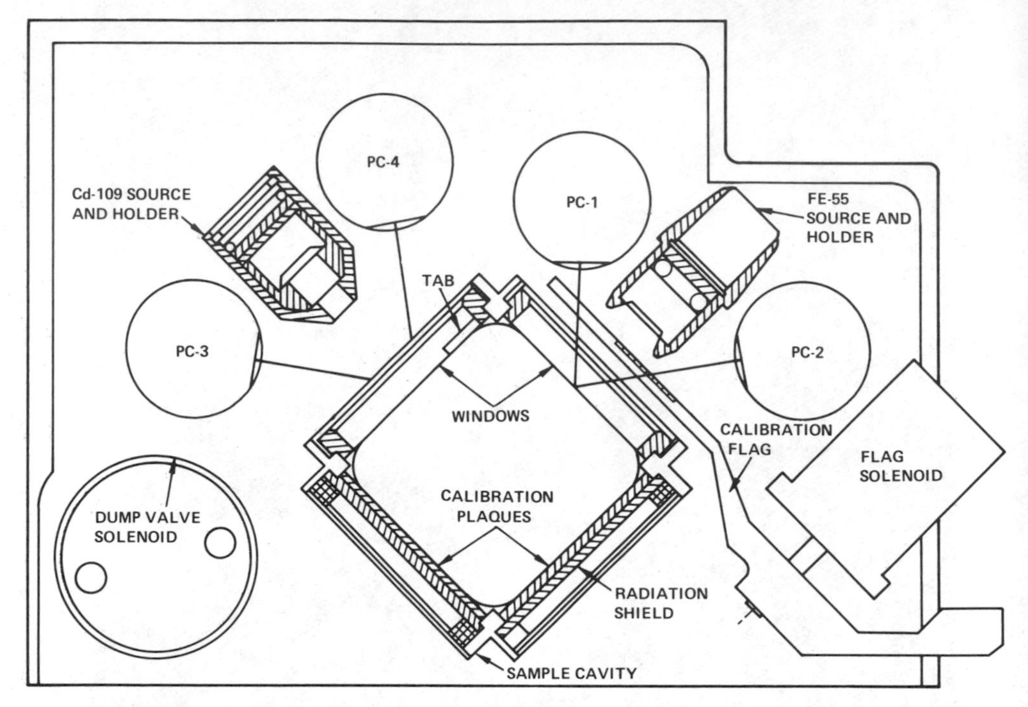

Figure 11.2. Analysis section of the XRFS. Shown are the radioactive sources, the proportional counter detectors, and Sample cavity. The flag calibrator is shown in the calibrate position. Source: Clark et al. (1).

sion was beyond the state of the art in proportional counter design. A study of the emission of the bonding epoxy showed that a partial decomposition occurred during the extended bakeout used for outgassing the counters. This required numerous changes in the manufacturing procedures to improve the gas purity of the sealed detectors. These procedures proved successful, and the temperature sensitivity was reduced to acceptable limits.

A second difficulty was encountered when during final testing one of the PC detectors suddenly failed to produce output, but revived after a few hours with the power turned off. This phenomenon was traced to the gradual charge-up of the thin epoxy surface on the inside of the counters, to the point where pinhole emissions were stimulated. The detectors were then redesigned in such a way that the thin windows were mounted outside the tube wall, rather than inside. With this done, the detectors then behaved in a satisfactory manner and no failures were observed.

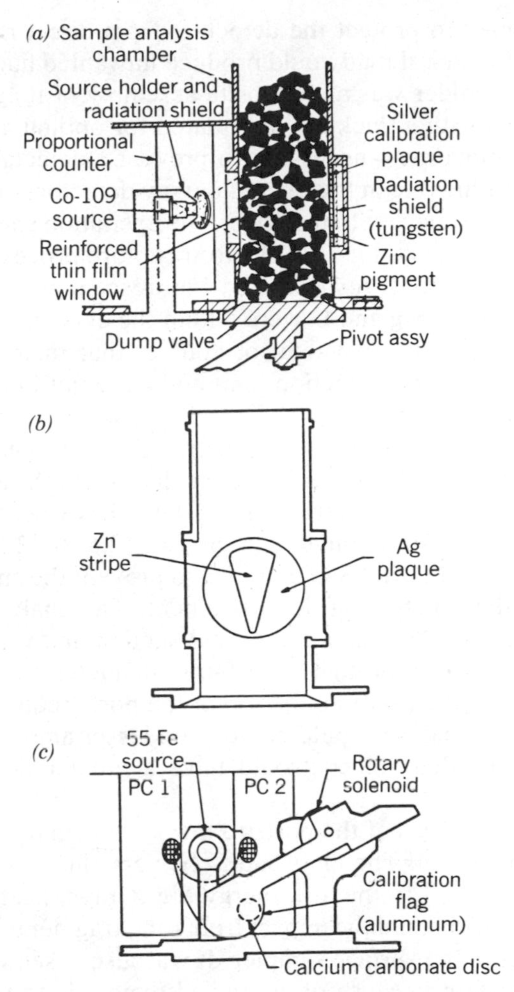

Figure 11.3. Details of the sample cavity and calibration targets. (*a*) Sample cavity filled with lithic and crustal fragments. (*b*) Plaque target for the ^{109}Cd. (*c*) Flag target for the ^{55}Fe source. Source: Clark et al. (1).

The radioactive sources ^{55}Fe and ^{109}Cd were fabricated *from* accelerator grade material with specific activities of $\geq$ 500 Ci/g. After additional purification, they were electroplated onto suitable backing foils, heat diffusion bonded, and sealed into a beryllium-windowed capsule by the use of metal sealing techniques. The radioisotope size was kept below 5 mm diameter. A 180-μm foil filter attenuated the Cu radiation (from the brazing alloy used in the ^{109}Cd source). Source holders and radiation collimators were designed to protect the detectors from direct radiation and to avoid the use of material that would produce unwanted fluorescence radiation. The ^{109}Cd holder was a composite assembly of tungsten alloy and high-purity silver with a thickness calculated for optimum attenuation of the 87.7-keV gamma ray—necessary to prevent a detectable increase of background counting rate in the ^{14}C radioassay detectors of the integrated biology instrument adjacent to the XRF. The excitation energies were 5.9, 22.2, and 87.7 keV due to the ^{55}Fe electron capture process and the ^{109}Cd nuclear gamma emissions, respectively. Fluorescent emission from all of the elements between Mg and U were within the excitation and detection range of the instrument, provided, of course, that their concentrations exceeded their maximum detection limit and were not interfered with by emission from other elements.

The XRFS was mounted within the thermally controlled compartment of the Viking Lander body. Samples were delivered to the analyzer by the surface sampler scoop in several ways. These have been described by Baird and associates (3). A convex screen with 12.5- × 12.5-mm openings was mounted inside the XRFS inlet funnel to prevent the entrance of large rocks or clods that might clog the instrument. The analysis sample was contained in a cavity 25 × 25-mm in cross section and was irradiated by the radioactive sources through two 20-mm diameter analyzing ports at right angles. To fill the cavity to the top of the ports required about 25 cc of material. While analyses could be done on lesser amounts of material, the concentrations determined would have a greater degree of uncertainty.

The thin film windows of the ports to the sample chamber represented a considerable problem. The films had to be very thin, so that they were reasonably transparent to the low-energy Mg X rays, heat resistant, and strong enough to withstand the impact from rock fragments falling into the sample cavity from the surface sampler. It was also essential that the film be radiation resistant because of continued exposure to the X radiation from the ^{55}Fe source. An estimate of the dosage was about 3×10^7 rads over the 2-y exposure time. Beryllium foil, the first choice, was ruled out because of susceptibility to fracture due to rock impact. The final choices were special organic films. Polycarbonate film was selected at the ^{55}Fe port, although the commercial film had to be stretched by radiatively heating it to the softening point under controlled tension. Specially cast polymide films proved to be acceptable for the ^{109}Cd port. Both films were supported by high-purity metallic grids (Ni for the ^{55}Fe window and Al for the ^{109}Cd window), (see Figure 11.3).

Operational stability was an important consideration, considering the duration of the mission. A considerable effort went into the design and testing of the XRFS electronics, with particular emphasis on the amplifiers and the high-voltage detector supplies. The bias supply for the detectors had a commandable output, in 5-V intervals from 75 to 1350 V, which permitted an adjustment of the PC detector gain over a range of 10,000 : 1, from the Earth. All four detectors were connected through signal isolation networks to a single high-voltage supply.

The procedure for the electronic data encoding and transmission was complex and beyond the scope of this text. Great pains were taken to assure that the dozen or so components in the data chain on Mars, in space, and on the Earth did not introduce unknown errors into the data. Clark and co-workers (1) have stated that the error detection/correction scheme developed resulted in at least 99.99% of the data being a perfect replica of the data generated by the XRFS instrument, and that all sources or error originated only in the instrument rather than in the data handling.

The calibration procedure selected was end-to-end, which determined the overall system response. This was accomplished by building calibration targets into the far walls of the sample analysis cavity (Fig. 11.3 *a* and *b*), so that with the sample absent the target elements would fluoresce. With the geometry of the system fixed, and since X-ray emission energies are independent of environmental parameters, the use of the fluorescence targets provided an absolute method for establishing the energy scale, element response function, and relative source strength level for each detector–source combination.

The targets were referred to as calibration plaques, the one opposite the ^{55}Fe was aluminum, and the one opposite the ^{109}Cd was a pure silver plaque. The silver plaque had a recessed triangular stripe of ZnO pigmented paint, as shown in Figure 11.3 *b*. The ZnO stripe was oriented downward so that as the material filled the cavity, the stripe became progressively hidden, and the Zn peak detected in PC3 and PC4 diminished to a negligible level.

The point of this particular scheme was that the Ag/Zn intensity ratio diminished as the chamber filled. Figures 11.4 and 11.5 are the calibration spectra for a clean cavity showing the Al, Ag, Zn, and the excitation source backscatter peaks. It was pointed out that the Ag line in Figure 11.4 in the PC1 spectrum is due to the PC1 being able to "see" the Ag plaque. The ^{55}Fe backscatter peak from the plaques is also seen in Figure 11.5. The signal reaches these detectors when the sample cavity is empty, by scattering of the plaques and interior walls of the chamber. The small Al tab shown in Figure 11.3 prevented a direct path of the ^{55}Fe source to the PC3, and also partially blocked the path of the Zn stripe radiation to the PC4, thus accounting for the lower Zn intensity shown by that detector.

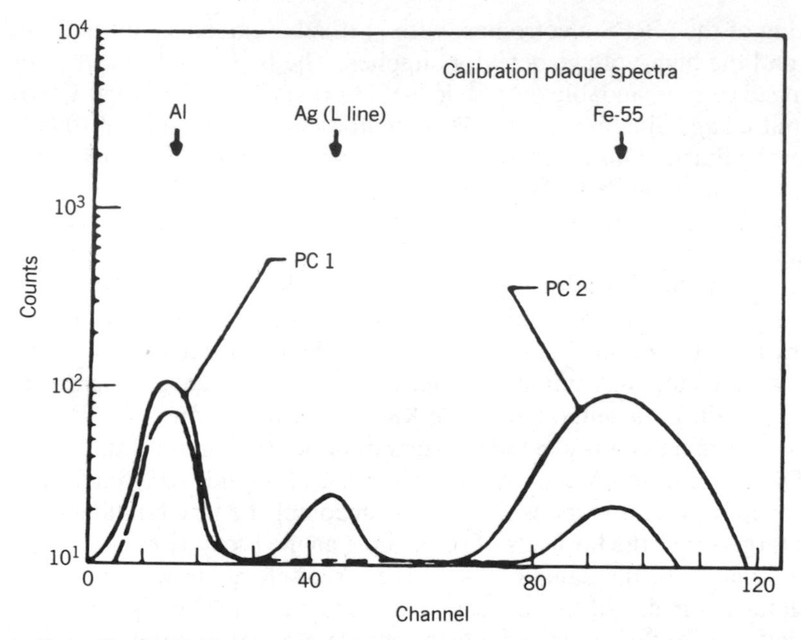

Figure 11.4. Typical calibration plaque spectra for an empty and clean sample cavity. Due to geometrical factors the detector PC1 "sees" the Ag plaque, while PC2 does not. Source: Clark et al. (1).

As an additional method of calibration for detectors PC1 and PC2, to improve the determination of Mg/Al and K/Ca ratios, a mechanism with Ca and Al targets could be inserted between the ^{55}Fe source and the sample by means of a solenoid activated lever. This "flag" calibrator had the following advantages: it could be used even when the cavity was filled with soil, it remained clean, and it produced a relatively high counting rate leading to adequate counting statistics in a single scan.

PRELAUNCH CALIBRATIONS

The constraints imposed by the nature of the Viking Mission made it necessary to devise many special approaches to preflight calibration. For example, reliability considerations required that all the Viking hardware be carefully protected from dust or other particulate contamination before flight. As a consequence, it was not possible to calibrate the XRFS with well-analyzed soil and rock powders. As an alternate, some 29 solid rock specimens were chosen and shaped by sawing and grinding to fit snugly into the sample cavity.

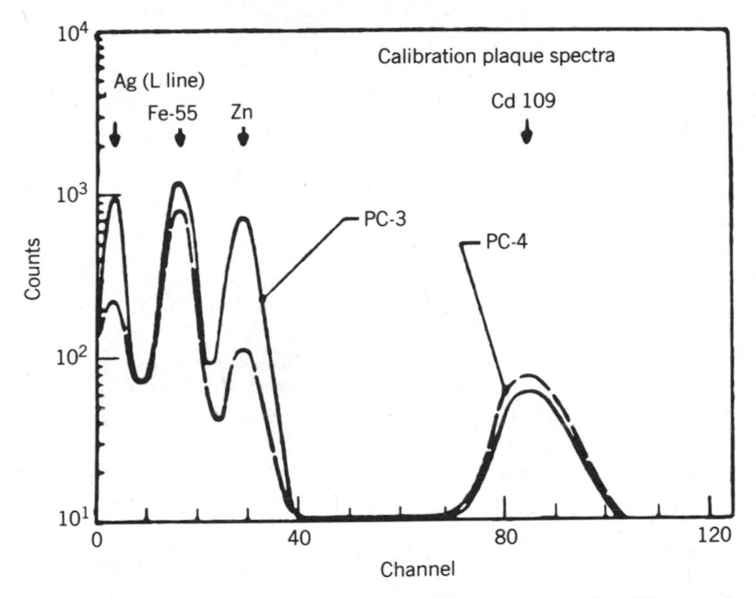

Figure 11.5. Calibration plaque spectra for the ^{109}Cd detectors. The ^{55}Fe signal reaches these detectors when the sample cavity is empty by scattering off the plaques and interior walls. The lower Zn peak for detector PC4 is due to a small Al tab which partially obscures the path from the Zn stripe. Source: Clark et al. (1).

Four such standard materials were selected to test the instrumental analytical capabilities. These ''performance standards'' consisted of well-analyzed rocks of the following types: a Columbia River Plateau basalt, an andesite and a quartz latite from Lake County, Oregon, and a rhyolite from Glass Mountain in California.

In addition, 25 standards covering various other rock types and materials were chosen so that the performance could be evaluated for a large variety of rock compositions. Pure element standards for Mg, Al, Si, S, K, Ca, Ti, Fe, Rb, Sr, and Zr were also developed either by machining pure elements or by mixing an appropriate powder with epoxy in a special mold. The elemental concentrations in the molded standards was adjusted to yield an overall counting rate of about 1000 counts per second in the detector being tested.

Typical response functions for the pure elements are shown in Figure 11.6. These functions were produced by a least-squares fit of Gaussian curves to the peaks and by hand-fitting over non-Gaussian regions. These data were used to extrapolate response functions for elements not directly measured using the fact that peak resolutions are to a first approximation proportional to the theoretical inverse square root of the energy law.

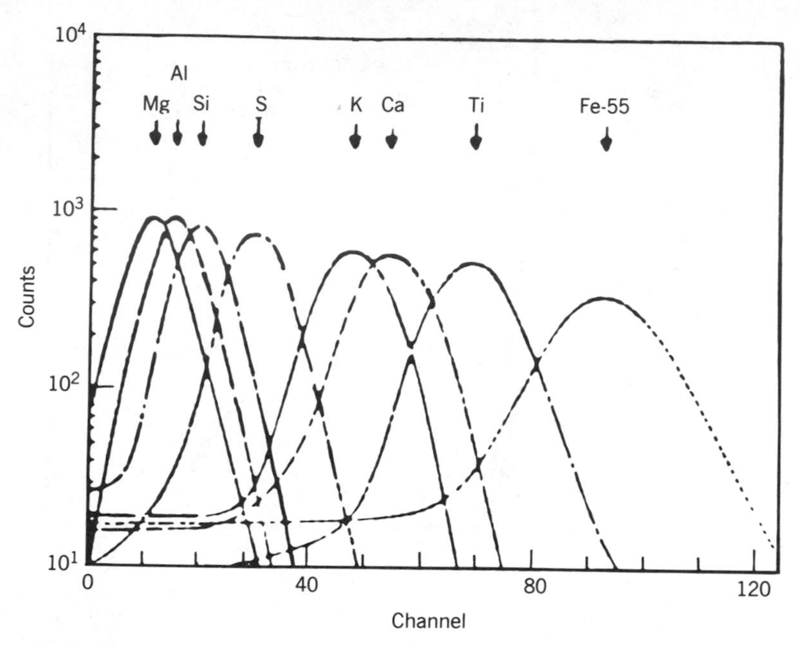

Figure 11.6. Pure element response functions for PC2. The overlap between adjacent elements mandates accurate gain knowledge. The ^{55}Fe backscatter peak provides an internal calibration for the gain. Source: Clark et al. (1).

On completion of the assemblies, they were tested by taking spectra of each of the 4 standards and reducing the data to determine if the analysis for some 10 elements were within acceptable error limits. An important follow-up test was for sterilization compatibility, because heat sterilization of the spacecraft was a vital requirement. Thus the XRFS was heated in a dry nitrogen atmosphere for 54 h at a temperature of about 112°C.

The instrument was also subjected to thermal cycles at a reduced CO_2 atmosphere, and, finally, it was retested against the four standards to see if it passed the performance specifications. Immediately before the final installation into the Lander, the radiations sources were replaced with freshly prepared flight level sources, and 225 spectra were recorded to characterize the response of the unit. These spectra included 25 solid rock standards, pure Ar, and the various plaques and flags described above.

QUALIFICATION OF THE XRFS

Clark (1) has described the intensive testing that the XRFS underwent during development. Numerous quasiblind, blind, and double-blind tests were performed. An early breadboard stage was evaluated using four powdered rock samples supplied by NASA. With the instrument at a spacecraft system level, samples of Pike's Peak granite, Pierre shale, and an unknown sample were analyzed during the science end-to-end test.

A good demonstration came when a deviation in the apparent granite composition was observed. This was traced to a particularly large microcline fragment, whose natural cleavage face covered a major fraction of the viewing port window. The unknown sample supplied by NASA was successfully identified as a rhyolite from a suite of five selected candidates of diverse composition.

In one test, a Chattanooga shale was found to contain relatively high concentrations of sulfur—conventional analysis had shown only a high ignition loss. Subsequent laboratory analysis confirmed the high sulfur (4.6%). With regard to trace concentrations, the analysis of a Riverside nontronite showed low but approximately equal amounts of Sr and Y. This was confirmed by later laboratory analysis, which showed 80 ppm Sr and 57 ppm Y. In further testing, a field portable version of the instrument was successfully used to identify a nontronite occurrence at Riverside, California, to detect a high Zr content in a Nevada rhyolite, and to identify sulfate salts in Utah lake samples.

ANALYSIS METHOD

From the 25 standards, 3 were selected for detailed calibration studies. Two of these, a syenogabbro and a biotite peridotite were, like the Martian soil, relatively high in iron, with a high Mg/Al ratio. The third sample, a Chattanooga shale, was important for quantitatively determining sulfur because it was the only preflight material containing appreciable amounts of this element. Figures 11.7*a*, *b*, and *c* show a comparison of the Mars spectra to the preflight spectra of the standards. The figures are for PC1, PC2, and PC4, respectively. It is obvious that none of the standards gives a close match to the Martian spectra. It has been found since that a comparison to several hundred different geological materials prior to and since landing has failed to yield a close match in all respects (3, 4). In fact, as Clark and others indicate (1), the natural material that most closely match are the first Martian sample from the Lander 1 site and the first sample at the Lander 2 site, although the two sites are separated by 6500 km.

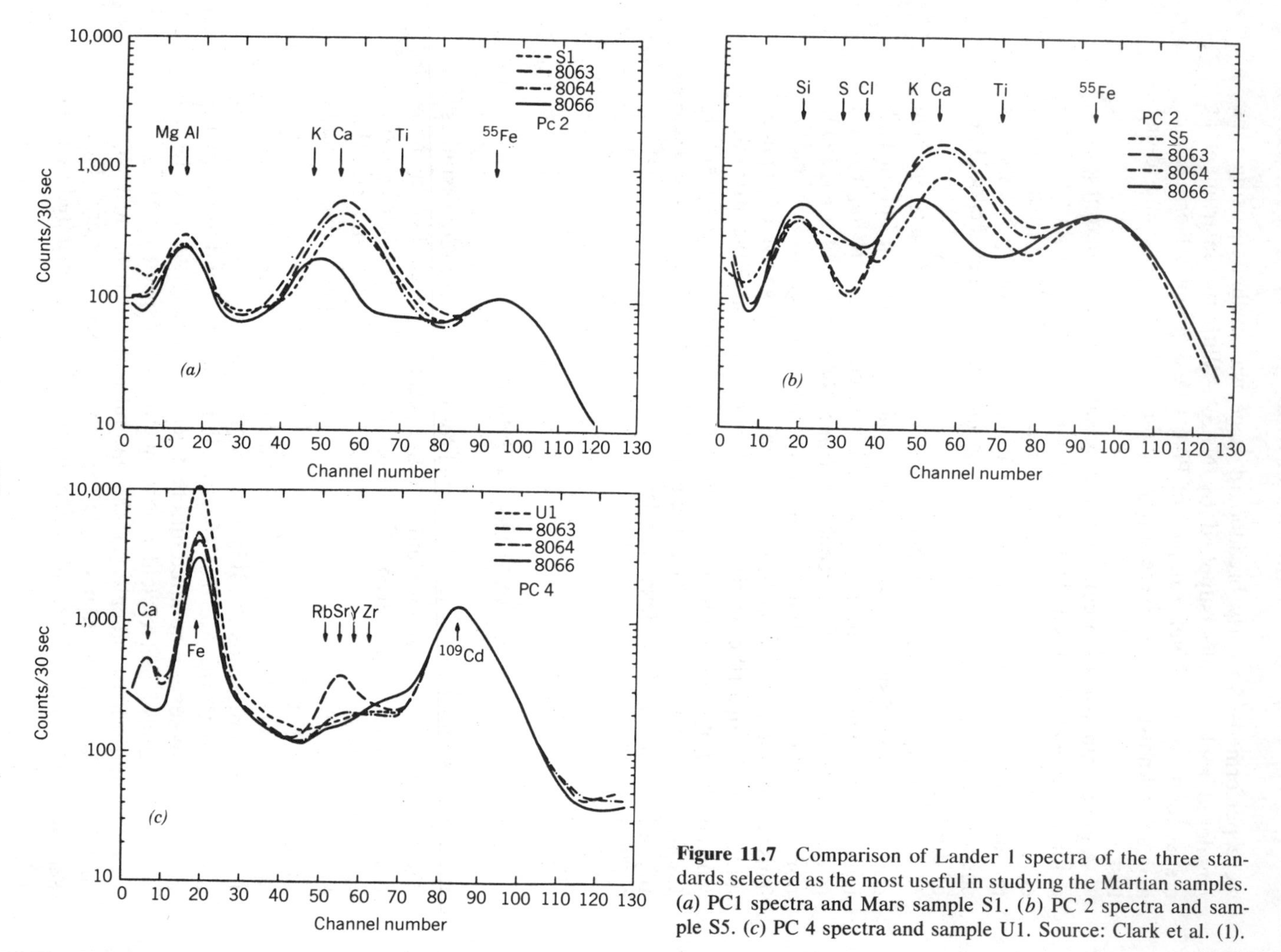

Figure 11.7 Comparison of Lander 1 spectra of the three standards selected as the most useful in studying the Martian samples. (*a*) PC1 spectra and Mars sample S1. (*b*) PC 2 spectra and sample S5. (*c*) PC 4 spectra and sample U1. Source: Clark et al. (1).

DATA REDUCTION

A detailed method of the methods of data analysis has been published by Clark and co-workers (1). A summary has been written in a more recent paper by the same research team (5). Each XRFS contains four proportional counters, with sensitivities optimized for different elements. Detector PC1 was optimized for Mg and Al; PC2 for Si, S, and Cl; both had good response for K, Ca, and Ti. The remaining two detectors, PC3 and PC4, were for the analysis of Fe and certain trace elements. In the initial report (1) K, Ca, and Ti were derived from the PC2 data alone. Based on the observed performance on Mars, it was found that the PC1 detector was superior to the PC2s both in terms of gain and resolution stability. Thus the final reported results were based on K, Ca, and Ti values derived from the PC1 spectra. The concentrations for these elements were also calculated from the PC2 data and found to be in good agreement. Final values for Si, S, and Cl were obtained from the PC2 data, while Fe was obtained from the PC4.

The spectra were adjusted to a standard energy scale by the use of a two-point energy calibration. The procedure, named STRETCH, is based on a linear model and applies gain and offset corrections so that the number of counts per unit energy increment is exactly preserved. For the PC1, the two energy points were the Al peak obtained from the built-in calibration flag and the radiation backscatter peak in the sample spectra.

For the PC2 detector, the calibration flag Ca peak was combined with the sample backscatter peak to get the gain-to-offset ratio. This ratio was then assumed constant and used with the backscatter peak location to establish the energy scale.

Thus the STRETCH'ed spectra were tied to key reference points. In the case of PC1, the results showed that Mg is detected in the samples because the composite peak was always shifted toward Mg from the Al-calibrator tic point. It was observed that the K/Ca peak was not shifted toward the K from the Ca-calibrator reference peak position as a result of the fact that the Ca/K ratio was very large for the Martian samples.

The processing of the data required correction for the background continuum produced by cosmic rays, the on-board radioisotope thermoelectric generators (RTG's), and induced radioactivity in the lander structure. The spectral form used was that of the dominant component, the RTG background. The increases in level with time were determined from analysis of the background measured by the ^{14}C detectors in the biology experiment and analysis of PC4 calibration plaque spectra. The absolute background intensity was assumed initially was the level predicted for the landing date, based on prelaunch measurements. Subsequently, the value was adjusted slightly to form a self-consistent data set between the earlier and later samples.

A final necessary correction involving the PC1 and PC2 detectors was for a stray radiation component due to the sample support structure. This component was determined from calibration plaque measurements. In the case of PC2 in Lander 1, a one-point empirical adjustment was made so that the Ca and Ti values were consistent with the PC1 results.

With the Martian spectra in hand, an extensive program was undertaken to simulate the results on Mars using the spare flight instrument. A large number of analogue samples were prepared from geologic materials and chemical oxides. The components were dried in a vacuum oven, weighed, and blended together in a roller mill. The components were then mixed intimately by grinding with a motorized pestle.

Significant effects were found on the relative intensity of the X-ray peaks for some of the elements, particularly the sulfur and chlorine, in response to the grinding time. For most of the elements, a grinding time of 8 h or more served to stabilize the intensities. The elemental concentrations calculated from preflight rock calibrations resulted in analogues whose spectra matched the Martian samples. Concentrations are reported based on the calibrations. The grinding experiments were used to establish the limits of absolute error due to matrix effects. An exception is the element aluminum. This element is difficult to deal with in the mathematical model because of uncertainties in the influence of the secondary fluorescence from the PC1 aluminum entrance window. Thus the final Al_2O_3 values were based on correlations with laboratory measurements of analogue samples run in the spare flight unit.

SAMPLING AND RESULTS

While a number of preliminary reports were written describing the XRFS experiments, the following represents a recent accounting by Clark (5). The camera observations, which were spectacular, disclosed that at both lander sites there were blocks, boulders, and cobble-sized rocks (to 1 m diameter) sitting in and on an apparently fine reddish material, either in distinct drift forms or as irregular hummocky masses. Most of the rocks gave the impression of being vesicular volcanics (3, 4), although there is terrestrial evidence of vesiculated blocks that are due to erosion to igneous or sedimentary material. An attempt was made to scratch or mar their surfaces with the sampler boom but the effort failed, leading the investigators to conclude that the rocks were not coated with soft weathering rinds.

The view of the surface also showed that pebble-sized particles as well as grit and coarse sand sizes (these last below the best camera resolution) appeared to be scattered about the surface at both lander sits. At some locations, these appeared to have high concentration. These materials were of great interest to the investigators because they might be samples of Martian rock that could be collected and analyzed by the XRFS. However, they turned out to be duricrust rather than rock the durifrost fragments were plate-shaped fragments, 0.5 cm or so in thickness, lying on the Martian regolith much as "caliches" on the Earth.

At the Lander 1 site these fragments had the same composition as the fines, except for a higher content of S and Cl, (this apparently due to the cementing salts that led to the durifrost). The experimenters were not equally successful in collecting durifrost at the second site. They have speculated that this was due to a weaker cementation, which caused the crust to break up during sampling. They have also concluded that pebble-sized rocks do not exist at the landing sites, for unknown causes.

The fine material collected for analysis came from a number of micro-sites at the lander locations. These resulted from various operations such as surface skims, direct penetrations of a few centimeters, trenches dug

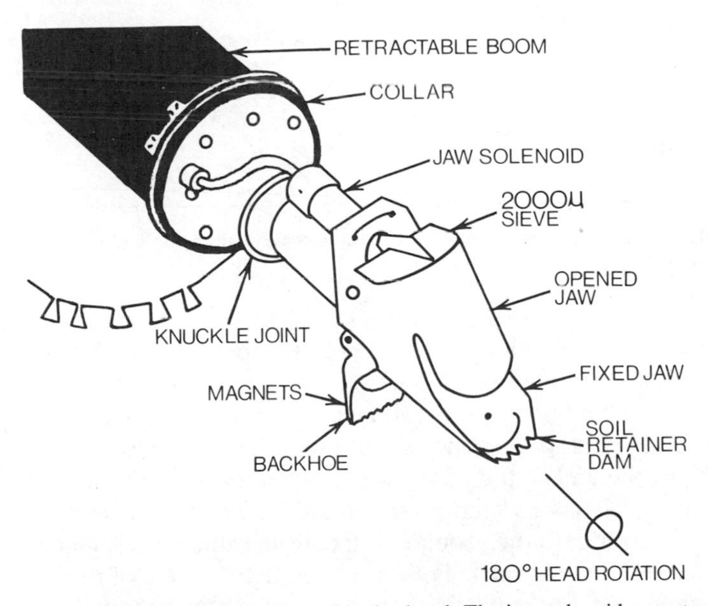

Figure 11.8. The Viking Lander surface sampler head. The jaw solenoid operates at 4.4 or 8.8 Hz. The backhoe is shown in its extended position. In its stowed position it folds backward toward the knuckle joint. Source: Baird et al. (2).

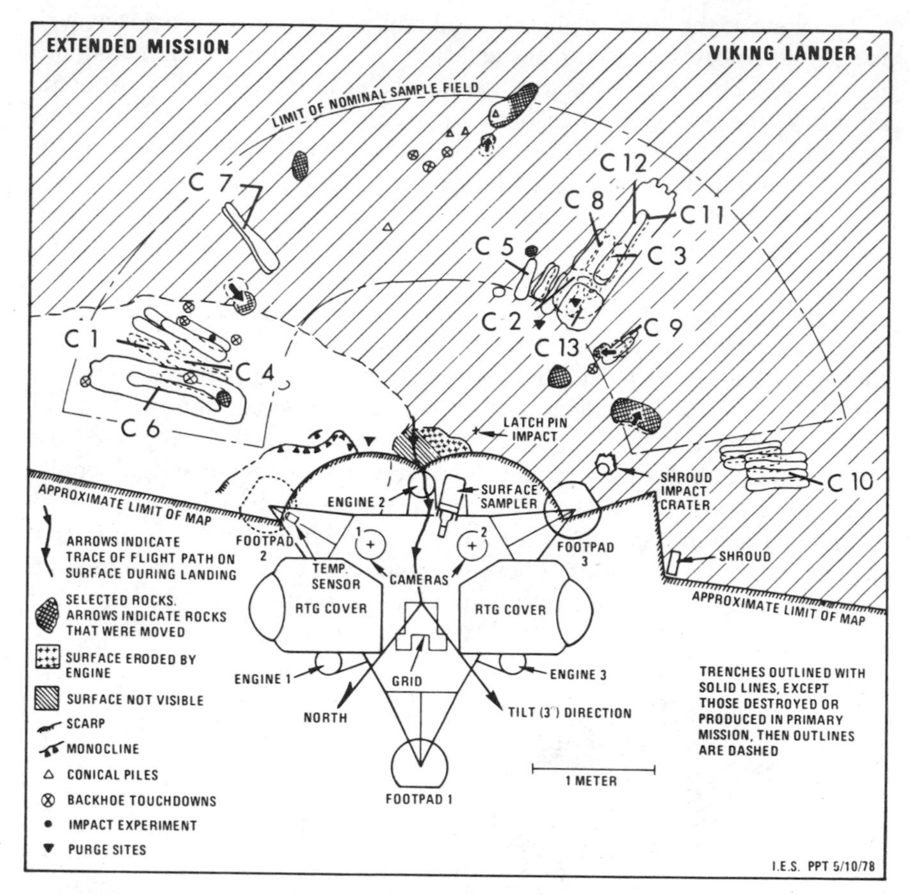

Figure 11.9. Sampling field and locations of samples taken for analysis in the XRFS. Source: H. J. Moore (7).

by the backhoe, tailings from the trenching operation, and from beneath rocks moved by the collector boom.

The retractable boom proved to be a remarkably flexible device. It was capable of sweeping a semicircular area about 3 m wide and 170° in azimuth. At the end of the boom was a collector assembly with one fixed and one movable jaw, which provided an opening of 4 × 2 cm and which could be inserted into the ground with considerable thrust. Once a sample was collected, it could be delivered to the XRFS in one of three ways: (1) upright collector, movable jaw (the upper) could be vibrated and the entire contents delivered to the XRFS; (2) inverted collector, the lowermost jaw vibrated and only material smaller than 2 mm was delivered

Table 11.2. Sampling Survey[a]

Sample No. Location	Sol	No. of Acquisitions	Acquisition	Sample Type	Depth (cm)	Delivery	Sample Level in Chamber
C-1 Sandy Flats	8	2	Two normal mode scoops into front of preexisting trench	Fines	0 to 7.5–9.5	Two fines deliveries: 60-s HF vibration through sieve	Full
C-2 Rocky Flats	34	2	Two normal mode scoops into undisturbed surface	Crust	0–6	Two deliveries of coarse material: 60-s HF vibration through sieve; delivered by 60-s LF vibration through jaw	Full
C-3 Rocky Flats	40	2	Two normal mode scoops into preexisting trench	Crust; fragments 2–5 cm across	0–6	Two deliveries of coarse material: 90-s HF vibration through jaw	Partial
C-4 Sandy Flats	91	1	One normal mode scoop into undisturbed surface	Fines	0–5	Spill of fines from biology experiment delivery 45-s HF vibration through sieve	Partial

Table 11.2. (*Continued*)

Sample No. Location	Sol	No. of Acquisitions	Acquisition	Sample Type	Depth (cm)	Delivery	Sample Level in Chamber
C-5 Atlantic City	177–180	4	One normal mode scoop into undisturbed surface; three normal mode scoops in same trench	Crust fraction of blocky material	0 to 6–7	Two deliveries of coarse material: 120-s HF vibration through sieve; delivered by 60-s LF vibration through jaw	Full
C-6 Sandy Flats (Deep Hole)	229, 250	2	Two skim scoops into bottom of deep hole after backhoeing	Fine material	22	Two fines deliveries: delivery 1 by 120-s vibration through sieve purge 1 by 60-s LF vibration through jaw; delivery 2 by 60-s HF vibration through sieve purge 2 by 60-s LF vibration through jaw	Full
C-7 Jonesville	285, 286	2	Two scoops into undisturbed surface	Fines	1–2	Two fines deliveries: 60-s LF vibration through jaw	Full

C-8 Rocky Flats	311, 312	2	Two normal mode scoops into undisturbed surface	Material	0–4	Two fines deliveries: 60-s HF vibration through sieve, 30-s LF vibration through jaw	Full
C-9 near Bashful Rock	378	1	One normal mode scoop from area behind Bashful Rock	Fines	0–3	One bulk delivery with jaw open	Full
C-10 Angie's Place	430, 431	4	Four skim mode scoops into undisturbed surface	Coarse material	0–2 or 3	Four deliveries of coarse material	Partial
C-11 Rocky Flats (Deep Hole 2)	468	1	One normal mode scoop after 23 backhoe strokes	Bulk	1–6	One bulk delivery with jaw open using rotating head	Partial
C-12 Rocky Flats (Deep Hole 2)	502	2	Two normal mode scoops after 24 backhoe strokes on SOL486	Bulk	1–6	Two bulk deliveries with jaw open using rotating head	Partial
C-13 Deep Hole 2 tailings	581	16	16 Normal mode scoops into tailings pile	Rocks and beneath	0 to 11–14*	16 Deliveries of coarse material with jaw open using rotating head after 90-s HF vibration through sieve to eliminate all but crust	Full

Table 11.2. (*Continued*)

Sample No. Location	Sol	No. of Acquisitions	Acquisition	Sample Type	Depth (cm)	Delivery	Sample Level in Chamber
U-1 Bonneville	29, 30	2	Two normal mode scoops into undisturbed surface	Fines	0 to 4–6	Two fines deliveries: 90-s HF vibration through sieve	Full
U-2 Notch Rock (under)	131	1	One normal mode scoop into undisturbed surface previously covered by rock	Fines	0 to 4–6	One fines delivery: 90-2 HF vibration through sieve; purge by 120-s LF vibration through jaw	Full
U-3 Spalling Valley	161	2	Two normal mode scoops into previously dug trench made by two normal mode scoops	Fines	0 to 3.5–4.0	Two fines delivery: 90-s HF vibration through sieve; purge to rock pile using 60-s LF vibration through jaw	
U-4 Badger Rock (under)	185	2	Two normal mode scoops into previously dug trench; sample from area originally under rock	Fines	0–3	Two fines deliveries: 90-s HF vibration through sieve; purge using 60-s LF vibration through jaw	Full

U-5 Physica Planitia	388	2	Two skim mode scoops into undisturbed surface. Each skim scoop make trench, first trench partly superposed on second	Bulk	0–2.3	Two bulk deliveries: 60-s LF vibration at 45° collector head position	Full
U-6 Spalling Valley (Intermediate Deep Hole)	417	1	One normal mode scoop into deep hole after 24 backhoe strokes	Fines	5.5–6.0	One fines delivery: 60-s LF vibration through sieve; no purge	Full
U-7 Spalling Valley (Deep Hole)	442	2	Two normal mode scoops into deep hole after 24 additional backhoe strokes (48 total strokes)	Fines	12	Two fines deliveries: 60-s LF vibration through sieve; no purge	
U-8 Bonneville	595, 596	10	10 normal scoops into existing U-1 trench	Bulk	6–8	10 Bulk deliveries	Full

[a] Sol, solar days at Mars, post landings; normal mode, insertion of jaw at angle into surface; fines, uncemented drift material; crust, cemented duricrust; bulk delivery, delivery mode 2, see text; coarse delivery, delivery mode 3, see text; HF vibration, high-frequency jaw vibration (8.8 Hz); LF vibration, low-frequency jaw vibration (4.4 Hz); skim: sampler subparallel ground surface; purge: sample dump outboard of lander.
Source: Clark et al. (5).

Table 11.3. Summary of Elemental Compositions (Expressed as Oxides)

	Type	SiO_2	Al_2O_3[a]	Fe_2O_3	MgO[a]	CaO	K_2O[b]	TiO_2	SO_3	Cl
Sample C-1	Fines	(43)	7.5	17.6	6	6	0	0.65	7	0.7
C-2	Crust	(42)	—	17.3	—	5.5	0	0.57	9	0.7
C-5	Crust	42	6.9	17.4	7	5.6	0	0.60	9.5	0.9
C-6	Deep fines	44	7.3	17.3	6	6.0	0.04	0.61	6.7	0.8
C-7	Fines	44	7.4	19.0	5	6.0	0	0.63	6.8	0.6
C-8	Fines	43	7.1	18.8	6	5.8	0	0.71	5.9	0.65
C-9	Bulk	45	7.5	18.9	5	6.0	0	0.71	7.2	0.8
C-11	Deep fines	—	7.2	17.7	6	5.4	0	0.64	—	—
C-13	Crust	(43)	7.0	18.2	7	5.4	0	0.59	9	0.9
U-1	Fines	42		18.9		5.8	0.03	0.60	8.4	0.3
U-2	Under rock	43		17.6		5.8	0.02	0.63	8.1	0.6
U-3	Fines	44		18.3		5.95	0	0.64	7.6	—
U-4	Under rock	44		16.9		5.7	0	0.52	7.9	0.45
U-5	Bulk skim	43		16.3		5.3	0	0.44	8.3	0.6
U-6	Deep fines	42		17.1		5.5	0	0.48	7.9	0.3
U-7	Deep fines	42		17.5		5.5	0	0.51	7.6	0.4
U-8	Bulk	41		—		5.6	0	0.47	(8.5)	—
Uncertainty estimates										
Instrument precision		±2	±0.4	±0.5	±1	±0.2	±0.15	±0.1	±0.7	±0.25
Calibration uncertainty		±3	±2.5	±2	±1	±1	±0.1	±0.1	±2	±0.2
Total uncertainty[c]		±6	±4	−2 to +5	−3 to +5	±2	±0.5	±0.25	−2 to +6	−0.5 to +1.5

[a] Magnesium and aluminum values for all Utopia samples are consistant with Chryse sample values, but details are not reported because of noise bursts in data.

[b] All K_2O values are as calculated form the data. Nonzero values are, however, all below the estimated intrinsic instrumental precision limit of ±0.15% and therefore should be considered as insignificant values.

[c] Absolute errors which could result if the Martian soil matrix strongly deviates from the assumed fine-grained homogeneity. Also includes precision and calibration uncertainties.

Source: Clark et al. (5).

through a screen in the jaw; (3) the vibration as described in mode 2 was performed outboard of the lander, then the collector was returned to an upright position and the delivery conducted as in mode 1, delivering only particles larger than 2 mm.

Figure 11.8 shows the details of the sampling arm. Figure 11.9 and Table 11.2 is a summary of the sampling that demonstrates the variety of sites, samples, and conditions. This represents in every way an extraordinary accomplishment considering the distances, the communication lag, and the duration of time over which these operations were carried out. The results of the chemical analysis are shown in Table 11.3. Clark et al. (5) tell us that, although the analyses were elemental, they are expressed as oxides. Most of the values cited are close to those reported earlier in an interim report (6).

Some of the data have been revised somewhat. The titanium content has been revised downward as a consequence of refinements in the knowledge of the electronic gain of the instruments. This was the result of the long-term operation on the Martian surface and more extensive use of the calibration flag later in the mission. Also, a result of better knowledge of the gain, laboratory simulation experiments, and a better understanding of the window fluorescence correction for detector PC1, it is now believed that the aluminum value is higher and the magnesium value somewhat lower than had been previously reported.

Table 11.3 shows concentrations that are reported only to the number of digits having relative significance. The numbers in parentheses are less certain because of temporary fluctuations in detector resolution. The missing values are due to missing data because of such factors as the servicing of the Voyager by the Deep Space Net. Several of the samples such as C3, C4, C10, and C12, were of inadequate volume to allow analysis to the precision limit of the instrument. However, all four of these samples were of the same compositional family as the samples listed.

ERRORS

The analytical uncertainties are obviously of importance. It is also clear that precision and accuracy in such undertakings, particularly where sample preparation is an enormously difficult problem, must generally be inferior to that of a carefully operated terrestrial analytical laboratory. These factors need to be considered in drawing conclusions about the relationships being sought. Clark and associates (1) have discussed the uncertainties of the XRF experiment.

Three levels of uncertainty are cited. These are tabulated in Table 11.3

and represent their best estimate of the 90% confidence limit for each error. Remarkably, the instrumental precision was such that the uncertainty was the lowest of the three categories. The instrument stability and ''repeatability'' as determined on Mars was found to be excellent. The exceptions were in the measurements of Mg, where the sensitivity was poor, and the determinations of Si, which was interfered with by the presence of unexpectedly large amounts of sulfur in the Martian samples.

The second level of uncertainty has arisen from the calibration procedures. We remember that only solid rock slabs were permitted in the instrument prior to flight, and also that none of these calibration standards were close in composition to the Martian samples. The uncertainty for the aluminum values were high because of the detector-window fluorescence effects.

Finally, matrix uncertainties were a source of absolute error in the results because of the possible heterogeneities that could lead to absorption or enhancement effects in the measured X-ray intensities. The heterogeneities could involve different particle size distributions for different mineral constituents, large particles, and discontinuous coatings on the particles. It has already been stated that Clark and co-workers (5) found marked changes in the S and Cl peaks on extended grinding.

In addition, they observed a moderate increase in the Al and moderate decreases in the Si, Ca, and Fe peaks. It took 6–8 h of mortar–pestle grinding before no further changes were observed. They also found that much more dramatic effects were observed optically; only 1% of a strongly colored material such as hematite could dominate a natural soil color when the two were ground. On the other hand, tumbler mixing required 10 times as much hematite to produce the same effect. Thus many of the effects observed could be traced to a superficial coating of larger grains by smaller grains of different composition.

In the initial mixtures, the relatively coarse crystallites of the S and Cl salts were apparently obscured by a coating of superfine SiO_2 and Fe_2O_3 powders. As a consequence, the S and Cl peaks were barely visible in the X-ray spectra of these mixtures. Translated to the Martian situation, if these were relatively coarse salt grains within the Martian soil samples, then the true salt content could be higher than the nominal value (5). Therefore the SO_3 content reported may be a lower limit, and the matrix uncertainty was given limits of −2% to +6%. On the other hand, if the S and Cl minerals were sufficiently fine, they could coat the other grains, and the estimates for these samples would be high.

The errors assigned to the MgO and the Al_2O_3 values due to matrix effects were based on the following: Even thin, superficial coatings (for example, iron oxide stains) could suppress the low-energy X rays emitted by the Mg and Al. If Fe was present as coarse particles, about 100-μm in diameter, the true Fe content could be as high as 30% Fe_2O_3. This was not consistent with the total summation. The investigators felt that the Martian soil had been subjected to an extensive abrasion history, leading to fine-grained material which was homogeneously distributed and free of a uniform stain. If this were so, then the absolute errors could be smaller than stated.

The X-ray spectra yielded no unambiguous evidence for potassium. Based on various limitations, the experimenters assigned an upper limit of 0.5% K_2O. While their 1982 paper (5) does not report on trace elements, they state that except for ultramafic minerals, most terrestrial and lunar igneous rocks contain far larger abundances of the trace elements Sr, Zr, Y, and Rb then any of the Martian samples.

SUMMARY

Clark (5) concluded that the results of the XRFS are important in that they at least demonstrate what Mars is not chemically. A number of pre-Viking concepts, which included such ideas as a color due to pink feldspars, limonite beds, and 60% SiO_2 (based on orbital infrared observations) have to be discarded. The element profile determined, is not consistent with highly differentiated source material, such as continental siliceous igneous rocks on the Earth, for example. It is conceded that the deduction of a unique, exclusive model for the origin of the Martian fines may not be possible from the X-ray data alone. Nevertheless, conclusions can be reached, no matter which model is selected for interpretation.

One of the more remarkable observations is the "striking" similarity in composition of the samples at two widely separated sites. The C-9 sample at Chryse was observed to be nearly indistinguishable from the U-3 sample at Utopia for all the elements. The averages for the samples from deep trenches or from under the rocks, referred to as the protected samples, were also close to being identical for the two landing sites. This is demonstrated clearly in Table 11.4. The only detectable differences were the S, Cl, and Ti, and here even the assumed salts differed by only about 10% on a combined anion basis ($Cl^- + SO_3^{2-}$). If the salts entered the soil by a mechanism such as aqueous transport, then it becomes very difficult to explain how soils such great distances apart would have become enriched to the same degree. A second requirement is that the soil had been globally mixed and homogenized in the silicates before invasion by the salt. It has been proposed that eolian transport has either blended the various mineral components (including the salts), or by its abrasive action has

Table 11.4. Average Composition of "Deep" Samples

Samples	Chryse: C-6, C-11	Utopia: U-2, U-4, U-6, U-7
SiO	44	43
AlO	7.3	(7)[a]
FeO	17.5	17.3
MgO	6	(6)*
CaO	5.7	5.7
K_2O	< 0.5	< 0.5
TiO	0.62	0.54
SO	0.8	0.4
Cl	0.8	0.4
Other[b]	2	2
Total	91	90

[a] Mg and Al assumed as at Lander 1 site.
[b] Includes such elements as P, Mn, and Na, none of which could be unambiguously detected by the XRFS.
Source: Clark et al. (5).

produced homogeneity down to and below the scale of transportable grains.

One of the consequences of the X-ray analysis, when combined with spectroscopic observations of widely distributed bright regions on Mars (7), is the possibility that much or all of Mars has a blanket of fine material of uniform composition. It is not likely that this is the result of igneous activity that has been so homogeneous in a chemical sense. It appears rather that the "universal" fines are a composite of the original material or the products of chemical weathering of several different igneous rocks that have been widely distributed by Martian eolian processes.

The concentrations of Si, Fe, and Ca in the protected fines were constant for both sites, but the Viking 1 landing site appeared to be slightly richer in Ti and Cl and to contain somewhat less S than the Viking 2 site. Four crust samples from site 1 (C-2, C-3, C-5, and C-13) all contained 50% more S than the loose fines. These were the duricrust samples which were interpreted as sulfate salt-enriched and cemented forms of the fines material. By contrast, the sample C-8, of fines collected in the same area where the C-5 crusts were samples, contained less S than any other sample, at either site.

The XRFS team has sought correlations in the data but has offered the following cautions: (1) the samples were similar; (2) there were only 17 samples for which high-precision analyses were done; and (3) the sampling technique was not appropriately designed. Despite the above, they feel that there were some trends:

1. The protected fines contained Fe_2O_3 in the 17% range, but the exposed surface fines contained 18%, except for C-1 and U-5. This could perhaps indicate a slightly higher iron content for the surface fines. However, statistical limitations in the number of samples collected and the measurement precision of the instrument (±0.5% concentration by weight for the iron) prevent a firm conclusion.
2. There was no apparent correlation between the sulfur content and depth at which the sample fines were taken. Sample C-6, collected at the greatest depth (23 cm), contains about the same sulfur as neighboring surface materials, the samples C-1 and C-7.
3. Three candidate cations Ca, Fe, and Al are uncorrelated or negatively correlated with sulfur. Mg trends in the same direction as the S, but the instrument precision is too poor to establish this with any certainty (this bears on any conclusion about the salt type).

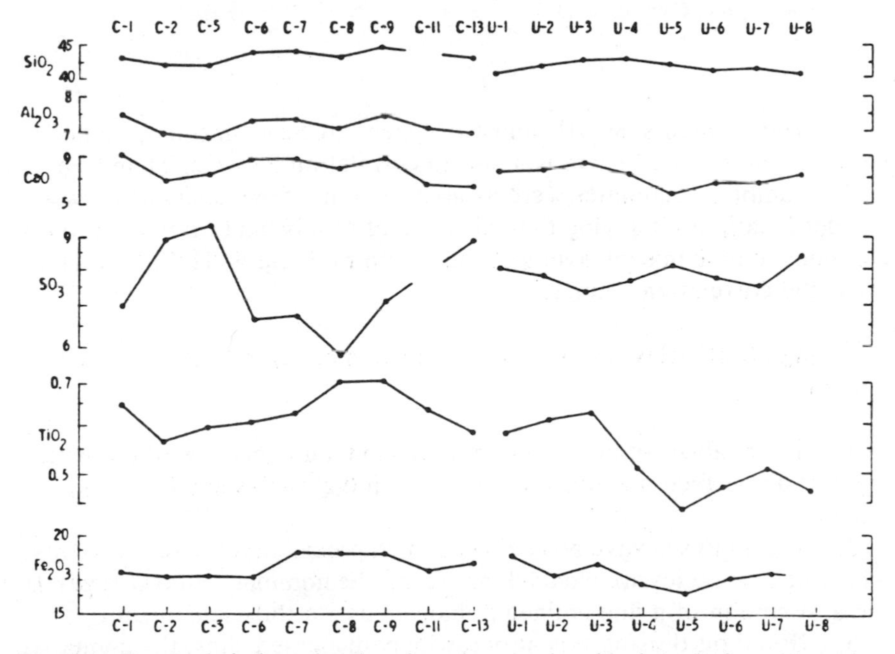

Figure 11.10. Analytical results plotted on the same logarithmic scale for oxide against sample numbers; C: samples from Chryse Planitia; U from Utopia Planitia. Source: Clark et al. (3).

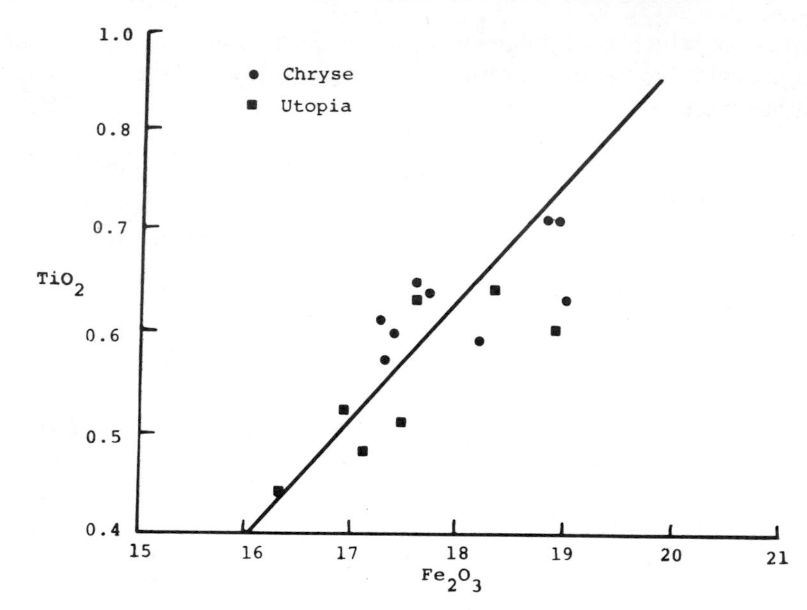

Figure 11.11. Correlation of Ti versus Fe. Data from both landing sites show a positive correlation. Source: Clark et al. (3).

4. The elements Si, Al, and Ca appear to be connected, and to a lesser extent the Fe. There is a strong correlation between Ti and Fe.

5. Sample variabilities were as great within a few meters as between Lander locations, implying the existence of a universal Martian regolith component of constant average composition. Figures 11.10 and 11.11 show the correlation studies.

Finally, the XRFS results have been summarized by the investigative team:

1. The analysis of major and minor element chemistry of the samples of Martian surface materials during the Viking missions has been completed.

2. The analyses have been made at high precision for many elements. Absolute accuracies are model dependent. The nominal results depend on the assumption of a fine-grained homogeneity on the scale of μm.

3. Based on the analysis at two widely dispersed sites, the compositions seem identical. Thus there appears to be a widespread uniform soil unit.

4. One cannot provide a unique interpretation for the origin of the fine material, but it appears to require that the source materials be predominantly mafic to ultramafic in nature.

REFERENCES

1. B. C. Clark, A. K. Baird, H. J. Rose, Jr., P. Toulmin III, R. P. Christian, W. Kelliher, A. J. Castro, C. D. Rowe, K. Keil, and G. Huss, *J. Geophys. Res.,* **82,** 4577 (1977).
2. A. K. Baird and B. C. Clark, *Geochemical Investigation of the Surface Composition of Mars,* 54, NASA Rept. (1969).
3. A. K. Baird, A. J. Castro, B. C. Clark, P. Toulmin III, H. J. Rose, Jr., K. Keil, and J. L. Gooding, *J. Geophys. Res.,* **82,** 4595 (1977).
4. P. Toulmin III, A. K. Baird, B. C. Clark, K. Keil, H. J. Rose, Jr., R. P. Christian, P. H. Evans, and W. C. Kelliher, *J. Geophys. Res.,* **82,** 4625 (1977).
5. B. C. Clark, A. K. Baird, R. J. Weldon, D. M. Tsusaki, L. Schnabel, M. G. Candelaria, and P. Magell, *J. Geophys. Res.,* **87,** 10059 (1982).
6. B. C. Clark, A. K. Baird, H. J. Rose, Jr., P. Toulmin, K. Keil, A. J. Castro, W. C. Keliher, C. D. Rowe, and P. H. Evans, *Science,* **194,** 1283 (1976).
7. E. Guiness, *J. Geophys. Res.,* **86,** 7983 (1981).

CHAPTER

12

VENUS—THE CHEMISTRY OF ITS ATMOSPHERE AND SURFACE

For many years prior to its exploration, Venus had been called the Earth's twin. This was due, in large part, to the fact that Venus is almost the same size as the Earth. Eternally cloud covered, Venus represented, until recently, a great scientific mystery that has now begun to yield to the scientific probes of the United States and the U.S.S.R. We now know that Venus is anything but a sister planet of the Earth. To quote Bevan French of NASA from his article in the volume *A Meeting With the Universe* (1),

> Venus is a strange and hellish inferno. Its thick atmosphere is composed mostly of carbon dioxide with little nitrogen and traces of water, oxygen and sulfur dioxide. At the surface the pressure of Venus's atmosphere is 90 times that of the Earth's, equal to the pressure a half mile down in the ocean! The thick atmosphere traps the Sun's heat producing a ''green house effect'' that keeps the surface of Venus at a scorching 480°C, literally hot enough to fry eggs.

It is these conditions that have made probing Venus such a difficult undertaking. Developing scientific probes and their associated electronics that could survive such extreme conditions has been a considerable task. Yet both the United States and Russia have succeeded. The Russian Venera spacecraft have actually made it to the surface and survived long enough to send back pictures of a rugged rocky surface from two different sites. Furthermore, a cleverly designed X-ray fluorescence device supplied the first chemical analysis of the surface of Venus. These efforts will be described in the following sections, beginning with the atmospheric studies and going on to the surface investigations.

THE PIONEER VENUS PROGRAM

The Pioneer Venus program (2) involved two scientifically related spacecraft missions, an orbiter, and a multiprobe bus, managed by NASA's

Ames Research Center. Both missions achieved a successful encounter with Venus in December 1978. A brief review of the effort follows:

Both Pioneer flights, launched within a short time of each other, were intended to explore the planet's atmosphere, to study its surface by radar, and to examine a number of geophysical parameters. The objectives called for 8 months or more of direct and remote sensing measurements. One spacecraft was a Pioneer Venus Multiprobe spacecraft designed to separate into five atmospheric entry craft about 12.9 million km from Venus. Each probe craft was intended to make measurements of the atmosphere from its highest region to the planet's surface. The second spacecraft, the orbiter separately launched, was scheduled for an elliptical orbit with a 24-h period and oriented about 75° to the equator.

The four probes were not meant to survive an impact with the surface. On arrival at Venus, the large probe was scheduled to take about 55 min to descend to the surface and the three small probes about 57 min. The bus (carrier for the probes) was scheduled to follow the probes into atmosphere about 80 min after. While the probes were equipped with heat shields, the bus was not. Thus, unlike the probes, it was designed to supply only data on the highest part of the atmosphere.

Both missions achieved a successful encounter with Venus. There were numerous findings, which have been published in various publications and summarized in a special issue of the *Journal of Geophysical Research* under the title "Pioneer Venus" (Vol. 85, No. A 13, Dec. 30, 1980). Some of the results of the compositional studies will be described here.

L. Colin (2) has summarized many of the key scientific questions that spurred the mission. Those related to chemical studies are listed here:

1. In addition to the CO_2, what does the lower atmosphere consist of and how are these constituents distributed?
2. What are the clouds made of?
3. What does the atmosphere tell us of the planet's surface and interior?
4. Why is the lower atmosphere so hot?
5. What role do phase changes play in the thermal structure?
6. Why does the atmosphere of Venus differ so much from that of the Earth?
7. How deep do the H_2SO_4 clouds extend?

These are but a small fraction of all the questions, and I shall here deal with only a few of these.

THE LOWER ATMOSPHERE OF VENUS

J. H. Hoffman, V. I. Oyama, and U. von Zahn (3), based on results from Pioneer Venus and the Venera 11 and 12 space craft, have obtained data on the chemical composition of the Venus atmosphere from 700 km to the surface. These investigators point out that since 1978 no less than seven gas analyzers provided in situ measurements of the chemical composition of the Venus atmosphere. Three mass spectrometers and two gas chromatographs have sampled the lower atmosphere (0–65 km), and two mass spectrometers have sampled the upper atmosphere above 130 km. The result has been a wealth of information about the atmosphere's composition. Table 12.1 is a summary of the instrumentation.

Hoffman and co-workers have reenforced an earlier statement about the difficulties of the exploration of Venus.

> The study of the composition of a planetary atmosphere a quarter of an A.U. [astronomical unit] away from the earth necessarily imposes constraint on a system employed for measurements, and when the atmosphere itself is as hostile as that of Venus, the task becomes a double challenge.

The Venus experiments succeeded in producing data on the composition of the atmosphere from about 700 km to the surface itself. Over this altitude the instruments experienced a pressure change of 17 decades. The pressure range samples by the instruments varied from 0.1 to 100 bars. This produced special problems in sampling the atmosphere for the mass spectrometric determinations. While the sampling of the upper atmosphere was direct—the atmospheric gases entering directly into mass spectrometer ion source—the measurements of the lower atmosphere required a pressure-reducing device, such as a calibrated leak. This was used to reduce the atmospheric pressure by 8–10 orders of magnitude (the mass spectrometer operated at about 10^{-5}–10^{-6} mbar.) It was, for example, essential to avoid alteration of the chemical composition of the sampled gas stream as it passed through the leak. By contrast, the GCs were able to sample the atmosphere directly. The various experiments will be discussed below.

NEUTRAL GAS COMPOSITION OF THE UPPER ATMOSPHERE

The composition of the upper atmosphere was investigated by a team headed by H. B. Niemann. The results were reported in 1980 (4). The experiment involved the use of a quadrupole mass spectrometer as part of

Table 12.1. Pioneer Venus and Venera 11 and 12 Gas Analyzers[a]

	LMNS	LGC	VNMS	VGC	BNMS	ONMS
Instrument type	Mass spec.	GC	Mass spec.	GC	Mass spec.	Mass spec.
Mission	NASA	NASA	USSR	USSR	NASA	NASA
Spacecraft	Pioneer venus sounder probe	Pioneer venus sounder probe	Venera 11 and 12 landers	Venera 12 lander	Pioneer Venus multiprobe bus	Pioneer Venus orbiter
Sampling	LA	LA	LA	LA	UA	UA
Altitude range (km)	0–60	22–52	2–23	0–42	> 130	> 145

[a] LA, lower atmosphere; UA, upper atmosphere; LNMS, large-probe mass spectrometer; LGC, large-probe gas chromatograph; VNMS, Venera mass spectrometer; VGC, Venera gas chromatograph; BNMS, bus neutral mass spectrometer; ONMS, orbital neutral mass spectrometer.
Source: Hoffman et al. (3).

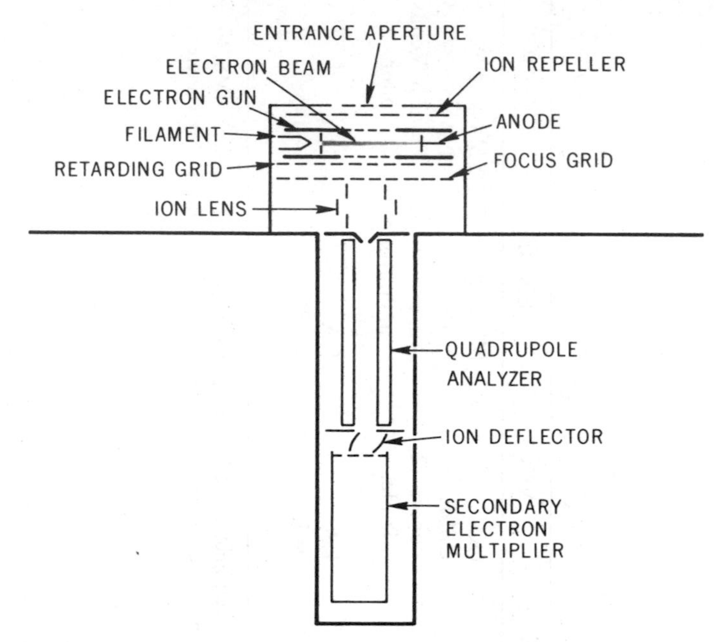

Figure 12.1. Schematic of the Venus Pioneer Orbiter mass spectrometer. Source: Niemann et al. (4).

the instrument payload of the Pioneer Venus Orbiter. Inserted into the Venus orbit on Dec. 4, 1978, the orbiter has continued to operate and supply scientific data. Initial data on the upper atmosphere were obtained once a day near periapsis at an altitude of 150–200 km above the surface, although as time goes on and due to the abandonment of orbit adjustments in an effort to save fuel, the periapsis has been rising due to solar gravitational effects.

A schematic view of the instrument is shown in Figure 12.1. The major components are the ion source, the quadrupole analyzer, and the secondary electron multiplier used as an ion detector. The ambient gas is introduced into the ion-source chamber through the entrance aperture, where it reaches thermal equilibrium with the surfaces through multiple collisions before being reemitted through the aperture to the atmosphere. The analyzer section and the secondary electron multiplier enclosures are vented to the atmosphere by a separate path. The connection between the two chambers is through a tiny aperture in the injection nozzle. Because of the speed of the spacecraft near periapsis was approximately 10 km/s (much higher than the thermal velocity of the gas), the gas density in the ion source greatly exceeded the ambient density. For CO_2 this density enhancement factor was about 100 when the aperture was pointed in the direction of spacecraft motion. The problems of making such measurements involving the relationship between ambient gas density and chamber density and the techniques employed for measurements from high-velocity spacecraft had been worked out previously.

Ions are formed by electron impact, and the energy of the ionizing electrons could be selected by ground control to be either 70 or 27 eV. The ionizing volume is located behind the entrance aperture, so that the entering gas could be ionized directly without prior surface collisions. The number of particles ionized directly was small relative to all the ions produced, but they were separable, because of their large velocity, from the ions produced from thermalized particles by means of a retarding grid (see Fig. 12.1). Incoming reactive gases that might be adsorbed on the walls or react with other adsorbed species could be measured directly.

The retarding mode of operation was under ground control. Measurements in the retarding mode were optimized by mounting the instrument on the spacecraft (which was spin stabilized) so that at the point of measurement, in the periapsis region, there was a minimum angle of attack for the aperture. The ions formed in the ion source were focused into the quadrupole mass filter. Mass peaks produced by the mass filter have flat tops, allowing a single measurement per peak without scanning. The instrument could be directed to operate in a stepping mode, where as many as eight masses between 2 and 46 amu are selected, or the instrument could be commanded to step sequentially 1 amu at a time through the full mass range from 2 to 46 amu.

RESULTS

By the time of the 1980 publication (4), data from 270 periapsis passes had been processed. For most of the passes, the masses 4, 14, 28, 44, 16, 32, and 30 amu were chosen with measurements being made in both the retarding and nonretarding modes. On nearly every seventh pass, the complete mass range from 2 to 46 amu was measured in unit-mass steps. On special occasions, diagnostic modes were selected. Some typical pulse-counting data, obtained in the nonretarding mode, for about 2.4 min at periapsis, are shown in Figure 12.2.

The instrument was operated in unit-mass stepping, and so the full mass range was sampled sequentially, requiring 8 s. The maximum pulse-counting rate is at periapsis with a drop on both sides (periapsis representing the highest atmospheric density). The modulation seen on the peaks is due to spacecraft rotation. The peaks seen, which come from major atmo-

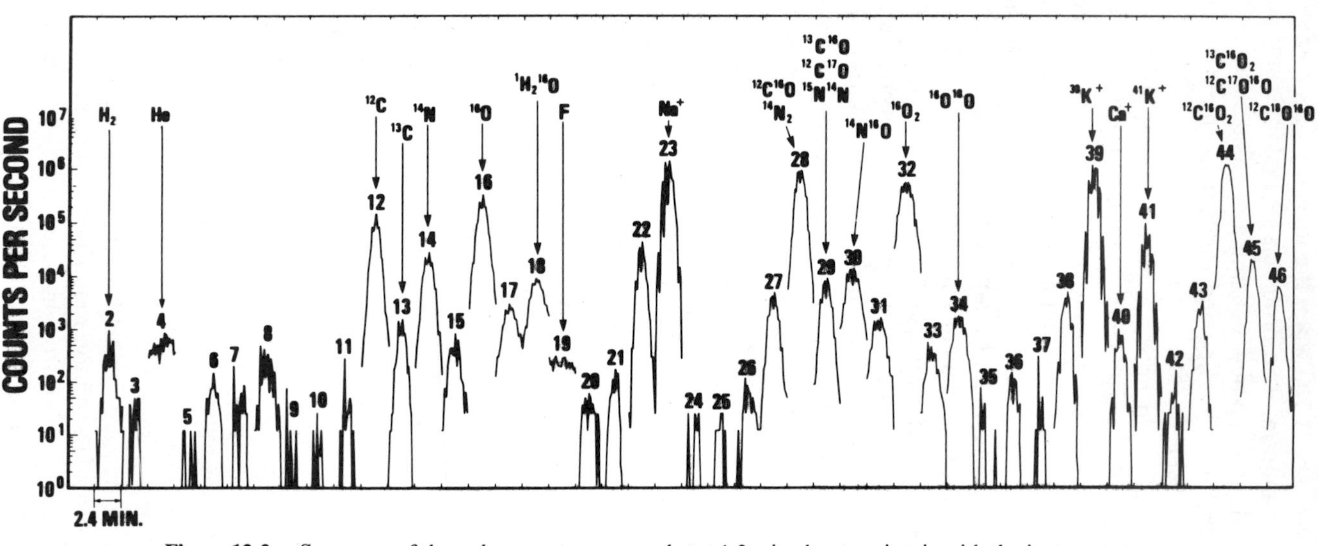

Figure 12.2. Summary of the pulse counts measured at ±1.2 min about periapsis with the instrument in the peak stepping mode. Starting at the left of each 2.4-min time interval marked on the abscissa, the altitude is 161 km on the ingoing leg. After passing through periapsis altitude at 141 km, the interval end at 160 km on the outgoing leg. The mass numbers are shown above each trace, and some constituents are noted. Source: Niemann et al. (4).

spheric constituents, are mass 4—He; mass 28—the sum of the ambient molecular N_2 and Co; mass 32—atomic oxygen recombined to molecular oxygen on the ion-source surface; and mass 44—ambient CO_2 and some contribution of surface-recombined CO and atomic oxygen. The other peaks are identified in Figure 12.2.

In general, the principal gases encountered were CO_2, CO, N_2, O, N, and He. Atomic oxygen was the major constituent above 155 km on the dayside and about 180 km on the nightside, where He became the major constituent. The average value of CO_2, CO, N_2, O, and N were found to remain nearly constant during the day, but an abrupt change was observed across the terminator from a high dayside value to a low nightside value. The density was observed to vary in the opposite way, with a distance bulge at night near the morning terminator. Large variations in density on the nightside, from orbit to orbit, suggested possible strong turbulent motion in the atmosphere below.

COMPOSITION OF THE LOWER ATMOSPHERE

In the following description of the experiment it will be obvious what a difficult undertaking it was. The study of the lower atmosphere was performed with a neutral mass spectrometer designed by Hoffman and associates (5), and flown as one of the instruments on the Pioneer Venus sounder probe. One of the major problems was the sampling of the atmosphere over a pressure range of 0.1–100 bars while measuring concentrations of the order of 1 ppm of the dominant gas, CO_2. Because the instrument was operated in a pressurized spacecraft, a gas removal or pumping system had to be included with the instrument to maintain the mass spectrometer at its operating pressure of $< 10^{-5}$ torr and to remove the gases previously sampled to avoid contamination. A constraint on the sampling system was to minimize alteration of the chemical forms of the gas species sampled from the atmosphere.

The sampling of the atmosphere was done through two small microleaks, each a 3.2-mm diameter passivated tantalum tube whose outer end was forged to a flat plate configuration with a conductance of about 10^{-7} cm^3/s for the primary leak and 10^{-6} cm^3/s for the secondary leak. These leaks extended from the sounder probe into the atmosphere.

The analysis of the gas flowing in through the leaks was done by a miniature magnetic sector field mass spectrometer that covered a mass range of 1–208 amu with a dynamic range of over 6 orders of magnitude. The removal of the atmospheric samples was done by a combination of chemical getters and a sputter ion pump. In order to preserve the large dynamic range of the instrument (to enable measurements of substances in the Venus atmosphere at the 1 ppm level of CO_2, as the atmospheric pressure varied over the 3-decade range while the probe descended), the larger conductance leak was closed at about 47 km. A variable conductance valve, between the instrument ion source and the primary getter pump, was opened pneumatically by the increased atmospheric pressure, causing the conductance from the ion source to the getter pump to be increased at a rate that approximated the increased throughput of the atmospheric sample leak, thus maintaining a nearly constant pressure in the instrument ion source.

A detailed description of the mass spectrometer has been given by Hoffman (5). The instrument had a high mass resolution capable of separating many mass doublets such as ^{40}Ar and C_3H_4 at mass 40, where the mass difference was one part in 580. Hoffman has reported that one of the problems of the operation of a mass spectrometer for the analysis of complex gas spectra is the overlapping of ion peaks, either as parent ions or as ion fragments formed in the instrument by the dissociative ionization of complex molecules. Thus the high resolution capability proved useful in separating many nearly overlapping peaks, for example, doubly charged ^{38}Ar from ^{18}OH at the mass 19 position; $C^{18}O$ and C_2H_6 at mass 30. In addition to the instrument's high resolution, another feature that helped in the identification of species was the availability of three different electron energies for ionization. The normal electron energy was 70 eV but, three times in flight during the descent, the ionization energy was cycled on successive scans of the mass spectrometer to 30 and 22 eV. This technique was used to remove all doubly charged ion species from the spectrum, and the fractionation patterns of complex molecules were significantly altered, making their identification simpler.

The mass spectrometer was controlled by an on-board microprocessor. The data acquisition involve a peak top-stepping routine and a data compression algorithm that led to a reading of the mass spectrum from 1 to 208 amu, obtaining a set of 232 measurements in 64 s (including a 5-s period for reading backgrounds and returning of the peak stepping routine). Only 40 bits/s were required to return the data to Earth. There was no scan or sweep of the mass spectrometer, as is the mode in conventional mass spectrometric operation. The use of the microprocessor in the peak stepping routine eliminated the time that is usually spent sweeping the valley between peaks. Information stored in the microprocessor defined positions in the mass spectrometer of a preselected set of mass peaks. These included several sequential positions around key mass numbers such as 15, 28, 40 and 136 amu, and mass defect positions for all known gas species to be measured. Two or more peak positions were set at some mass numbers. These were 16 (O and CH_4), 32 (O_2 and S), and 34 (^{34}S and H_2S). The microprocessor directed the instrument to measure only the amplitudes of the preselected peaks.

In practice, 45 scans were completed after entry into the Venus atmosphere. Surface impact came during the 46th scan. The instrument provided four scans prior to entry, which were used to obtain background information. The mass spectrometer's high mass resolution made it necessary to do the measurements with closely controlled ion acceleration potential—the tolerance was about 0.02%. To do this required a periodic finetuning of the program parameters, done by the positions of spectral peaks from two calibration gases that produced peaks at the following amu's: 136($^{136}Xe^+$), 68($^{136}Xe^{2+}$) and 15(CH_3^+).

The preflight calibration of the flight instrument was carried out in a high-pressure–high-temperature chamber called the Venus Atmospheric Simulator (VAS). The chamber was capable of reproducing the Venus descent profile, temperature, and pressure conditions down to surface values of 740°K and 90 atmospheres. The instrument inlet was inserted into the VAS chamber to sample its gases, simulating the Venus operation. In the VAS mixtures of gases (e.g., N_2, SO_2, COS, Ar, Xe, Hg, C_2H_6, HCl) with CO_2 as the principal gas were admitted in a stepwise manner in accordance with a model Venus atmospheric profile, while measurements were made with the mass spectrometer.

RESULTS

In order to interpret the results properly, the behavior of the instrument had to be taken into account. A consequence of the gas-pumping system of the mass spectrometer was that chemically active and chemically inert gases were pumped by different mechanisms. The gases flowed through different paths, with different conductances within the instrument envelope. The active gases flowed principally through the variable conductance valve to the main chemical getter, which maintained a vacuum in the ion source.

It had been mentioned previously that the throughput of this pneumatically operated valve continually increased with increasing pressure. The chemically inert gases were not pumped by this getter, and their only path of evacuation was through the narrow ion optics slit that connected the ion source with the mass analyzer. The difference in conductance, slower for the inert gases, resulted in the partial pressure of the inert gases increasing more quickly than the active gases, thus producing an artificially larger signal for the inert gases. As an example, the ratio of the Ar

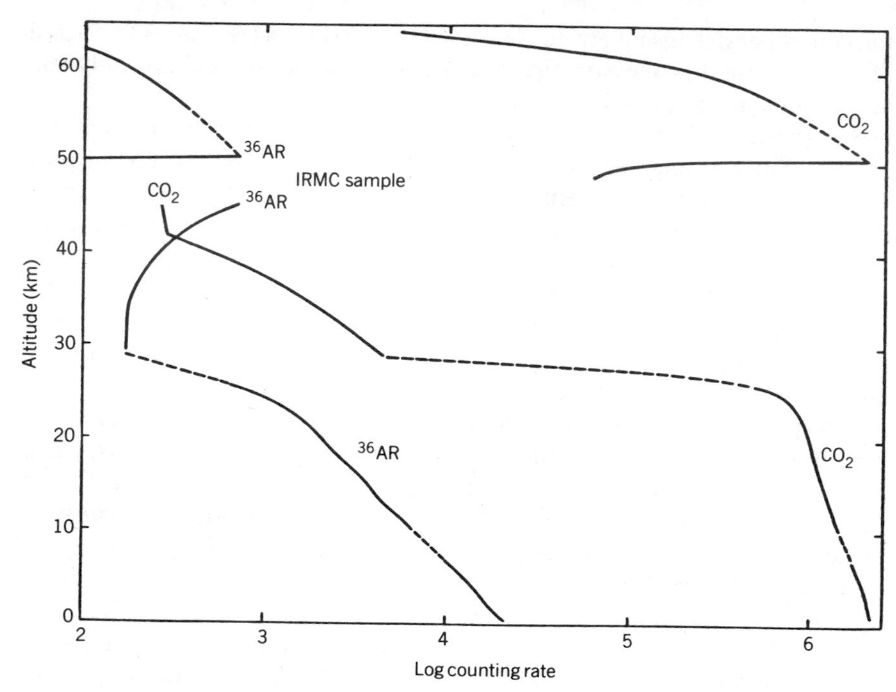

Figure 12.3. Altitude profile of ^{36}Ar, a nongettered gas, and CO_2, a gettered gas. The decrease in counting rate at 50 km is due to the leak blockage by H_2SO_4 droplets. The leaks reopened near 28 km. The dashed portions of the curves show the times when low-energy data were taken (not plotted). Source: Hoffman et al. (6).

peaks to the CO_2 peaks continually increased with time and was always greater than the atmospheric mixing ratio.

Figure 12.3 shows a flight profile of CO_2 and ^{36}Ar from the point at which the probe began making measurements. This occurred at an altitude of 62 km and after the heat shield had been ejected and the parachute deployed. The observation was made that, on initial entrance of the atmosphere into the mass spectrometer, the ratio of ^{36}Ar to CO_2 decreased. This was traced to the fact that CO_2 was pumped by the clean walls of the vacuum envelope and the electrodes faster than the Ar. Thus the ratio of ^{36}Ar and CO_2 decreased until the surfaces came to equilibrium inside the instrument, at about 60 km.

It was also observed that there was a sudden decrease in density (see Fig. 12.3) at about 50 km. The suspected cause was a blockage of both inlet leaks by sulfuric acid droplets from the middle or low cloud region in the atmosphere. The secondary leak had been programmed to be valved off at about 3 km below where the blockage occurred. Thus the inflow of ambient gas was effectively stopped, at which point the investigators feel the overcoating of the primary leak evaporated and the flow of ambient gas began again.

Beyond this point and to the surface of the planet, the profiles with altitude were characteristic of nongettered and gettered gases. The differences observed in the slopes were instrumental because of the pumping phenomena described above. Hoffman states that any gas having a constant mixing ratio over this altitude would have a profile lying between the two curves as limits. To determine the mixing ratio of the inert gas relative to that of the CO, it was necessary to know the conductance paths for each type of gas in the instrument. The mixing ratio was determined by the following relationship:

$$M_X = \frac{R_X}{R_{CO2}} \rho \frac{C_A}{C_V + C_A} \frac{[CO_2]}{[CO_2 + N_2]}$$

where M_X is the volume mixing ratio of gas X, R_X the counting rate for gas X, ρ the instrument sensitivity ratio of CO_2 to gas X C_A the conductance from ion source through instrument ion optics slits to analyzed ion pump, C_V the conductance of variable conductance valve, a function of atmospheric pressure, $[CO_2]/[CO_2 + N_2]$ the factor that converts M_X to a mixing ratio since N_2 is a nonnegligible part of the Venus atmosphere. Because C_A/C_V is very small (about 10^{-2}), the counting rate ratio of Ar to CO_2 as shown in Figure 12.3 was much larger than the mixing ratio.

The measurement of C_A and C_V were straightforward and were done in the laboratory. The uncertainties in these values came from the unknown pumping speeds of the getters and pumps.

The inlet valves, blocked by the overcoating of sulfuric acid droplets caused a loss of data from about 50 to 30 km. In this period, the peaks given by the mass spectrometer at 64, 48, and 18 amu showed a marked increase, which indicated that the material on the leak was forming SO_2 and H_2O vapor, apparently due to the dissociation of sulfuric acid. No parent molecules of sulfuric acid were observed, and SO_3 was not seen either. Laboratory tests confirmed that SO_3 would not pass through the inlet leaks without dissociation. In order to determine how H_2SO_4 dissociates during flight conditions, a laboratory simulation of the leak blockage was performed. The inlet leaks were coated with a solution of 85% H_2SO_4 and water until the flow of gas was blocked. The leaks were then inserted into the Venus Atmosphere Simulator, the chamber filled with CO_2, and the temperature in the chamber raised in correspondence with the temperature profile observed on the Venus desent. It was observed that the leak reopened and the flow of CO_2 began at about the same temperature

Table 12.2. Volume Mixing Ratios of Gases Found in the Venus Atmosphere

Gas[a]	Venus Atmosphere Mixing Ratio (ppm)	Earth Atmosphere Mixing Ratio (ppm)
^{20}Ne	9	16
^{22}Ne	1	2
Total Ne	10	18
^{36}Ar	30_{-10}^{+20}	31
^{38}Ar	6	6
^{40}Ar	31	0.93%
Total Ar	40–20	0.93%
^{84}Kr	< 0.2	0.5
N_2	4 ± 2%	78%
O_2	< 30	21%
SO_2 55 km	< 10	
Below 24 km	< 300	
COS Above 24 km	< 3	
Below 20 km	< 500	
H_2S	3	
C_2H_6	2	
H_2O	< 1000	
Cl	< 10	
Hg	< 5	

[a] In those rows where the altitude is unspecified, result is the average value from 24 km to the surface.

(250°C) as in flight. Furthermore, during the period of blockage, water vapor, SO and SO_3, and several reaction products of H_2SO_4 and vacuum sealant were observed in the mass spectrum in approximately the same ratios seen in the Venus data. Based on these studies, the team concluded that it was clear that the blockage material in flight was H_2SO_4 from the Venus clouds.

A summary of the results of the determination of the atmospheric composition from the mass spectrometric data is given in Tables 12.2 and 12.3. Table 12.2 gives the volume abundances of gases relative to CO_2. In many cases the uncertainties of measurement relative to CO_2 are large—traceable to the pump speed problem already mentioned. Relative isotopic abundances of some elements are shown in Table 12.3. For a detailed discussion, element by element, the reader is referred to Hoffman and associates (6).

Table 12.3. Isotopic Ratios of Some Substances Found in the Venus Atmosphere

Gas	Venus Atmosphere Isotopic Ratio	Earth Atmosphere Isotopic Ratio
$^{3}He/^{4}He$	$< 3 \times 10^{-4}$	1.4×10^{-4}
$^{22}Ne/^{20}Ne$	0.07 ± 0.02	0.097
$^{20}Ne/^{36}Ar$	0.3 ± 0.2	0.58
$^{38}Ar/^{36}Ar$	0.18 ± 0.02	0.187
$^{40}Ar/^{36}Ar$	1.03 ± 0.04	296
$^{13}C/^{12}C$	$\leq 1.19 \times 10^{-2}$	1.11×10^{-2}
$^{18}O/^{16}O$	$2.0 \pm 0.1 \times 10^{-3}$	2.04×10^{-3}

Source: Hoffman et al. (6).

In summary, the major constituents reported by Hoffman are CO_2 and N_2, where N_2 is about 3% by mass. Because of the excessive amounts of Venus atmosphere compared to the Earth, the amount of atmospheric nitrogen on Venus is about 3 times that in the Earth's atmosphere. However, if one adjusts the calculations to account for the 2 or 3 atmospheres of nitrogen fixed in the Earth's crust, then the total mass mixing ratio of nitrogen becomes similar for both planets. All the other constituents of the Venus atmosphere have volumetric mixing ratios of less than 100 ppm except for water vapor, COS, and SO_2, for which higher upper limits are given. A particularly significant finding is a hundredfold excess of nonradiogenic Ar and Ne on Venus relative to terrestrial abundances. There is also a small deficit of radiogenic ^{40}Ar. Other minor constituent are He and ethane. "There is strong but inconclusive evidence for H_2S" (6).

PIONEER VENUS GAS CHROMATOGRAPHS

The sounder probe of the Pioneer Venus mission also carried a gas chromatograph (SPGC) to measure the chemical composition of Venus' atmosphere in the altitude range of from 64 to 22 km. Like the mass spectrometer, the instrument was mounted in the pressure vessel with a sampling probe extended outside the vessel and beyond the aeroshell. The experiment and results have been described by Oyama and associates (7). The investigators state that the special attributes of the GC make it an ideal instrument for obtaining accurate in situ measurements of the chemical constituents of the atmosphere of Venus, as well as other planets. One of the important aspects of the instrument is that it is rugged and uncomplicated, and in many laboratories GC instrumentation has been used constantly over periods of years without service downtime.

Some of the unique analytical features that make GC so useful for planetary missions is that it offers a direct method for simultaneously identifying and determining the abundances of each molecular species of interest. Further, the capability of accepting relatively large samples at the ambient pressure and temperature of Venus directly into the sample path reduces contamination of the sample. In addition, the SPGC was engineered to provide a continuous flow of atmosphere through the sample path. It takes advantage of the increasing pressure during descent to keep the sample path constantly equilibrated with the atmosphere being sampled, thus preventing selective sorption, which may occur in low pressure devices.

GC has a marked advantage in that it is possible to measure distinct molecular species without the need for any disruption of the original sample component. As a consequence, there is no need to do calculations on the contribution of fragments to a total measurement. Finally, the GC typically does not separate isotopes. However, it readily separates isobars which, for example, permit the resolution and measurement of CO and N_2 while providing a total value for all Ar isotopes.

Figure 12.4 is a schematic of the mechanical subsystem of the SPGC. This instrument is a modified version of the GC used in the Viking gas exchange experiment (see Chapter 11). The figure shows the atmospheric sample path, temperatures to be obtained for the sampling assembly before each sample point, retention time internal standard, helium carrier gas flow path, gas sample valves, columns, and detectors.

In detail, the qualitative separations of gases were carried out by two sets of columns arranged in parallel to the carrier gas stream. The long column pair (stainless steel reference and sensing columns, 15.85-m long, 1.1-mm i.d.) were packed with Porapak N 100/120 mesh material. The columns were maintained at 18.3°C. Helium carrier gas pressure was maintained at 17 bars by means of a pressure regulator referenced to probe pressure, which maintained a flow of 35 cm^3/min.

The tubing for the short column set was 2.13-m long, 1.1 mm-i.d., and packed with 180/220 mesh polydivinyl benzene porous polymer beads, prepared specifically for the mission. The column was maintained at 62°C in an insulated shell. The flow through the column at the same head pressure was 45 cm^2/min.

Quantitative analysis was carried out by two thermal conductivity detectors using matched thermistor beads, one in the sensing column flow path and the other in the reference column flow path. The detectors were maintained at temperatures slightly above those of their associated columns.

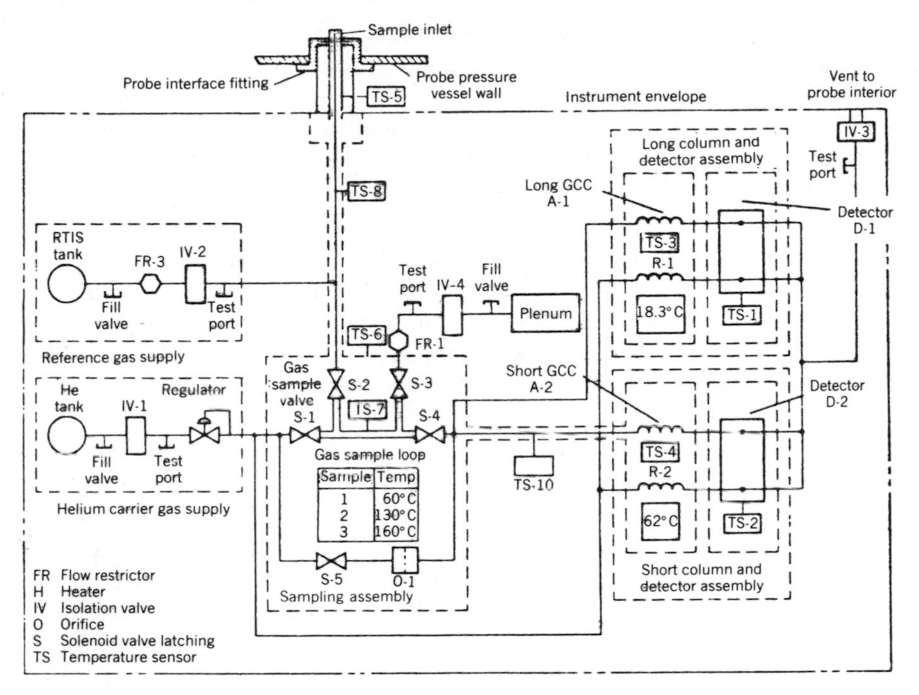

Figure 12.4. Schematic of the mechanical subsystem of the sounder probe gas chromatograph (SPGC). Source: Oyama et al. (7).

Just before entry into the Venus atmosphere, the columns were rejuvenated by priming them with 1% by volume of O_2 and SO_2 in helium. This was necessary for the sensitive measurement of these gases. Preflight tests had shown that long-term storage under vacuum caused trace amounts of injected O_2 and SO_2 to be irreversibly adsorbed on the long and short columns. The sampling plenum provided a way of storing this gas mixture. By proper sequencing of the valve operations, the rejuvenating gas was introduced into the head of each column, followed by a helium purge to sweep the sampler and columns of loosely held residual gases. The rejuvenating gas that remained in the plenum was exhausted to the atmosphere through the inlet port. As the probe descended, the atmospheric pressure began to exceed the plenum pressure and the flow was reversed, moving the ambient atmosphere through the gas sample path.

The sampling has been described in detail by Oyama (7). The first sample acquired by the SPGC occurred at about 9 min after entry. This involved the closing of the valves S-2 and S-3. A second later, S-1 and S-4 were opened simultaneously with the closing of S-5, diverting the gas flow through the sample loops into the splitter, which divided the gas sample into the long and short sensing columns. Seven seconds later, the valves were switched to their sample-receiving positions with S-1 and S-4 closed and S-5, S-2, and S-3 open.

The second and third samples were collected at 18:35 min and 29:05 min after entry. Immediately before the third sample acquisition, the thermal isolation valve (IV-2) was opened and the contents of the retention time internal standard (RTIS) entered the flow path. The RTIS was a mixture of Freon 14 and 22 in helium. Freon 14 was used to provide a reference peak for the long column, giving a retention time for a known material. Freon 14 and 22 provided markers for the short column. The choice of these two materials was dictated by the fact that they offered peaks well separated from other gases in the Venus atmosphere and were themselves not expected to occur naturally.

Calibrations were done before flight on the actual flight and spare instruments, using certified primary gas standards over the expected flight ranges. Calibrations involving water vapor were performed using permeation tubes, controlled temperatures, and flow conditions. The water vapor content loss from these permeation tubes was determined gravimetrically. Calibrations were also done for H_2S, COS, and SO_2.

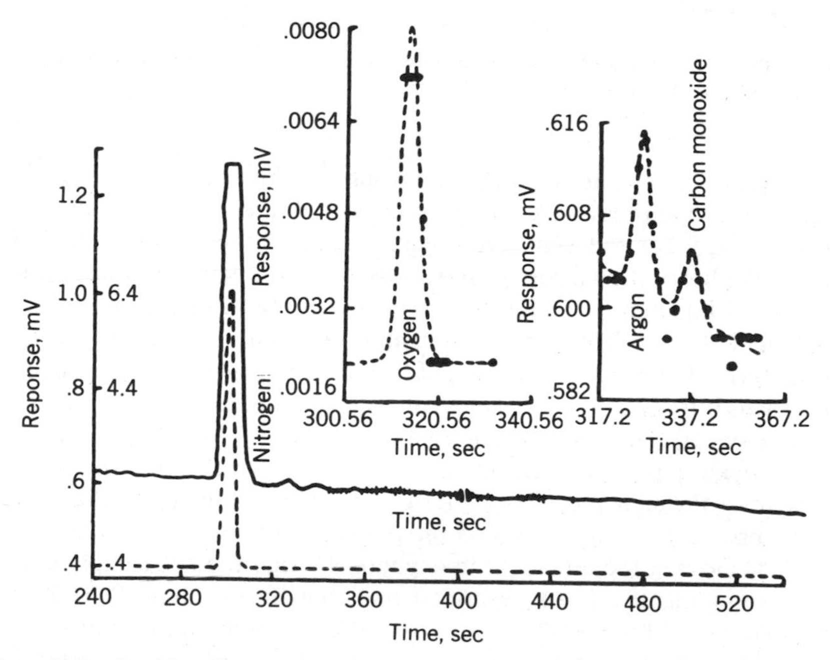

Figure 12.5. Raw data chromatogram of the first sample (long column), showing composite N_2, O_2, Ar, and CO peaks.

The actual chromatograms obtained by Oyama are shown in Figure 12.5–12.10. The blocked-out portions of the chromatographs show the peaks on an expanded scale. Figures 12.5 and 12.7 show the reconstructions of the O_2 peaks used to obtain the quantitative results.

The identification of peaks in the reconstructed chromatograms was based, first, on retention indexes and, second, on peak shape characteristics. The confidence in these techniques was based on

1. the limited number of small-molecular-weight species,
2. the complete chromatographic separation of the gases, and
3. the degree of redundancy provided for some gases by the dual column approach.

Once the raw data were obtained, absolute retention times were used for preliminary identification of the constituents in the lower atmosphere. While these data were considered reliable (the absolute retention time data of calibration gases closely matched the flight data), further analysis showed that the best fits could be achieved by using the relative retention data. These were based on the internal retention time standards, Freon 14 and 22, released into the third sample (see Figs. 12.9 and 12.10).

There was a deviation in the absolute retention time data from the prediction that was based on the calibration retention time for all gases. This resulted from the overall higher pressure and mass flow rates in the GC due to higher than expected internal probe pressure. The effect was a consequence of an early design decision that has been documented by Oyama as follows: Due to the need for the prevention of high-voltage

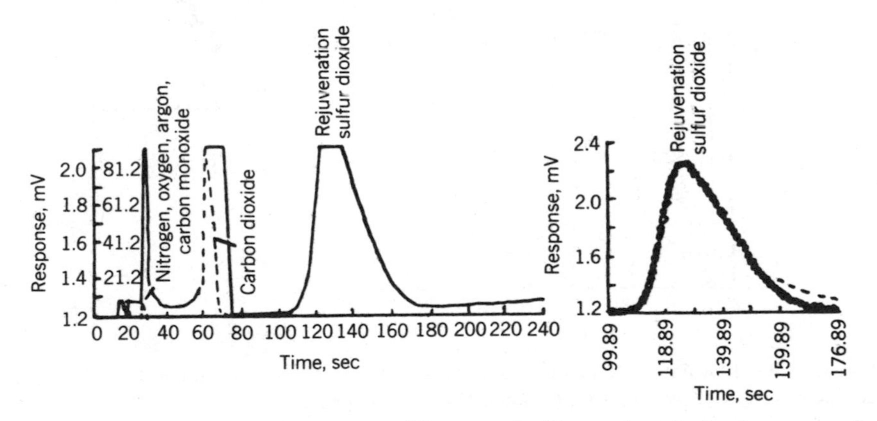

Figure 12.6. Raw data chromatogram of first sample (short column), showing composite N_2, O_2, Ar, and CO, CO_2 and anomalous rejuvenation SO_2 peaks.

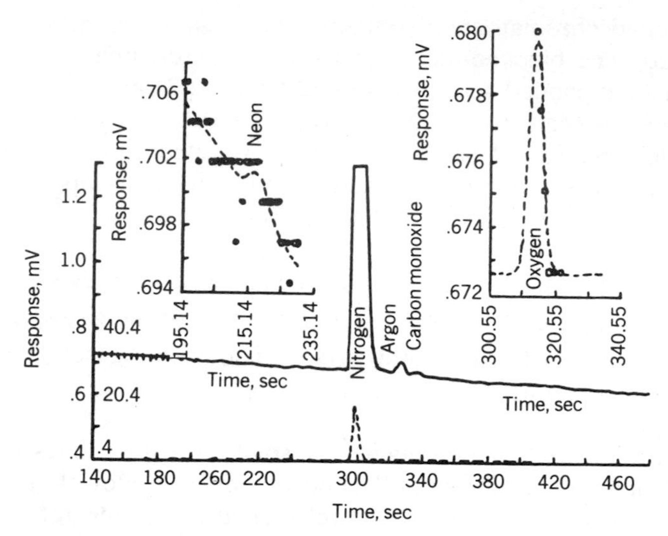

Figure 12.7. Raw data chromatogram of the second sample (long column) showing Ne, N_2, O_2, Ar, and CO peaks. Source: Oyama et al. (7).

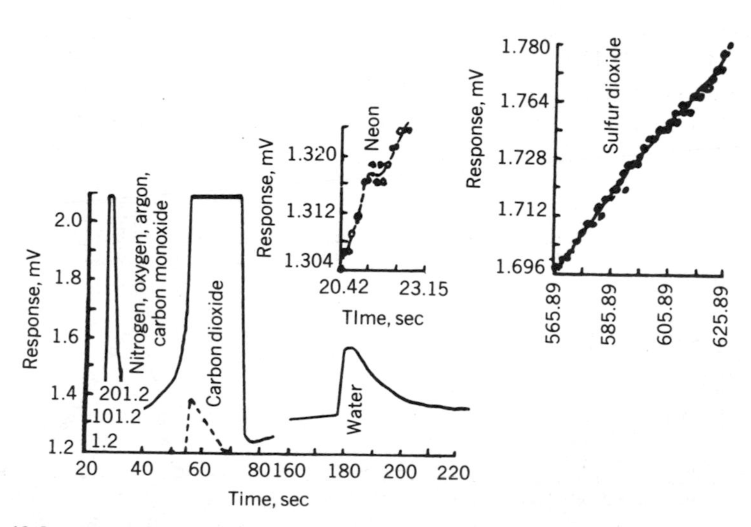

Figure 12.8. Raw data chromatograms of the second sample (short column) showing Ne, a composite N_2, O_2, Ar, and CO, CO_2 and H_2O and SO_2 peak.

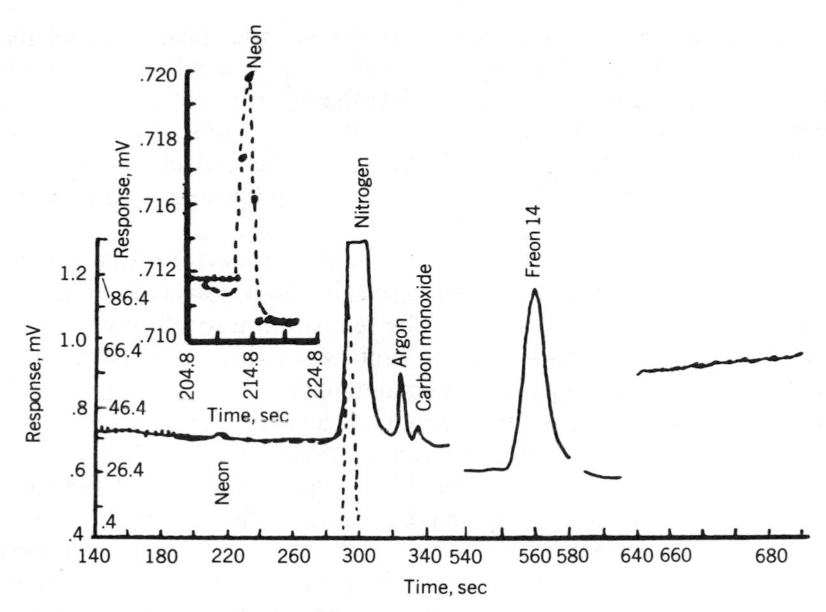

Figure 12.9. Raw data chromatogram of the third sample (long column) showing Ne, N_2, Ar, and CO and Freon 14 (internal standard) peaks. Source: Oyama et al. (7).

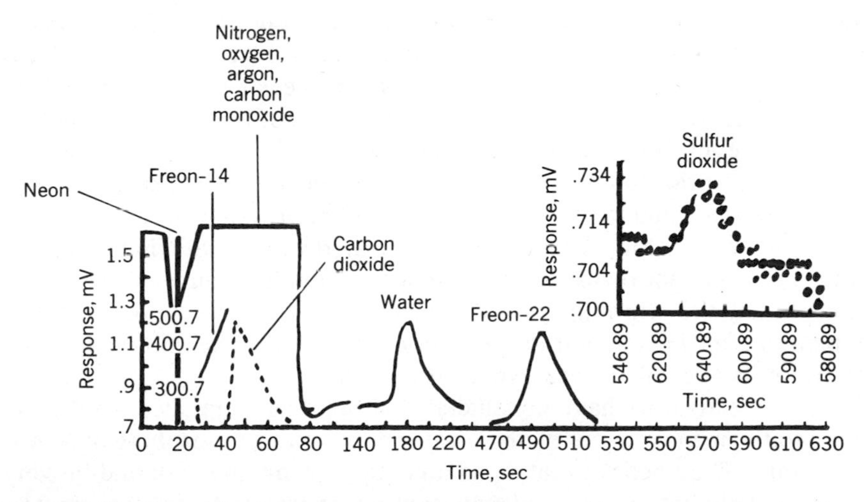

Figure 12.10. Raw data chromatogram of the third sample (short column) showing Ne, a composite N_2, O_2, Ar, CO, and Freon 14 (internal standard) CO_2, H_2O, and Freon 22 (internal standard) and SO_2 peaks. Source: Oyama et al. (7).

breakdown in the electrical portion of the sounder probe neutral mass spectrometer, the Pioneer Venus project office selected one atmosphere of N_2 as the initial probe pressure inside the pressure cell that housed the instruments. However, the likelihood that the probe pressure vessel might lose pressure during deep space transit was considered, and so an additional source of N_2 was provided to insure at 0.6 atmosphere by entry time.

Because the pressure vessel maintained its pressure, it actually arrived at entry with an added N_2 pressure so that the total was 1.3 atm. This pressure rose to 1.8 atm as the probe temperature rose during descent. The helium regulator and detector vents were referenced to the probe interior, and the pressure drop was maintained by the regulator. Thus the final pressure in the GC columns produced a larger carrier gas mass flow. Because the partition coefficient remains relatively constant over this pressure change, gas solute molecules would be dissolved in the carrier stream to a greater extent and arrived sooner at the detector.

There was an anomaly (Fig. 12.6) in that a large peak was observed from the short column. Its shape was identical with SO_2, but it appeared at 126 s after the injection of the first sample and so, because of bandwidth-retention time, relationships could not be associated with SO or any other constituent in the first injection. The investigators now believe that this peak is associated with the rejuvenation gas retention in one of the valves during the warmup sequence just prior to entry. The argument they present is that the first injection occurred 545 s after entry, and it is assumed that a total of 671 s passed before the peak emerged. This is 90 s later than the SO_2-absorbed retention time at an isothermal temperature of 62°C. However, the temperature of the column, after the end of the warm-up sequence prior to instrument turnoff, is expected to be less than 62°C, with power off and with flow terminated. This resulted in an extended retention time for SO_2. The O_2 peak associated with the SO_2 in the rejuvenation gas was not recorded from the long column because of the loss of data during the warm-up period. Since no measurable amount of SO_2 appeared after injection at the retention time for this gas, no further SO_2 retention in the valves was expected to occur.

The investigators have concluded that GC has been successfully applied to the measurement of chemical species in the atmosphere of Venus and Mars. They believe that as an analytical technique it should be very useful in the study of other planetary atmospheres and cometary comas. In fact, they strongly recommend its use for measuring the chemical constituents of the hydrogen-rich atmospheres of the outer planets. It is likely, they feel, that trace amounts of organic species other than methane are likely. Such trace quantities can be measured to an accuracy of parts per billion with the use of sensitive ionization detectors.

X-RAY FLUORESCENCE SPECTROMETRY ON THE SURFACE OF VENUS

One of the remarkable efforts to come out of planetary exploration has been the performance of X-ray fluorescence spectrometry on the surface of Venus (8). In March 1982, a descent module launched by the Venera 13 touched down on the surface of Venus. Its primary purpose was to perform the first analysis of the rocks on the surface of Venus. Although Russia had hoped for about 30 min of data collection before the instrument failed under the unbelievably harsh conditions, the craft actually continued to transmit signals for over 2 h. A second probe from Venera 14 landed about 600 miles from the first site, where it too performed analytical and physical testing.

These probes were asked to operate at temperatures of over 500°C and at pressures of 90 atm. Obviously, delicate analytical equipment could not operate under such harsh conditions; therefore, the samples were transported into the landing module where they were analyzed by the analyzer under conditions which were relatively normal. The problems that had to be dealt with were: (1) the absence of suitable rock standards, (2) the absence of a fixed geometry for analysis, and (3) the short time period for the analysis. A description of the experiment has been furnished by Surkhov and associates (8).

While X-ray spectrometry has been used on other occasions for the analysis of extraterrestrial materials, the use of X-ray fluorescence for the analysis of rocks on Venus required the production of a fundamentally new device and program of testing under conditions approximating those on the Venus surface.

INSTRUMENTATION

The surface sampling was done by means of a sampling device mounted in the shock-absorbing torus of the landing module. The claim of the investigative team was that "like a miniature drilling rig, the device is capable of drilling rock of practically any hardness corresponding to terrestrial types of rocks." As for the drilling depths and the mass of rock sample collected, it depended on the prevailing conditions under which the device

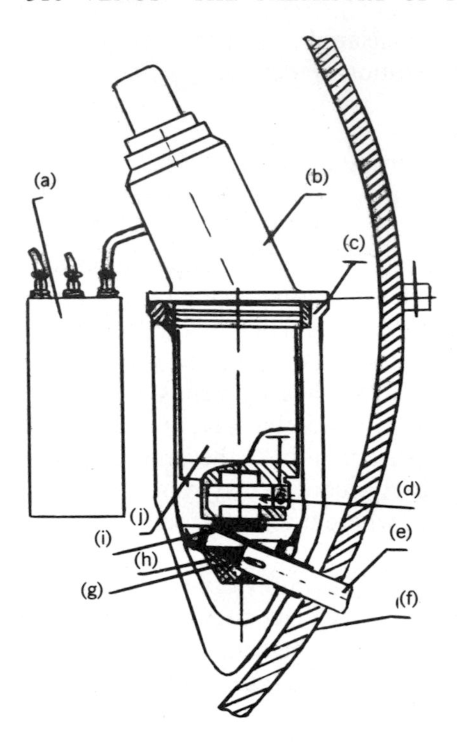

Figure 12.11. Location of X-ray fluorescence spectrometer on the landing module of Venera 13 and 14: (a) pulse height analyzer, (b) lid of detection unit, (c) casing of the detection unit, (d) counters, (e) air lock, (f) hull of the landing module, (g) sample under study, (h) rock receiver, (i) radioisotopic sources, (j) electronic section. Source: Surkhov et al. (8).

operated. Experiments performed under conditions approximating those on the planets surface showed that the sampling procedure always resulted in a sufficiently representative sample adequate for the accurate determination of the elemental content.

The location of the X-ray fluorescence spectrometers on the Venera 13 and 14 landing modules, as well as some of the details of the instrument are shown in Figure 12.11

The instrument consisted of a detection unit and a multichannel pulse-height analyzer. Both of these units were located inside a sealed and thermostatically controlled portion of the landing module. The analysis sample taken by the drilling device was fed into the chamber, which was then isolated from the external medium. Following this, the Venus atmosphere was pumped out and the sample carried through the airlock directly into the measuring cell of the detection unit. Once inside, the sample in the rock receiver (see Fig. 12.11) was excited into X-ray fluorescence by the radioactive sources. The fluorescent X rays were measured by the detectors and the amplified signals processed by the pulse-height analyzer. The accumulated data were periodically sent to

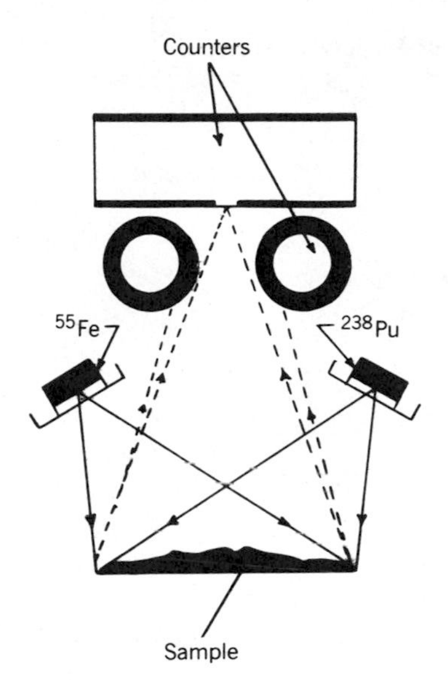

Figure 12.12. Diagram showing the relative location of the sample being analyzed, the radioisotopic sources, and the proportional counters. Source: Surkhov et al. (8).

telemetry for transmission to the Earth. The detection unit was built of a double titanium body designed to withstand the pressure of 100 atm. For temperature control, there was a coolant liquid between the walls of the body. The high-strength titanium was used to guarantee adequate performance of the analytical instrument in the event of the hot atmosphere breaking through the air-lock channel.

The rock receiver was built entirely of fluorocarbon plastic, as were all the walls of the measuring cell. The detection unit had a sealed body separated into compartments. The lower compartment housed the measuring cell, the middle compartment contained the electronics such as the preamplifiers, amplifiers, and power sources, and the upper compartment housed temperature-sensitive elements and pressure sensors. The arrangement of the sample, radioactive sources, and detectors is shown in Figure 12.12.

Surkhov and co-workers in describing their experiment, have stated that the radioactive sources and radiation detectors were the key elements of the experiment. Their selection and location determined the capability of the entire experiment. Because the basic rock-forming elements are in the atomic number range of 12–27, radiation sources were chosen to most efficiently excite the elements in this range and at the same time to minimize backgrounds. The radioactive sources were a ^{238}Pu (50 mCi) and two ^{55}Fe sources (250 mCi).

Table 12.4. Characteristics and Parameters of Radiation Detector

Mass	8 kg
Power consumption	9 W
Size of detection Unit	400 mm high × 107 mm max. diam.
Number of analyzer channels	256 (2 × 128)
Channel capacity	2^{16}–1
Range of energies	1.1–8 keV
Energy resolution	20–25% at 5.9 keV
Range of elements	Mg to Fe

Alpha radiation from the ^{238}Pu is very effective in exciting fluorescence from the lightest elements such as Mg, Al, and Si. Iron-55 was used to excite the heavier elements such as K, Ca, and Ti. The X rays from the ^{238}Pu were useful for exciting the elements with atomic numbers between 24 and 35. It has been shown that the yield of characteristic X rays for the Mg–Al–Si group is two to three times higher with the ^{238}Pu source, compared to that generated by the ^{55}Fe sources. Plutonium-238 alone would,

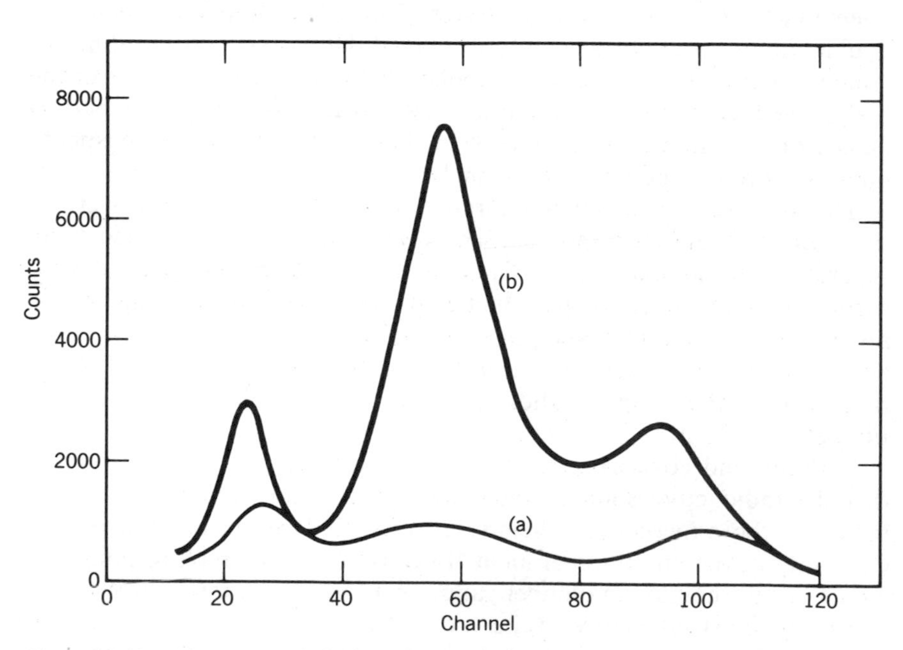

Figure 12.13. Fluorescent radiation of granite excited by *a*, ^{238}Pu and *b*, ^{55}Fe. Source: Surkhov et al. (8).

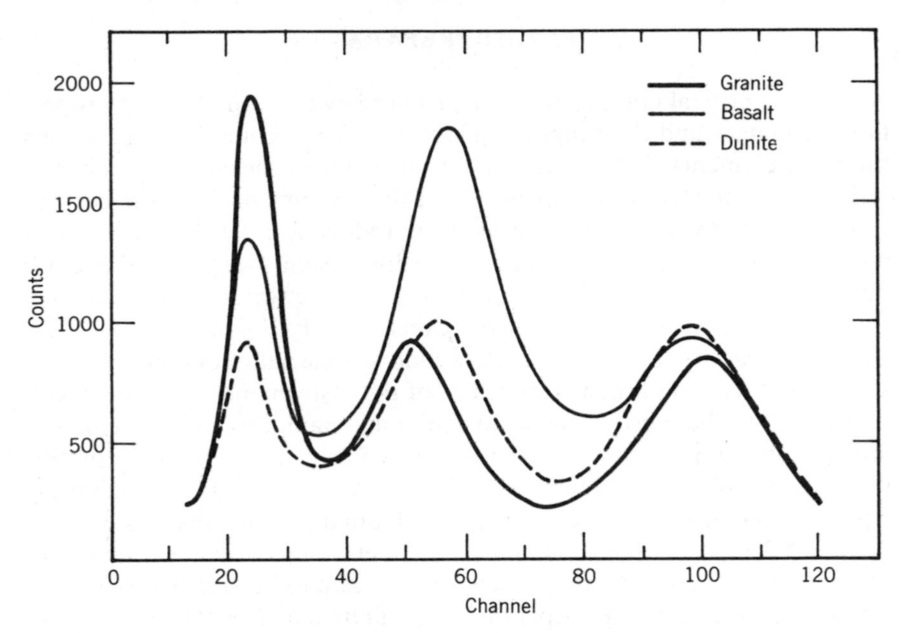

Figure 12.14. Spectra of fluorescent radiation of some Earth rocks as measured by the spectrometer. Source: Surkhov et al. (8).

however, not have allowed the determination of the elements in the K–Ca–Ti group for concentrations of the order of a fraction of a percent. To reduce backgrounds, collimators were mounted on the sources in order to reduce scattering, while maintaining a large area of irradiation on the sample.

The proportional counters were filled with 90% Kr and 10% CO_2 to a pressure of 220 mm of Hg. The beryllium window was 6 mm in diameter and the X-ray absorption path length was 15 mm. The energy resolution for the 5.9 keV Mn line was 20–25%. A summary of the characteristics and parameters are shown in Table 12.4.

Some examples of the instrument performance are shown in Figures 12.13 and 12.14. These were acquired with the Venera X-ray fluorescence spectrometer. In Figure 12.13, we see the spectra of granite excited by ^{238}Pu and ^{55}Fe. Note that the characteristic radiation of the light elements in channels 10–30 is weakly excited by the ^{55}Fe, but more intensely excited by the ^{238}Pu. The situation is reversed for the heavier elements in the higher channels (K, Ca, and Ti). Thus the use of both sources together made it possible to excite the fluorescent X rays more effectively in the range of rock-forming elements.

PREFLIGHT PREPARATION

Among the critical components that required evaluation were the proportional counters and their input windows, since the counters were among the main elements that determined the instrument measurement characteristics. To maximize the sensitivity of the instrument, it was necessary to increase the window transmission, by reducing the window thickness without compromising window strength. Instrument failure could occur if microholes were present in the windows, or if the windows inadequately resisted the large pressure changes. Consequently, a great deal of effort went into the choice of a suitable window material. Counter tests had shown that for long-time performance of the instrument the counter windows had to be absolutely leakproof and maintain their mechanical strength under pressure changes from 0.05 mm up to 1 atm. For this reason, a beryllium window of about 40-μm was selected, although it did cut the efficiency of the detector due to limited X-ray transmission.

The ^{238}Pu was another component that determined the performance of the instrument. The source's protective film had to be sufficiently thin so it would be adequately transparent to the alpha particles (for exciting the soft X rays, and yet radiation resistant to the environment). To check the safety of the source under conditions similar to those in space flight and on the surface of Venus, an attempt was made to simulate the chemical and physical effects of the environment to be encountered. The results of these tests were a more effective film and source construction. Similar tests were performed on the ^{55}Fe, although the requirements were far less demanding.

The detection unit that housed the rock powder receiver was filled with dry nitrogen at a pressure of 40-mm Hg in order to avoid the violent release of gas when the sample was introduced. The effect of this gas on the alpha-particle energy reaching the sample was very small. With regard to the attenuation of the low-energy X rays, this was accounted for in the instrument efficiency curve, as shown in Figure 12.15.

Before flight, comprehensive systems and subsystems tests showed the instrument to be stable over its normal operating range of -10 to $+50$°C. The gain factor variation was within 4%. Tests were performed under normal operating conditions, and in a chamber simulating the conditions on the surface of Venus. Tests of the instrument and the sampler were also done. These test were done in order to learn about the sample distribution in the rock receiver so that model experiments could be performed, and so that a library of rock standard spectra could be obtained. Tests in the chamber simulating the Venus surface made it possible to completely simulate rock sampling, the transportation of the sample to

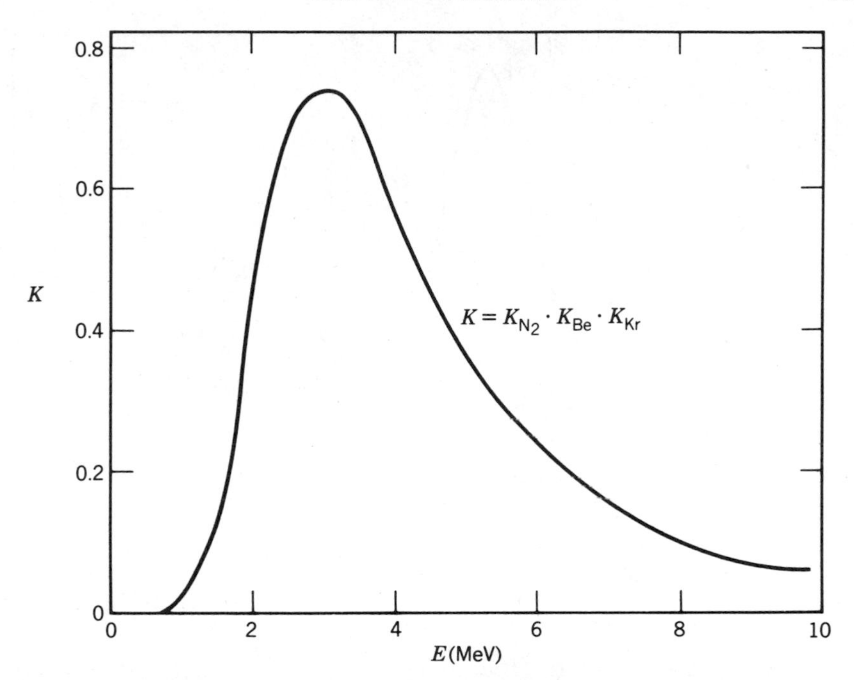

Figure 12.15. Efficiency (K) of the proportional counters for the fluorescent X rays as a function of energy. The factors attenuation by the nitrogen cover gas and beryllium window and absorption by the Kr counter gas. Source: Surkov et al. (8).

the rock receiver, the recording of spectra, and the determination of elemental concentrations.

SURFACE OPERATION

On the surface of the planet the X-ray fluorescence spectrometer was operated under commands from the on-board program/timing device. The first 4 min were used to obtain background, with the rock receiver empty. Simultaneously, information was acquired from the temperature-sensing elements, the pressure pickup, and the count-rate signal. Four minutes after the landing, the collected sample was transported to the rock receiver and a series of spectra obtained.

The X-ray spectrometers on both landing modules performed reliably during the active life of the module. Venera 13 survived for 127 min and Venera 14 for 53 min. Venera 13 delivered 38 spectra, as compared to 20 spectra from Venera 14. The accumulation, as reported by Surkhov and

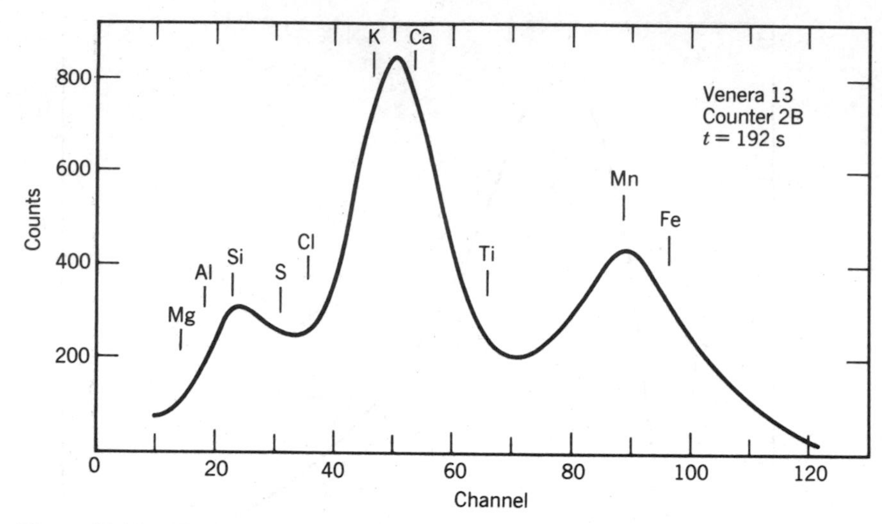

Figure 12.16. Fluorescent X-ray spectrum of venusian rock obtained by Venera 13. Source: Surkhov et al. (8).

others, was 384 s (counters 1B and 3D) and 192 s (counters 2B and 4D). The measured spectra for each of the four detectors on each instrument were reduced to a single scale and summed.

The results are shown in Figure 12.16 and 12.17. These are spectra of the Venus rocks. The spectra were processed on a computer by means of a special program (8) and a library of spectra from 200 rock samples. The

Table 12.5. Chemical Composition of the Venusian Rocks (Wt.%)

Elements (Oxides)	Venera 13	Venera 14
MgO	11.4 ± 6.2	8.1 ± 3.3
Al_2O_3	15.8 ± 3.0	17.9 ± 2.6
SiO_2	45.1 ± 3.0	48.7 ± 3.6
K_2O	4.0 ± 0.63	0.2 ± 0.07
CaO	7.1 ± 0.96	10.3 ± 1.2
TiO_2	1.59 ± 0.45	1.25 ± 0.41
MnO	0.2 ± 0.1	0.16 ± 0.08
FeO	9.5 ± 2.2	9.1 ± 1.9
Sum	95	96

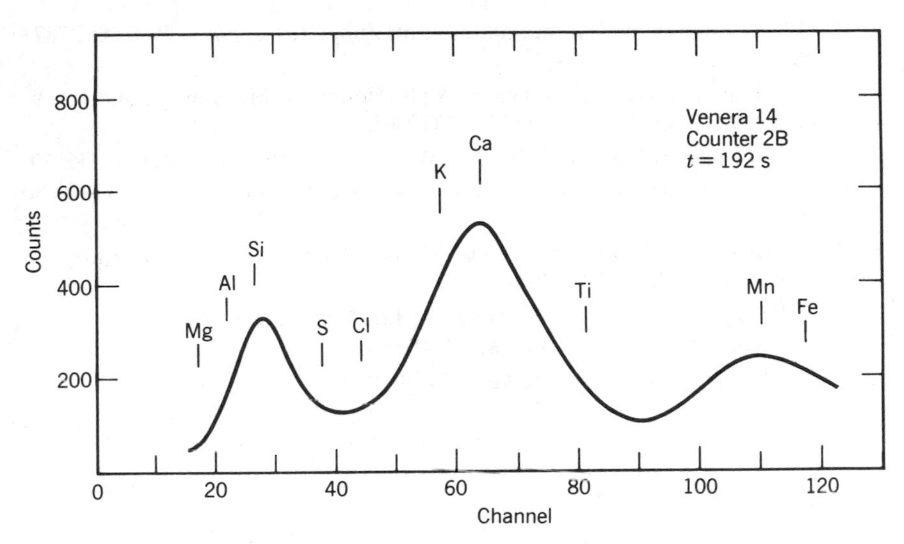

Figure 12.17. Fluorescent X-ray spectrum for venusian rock obtained by Venera 14. Source: Surkhov et al. (8).

investigators claim that it was possible to determine the concentrations of the basic rock-forming elements from Mg to Fe at both landing sites. Some of the results are shown in Table 12.5.

The investigators have drawn the following conclusions: The composition of the rocks from the different landing sites differed sufficiently to indicate different geomorphological provinces. The rock at the Venera 13 site is close in composition of potassium alkaline basalts on Earth. This is a type of rock not widespread on the Earth. It is found mainly on the oceanic islands and rift zones. The sample at the Venera is similar to the oceanic basalts of the Earth's crust.

In summary, this was truly a remarkable effort. Although the results and interpretations represent large-scale extrapolations from relatively small amounts of data taken under extreme conditions, they are to date the only compositional data from an in situ measurement on the surface of Venus. The accomplishment of this objective is a very impressive scientific act.

REFERENCES

1. B. M. French, *A Meeting With the Universe,* NASA EP-177 (1981).
2. L. Colin, *J. Geophys. Res.,* **85,** 7575 (1980).

3. J. H. Hoffman, V. I. Oyama, and V. von Zahn, *J. Geophys. Res.,* **85,** 7871 (1980).
4. H. B. Niemann, W. T. Kasparzak, A. E. Hedin, O. M. Hunten, and N. W. Spencer, *J. Geophys. Res.,* **85,** 7817 (1980).
5. J. H. Hoffman, R. R. Hodges, Jr., W. W. Wright, V. A. Blevins, K. D. Duerksen, and L. D. Brooks, *IEEE Trans. Geosci. Electron.,* **GE-18,** 80 (1980).
6. J. H. Hoffman, R. R. Hodges, T. M. Donahue, and M. B. McElroy, *J. Geophys. Res.,* **85,** 7871 (1980).
7. V. I. Oyama, G. C. Carle, F. Woeller, J. B. Pollack, R. T. Reynolds, and R. A. Craig, *J. Geophys. Res.,* **85,** 7891 (1980).
8. Yu A. Surkhov et al., *Anal. Chem.* **54,** 8 (1982).

CHAPTER

13

A BRIEF LOOK AT THE FUTURE

Although many exciting discoveries have occurred in our search for knowledge about the universe, there is promise of still further exciting scientific adventures. An examination of efforts to be made in the near future in connection with planetary exploration was mandated by the Solar System Exploration Committee (SSEC) of the NASA Advisory Council. Their recommendations were published in a comprehensive document, *Planetary Exploration Through the Year 2000*. The introduction to the report contains a well-written summary of what has already occurred.

> The exploration of the solar system by spacecraft has now spanned more than two decades and produced an avalanche of exciting discoveries and wealth of data. More than two dozen unmanned spacecraft have transformed our view of the planets from one of shimmering, telescopic images to one of crisp global perspectives.

In the past 20 years or so, spacecraft such as the Rangers, Surveyors, Pioneers, Mariners, Vikings, and Voyagers have visited every planet known to the ancients: Mercury, Venus, the Moon, Mars, Jupiter, Saturn, its rings, and many of its moons. There have been numerous surprises and as yet no Rosetta Stone.

In this volume we have examined a particular phase of the planetary exploration program, concentrating on the chemical studies of extraterrestrial materials and the implications of the findings. The early chapters dealt with materials studied in Earth laboratories, such as meteorites and returned lunar samples. The later chapters were devoted to the observations of materials at large distances by remarkable instruments and methods that have demonstrated extraordinary imagination and ingenuity.

Obviously, the account in this text is abbreviated. Actually the published information runs to many thousands of pages. It should be apparent that many of the developments described here will have a considerable impact on analytical chemistry.

But now, turning to the future, the SSEC was empowered to develop a mission strategy for Solar System exploration through the end of the 20th century. Based on scientific priorities established previously by the Space Science Board (SSB) of the National Academy of Sciences, and an assessment of technical capabilities as well as fiscal constraints, the SSEC has drawn the following conclusions. "In order to maintain U.S. leadership in solar system exploration and to realise any reasonable progress towards the scientific goals recommended by the SSB, NASA should immediately initiate a Core planetary program." (1)

The core program will be described in part. Since this text is devoted to chemistry, the portion of the core program that deals with compositional studies will be emphasized. An important aspect of the core missions is further exploration of the inner planets, so we shall examine some of the components of the core mission's program.

MARS GEOSCIENCE CLIMATOLOGY ORBITER (MGCO)

The MGCO is presently in an active planning stage with a possible launch in 1990. As we have seen, Mars, along with Venus and the Earth, are parts of a related triad of the inner Solar System planets that have an atmosphere. To understand their origin and evolutionary history is an important goal of the planetary exploration program. Knowledge of Mars is by now quite extensive, but there are still conspicuous gaps. The consensus of the SSEC is that there are a number of first-order questions that can be investigated from an orbital platform such as the MGCO can provide. Among the compositional questions to be investigated are the following:

1. What is the global elemental and mineralogical character of the surface?
2. What are the sinks for volatile materials and dust over a seasonal cycle?
3. What is the distribution of condensed or trapped volatile material? What is the character of the underlying residual polar cap, and is there at present a net transport of water between the hemispheres?
4. What is the escape rate of atomic hydrogen and how does the rate vary? What is the interrelationship between the atomic hydrogen, ozone, and water?

To answer some of these questions, the instruments to be used and the results expected are:

1. Gamma-ray spectrometer—elemental abundances of K, U, The, Fe, Ti, Si, O, C, and H.

2. Infrared radiometer—profiles of temperature, water, and dust.
3. UV spectrometer—ozone profiles.
4. UV photometer—atomic hydrogen column abundance.

The present plan for this mission is a launch in 1990 (other opportunities in 1992, 1994, and later). After a year's flight, the spacecraft would be inserted into a polar orbit around Mars, with observation times of about a year.

VENUS ATMOSPHERIC PROBE

The Pioneer Venus and Venera missions have raised questions about the Venusian atmosphere that will require an atmospheric probe instrumented for in situ analysis. It will be necessary to verify the Pioneer Venus findings of large Ne and ^{36}Ar abundances and large Ar/Kr, Ar/Xe, and D/H ratios. Precise values for $^{22}ne/^{20}ne$, $^{84}Kr/^{86}Kr$, and $^{132}Xe/^{129}Xe$ are needed to supply constraints for theories of the origin of the planetary atmospheres. Other major questions concern the oxidation states of the lower atmosphere, H_2 and H_2O abundances, and density profiles for sulfur compounds H_2S, COS, and SO_2.

The proposed instruments for this mission and the expected results are:

1. Neutral mass spectrometer: to yield the composition and physical state of the atmosphere as a function of altitude.
2. Gas chromatograph: to measure the profiles of trace constituents, including the noble gases (Ne, Ar, Kr), sulfur compounds such as H_2S, COS, SO_2, and H_2O.
3. Differential thermal analyzer: composition of the aerosols.
4. X-ray fluorescence spectrometer: composition of the dust within the aerosols.
5. Visual spectrophotometer: water abundances.

MARS NETWORK MISSION

The general objective of the Mars Network Mission is to establish a global network of seismic stations, meteorological stations, and geochemical and geophysical observation sites that would operate for an extended period of time. The concept calls for the release of a series of surface penetrators as the parent spacecraft approaches the planet. Each penetrator would be directed towards an appropriate preselected target. The projection is for the penetrators to enter behind a deployable heatshield and descend on parachute. Each probe would bury itself in the Martian surface to an extent that would leave some instrumentation and an antenna at the surface. Transmission to the Earth would be via an orbiting spacecraft.

Among the various questions to be answered are those related to the chemical composition of the Martian near-surface material. The plans for obtaining such compositional information include a gamma-ray spectrometer for the analysis of the naturally occurring radioactive elements such as K, La, Lu, Th, and U, and the elements activated by cosmic rays such as Fe, Mg, Ti, O, and Si. A second instrumental technique involves a pulsed neutron spectrometer for determining the major rock-forming elements, plus carbon, oxygen, and hydrogen. The nature of the method would permit analysis of material up to 1 m from the penetrator.

MARS SURFACE PROBES

The plans for Martian Surface Probes closely resemble those for the Mars Network missions. The spacecraft near Mars would launch penetrators that would be targeted for volcanic areas or ice caps identified on existing images. Location could be determined from tracking, nested entry images, or after body imaging. The objective is a determination of bulk chemical composition, including key trace elements. Information such as this is considered by the community of planetary scientists to have immense cosmochemical significance.

To achieve these objectives, the plans call for a gamma-ray spectrometer and a pulsed neutron spectrometer. As in the Mars Network mission, the elements called out for determination are the radioactive elements such as K, La, Lu, Th, and U, and those excited by cosmic ray interactions such as Fe, Mg, Ti, O, and Si. The pulsed neutron spectrometer should make it possible to analyze the major rock-forming elements plus C, O, and H up to distances of about 1 m from the penetrator.

LUNAR GEOSCIENCE ORBITER

Our present knowledge of the Moon is substantially more comprehensive than that about Mars. Our analysis of samples acquired directly from the Moon makes it unique among the Solar System bodies explored (except, of course, the Earth). Despite the intensive study of the Moon, there are still a number of compelling questions:

1. What is the compositional heterogeneity of the Moon, and what was the time sequence of lunar differentiation? What is the relationship between theories of lunar origin and its thermal history?
2. How do the detailed surface elemental and mineralogical phase compositions vary globally?
3. How do surface composition and gravity variations relate to magnetic variations?
4. Are there volatile materials such as water-ice trapped in polar regions of the Moon?
5. What is the nature and time distribution of explosive volcanism on the Moon?

In order to accomplish these scientific objectives, instruments such as a mapping spectrometer and gamma-ray spectrometer are being considered. The mapping spectrometer would be employed to determine surface mineral composition, while the gamma-ray instrument would be used to answer questions about cold trapped volatiles in the polar regions, and about surface elemental composition—Si, Al, Mg, Fe, Ti, Ca, Na, U, K, and Th. Both of the devices would be used for a global resource survey.

A Lunar Geoscience Orbiter could be launched at any time. Initially, the orbit would be an elliptical orbit with a low periapsis (less than 30 km) on the anti-sunward side for the Gamma-Ray Spectrometer calibration and magnetic measurements, followed by a 1-yr observation period at a 50–100-km circular polar orbit.

It is likely that identical instrumentation could be employed to study a near-Earth Asteroid. One possibility, which is attractive financially, is to send the Lunar Geoscience Orbiter to a near-Earth Asteroid after the completion of the lunar mission.

Having discussed proposed inner planet missions, let us now examine the plans for some small bodies missions.

COMET RENDEZVOUS

Comets are thought to be the most primitive bodies in the Solar System, and quite possibly carry some of the most important clues to the formation of the Solar System. The close-up study of a comet is one of the most important challenges in space exploration. The SSEC has outlined a mission of the following character. A comet rendezvous spacecraft will be launched using the shuttle–centaur configuration. After about a five-year cruise period, propulsive maneuvers will match the trajectories of the spacecraft and comet several months before the comet's perihelion passage. Following this, only a slight propulsive maneuver will be required for station keeping, and it will be possible to examine the comet from any selected distance and direction. Remote sensing measurements will begin immediately and as the comet approaches the Sun, the in situ instrumentation will measure the increasingly active expulsion of dust and gases. The mission would be expected to last for about 6 mo after initial rendezvous. Among the many questions to be answered are the following:

1. What is a cometary nucleus like? Is it, as often described, a dirty snowball? Does material boil off from all parts of the comet or only in discrete jets?
2. What atmospheric and surface changes occur as the comet approaches and then moves off from the Sun?
3. What is a comet made of? What are the abundances of the various elements in the nucleus? What materials and ices lie on the comets surface? What volatile molecules escape from the comet?
4. What are the physical structure and chemical composition of cometary dust grains? Are they all similar or are there different types? How abundant are dust particles of various sizes?
5. What is the generic relation of comets to interstellar dust grains, to meteorites, to Asteroids, and the planets? Are comets pristine samples of the solar nebula or have they undergone some processing? Are comets the building blocks of the outer planets? Did comets contribute measurably to the atmosphere of the terrestrial planets?

The instruments under consideration are X-ray and Gamma-ray spectrometers for determining the elemental composition of the nucleus; neutral and ion-mass spectrometers to analyze for elemental, molecular, and ion composition and density of the coma as a function of time and position; a dust collector and analyzer to provide elemental composition and physical character of the dusts; and finally an IR reflectance spectral mapper to study the mineral phase and composition of the nucleus and composition of the coma as a function of time.

An alternative mission scenario involves the collection of samples of volatile and nonvolatile material during a fast flythrough of a comet, with a return to Earth of the samples for subsequent analysis. The scheme calls for a launch of a shuttle-IUS, which travels on a ballistic trajectory and thus reaches the comet after about two years. Remote sensing equipment, if carried, would be turned on about 60 days before encounter, close to perihelion. The proposed flythrough would take place at 10–15 km/s and last only minutes, but would allow for the collection of atomized grains and gases. After encounter, the dust collector panels would be stored in an on-board capsule. By means of a small propulsive maneuver, the spacecraft would be placed on an impact trajectory with the Earth. The sample capsule would enter the atmosphere directly and parachute to the Earth for recovery.

Aside from a sample collection module, the possible instruments under consideration are a neutral and ion mass spectrometer, a dust collector, an IR reflectance spectral mapper and an imaging device. Such instruments have already been discussed.

OUTER PLANET MISSIONS

Among the outer planet missions for which plans are presently being drawn are those to explore Jupiter, Saturn, Titan, Uranus, and Neptune, although the time frame for some of these missions lies well into the future.

RETURN TO JUPITER—THE GALILEO MISSION

The amazing discoveries about Jupiter and the Jovian system that have flowed from the Voyager flybys have spurred a great interest in a return to Jupiter. Accordingly, NASA has begun the preparation of a Jupiter mission, called Galileo, tentatively scheduled for a 1986 launch.

The plans involve a multiple-vehicle system made up of two elements, a probe for the Jupiter atmosphere investigation, and an orbiter to explore Jupiter, its satellites, and magnetosphere. Flybys of the Jovian satellites will be used to alter the orbit of the Galileo spacecraft so that a tour of the Galilean satellites will occur and the moons will be examined at different geometries, and a deep penetration into the magnetosphere, in an as yet unexplored region in the space behind Jupiter, will take place.

The science to be done is divided into two groups of experiments: probe science and orbiter science. The main emphasis in the study of Jupiter is on direct measurements with the probe. If the mission succeeds, it will, for the first time, be possible to examine the atmosphere of this giant planet directly. The hope is that by measuring the temperature and pressure as the probe descends through the clouds, it will be possible to determine the structure of the atmosphere with higher precision than has been achievable by any remote measurements. In addition, the probe would be able to make direct measurements of the composition of the gases—in some cases, with the sensitivity of parts per billion using a neutral mass spectrometer.

Direct studies of Jupiter's clouds will be done by means of a nephelometer on the probe. Estimates will be made of the sizes and compositions of the individual aerosol particles. Infrared instrumentation will be used in the measurements of the cloud layers and the amount of UV deposited in different regions of the atmosphere.

The composition and structure of Jupiter's upper atmosphere as well as its moons will be done with a UV spectrometer carried on the Orbiter. A near-IR mapping spectrometer will be employed to determine the composition and structure of the upper atmosphere. In addition, there will be a large variety of other instruments for the measurement of a number of physical and geophysical parameters.

The Galileo Orbiter and Probe are scheduled to be launched with NASA's new Space Shuttle and Inertial Upper Stage. In order to maximize the payload carried to Jupiter, the mission will depend on achieving a boost from Mars. It is anticipated that the Galileo will approach Jupiter sometime in the mid-1980s. The initial trajectory will take the spacecraft within five Jupiter radii, at which point by firing its rocket engines it will shed velocity sufficiently to be captured by Jupiter. During this first pass there will be a close flyby of Io.

Because Jupiter is known to have an intense radiation belt, the Orbiter will not be permitted to spend too much time in the inner magnetosphere near Io's orbit. The extended exposure to the intense radiation could be destructive to the electronics and bring about a premature conclusion to the mission. Thus it is contemplated that an additional thruster firing during the first orbit would boost the periapsis to about 10 Jupiter radii. While no further close-up observations of Io would be possible, studies of this satellite could be made on each subsequent orbit with sufficient resolution to study the details of the volcanic eruptions and to observe those changes in the surface produced by the volcanic activity.

Galileo will also be programmed for a close flyby of one of the outer moons. Several passes each of Callisto, Ganymede, and Europa are possibilities. The description of the mission in the NASA document *Voyage to Jupiter* (NASA SP-439), states that the satellite tour does not need to be fully planned in advance. By adjusting the spacecraft's trajectory with small bursts of the thruster motors, navigation engineers can modify the orbits to meet scientific needs.

The total duration of the mission is planned for about 20 months. It will be possible to add to the basic mission if the spacecraft continues to function and the fuel reserves are adequate. By contrast to the orbiter, the Galileo probe mission would last only a few hours.

TITAN FLYBY/PROBE

The Voyager flyby has demonstrated that Titan is a particularly fascinating Solar System body. The atmosphere of Titan is uniquely interesting from the standpoint of planetary chemical evolution. The organic chemistry observed to be occurring on Titan provides the only planetary scale laboratory for a study of those processes that may have occurred on the Earth in the prelife terrestrial atmosphere. The compelling questions identified by the SSEC are:

1. How did Titan develop its present atmosphere?
2. What gases and aerosols exist at different heights in the Titan atmosphere?
3. What are the chemical processes taking place in the atmosphere? What organic molecules are present and what can they tell us about the origin of life on the Earth?
4. What are the nature and structure of the clouds in the Titan atmosphere?
5. What is the energy source for the UV dayglow?
6. What is the chemical composition of the Titan orange haze?
7. How much sunlight reaches Titan's surface, and what is the extinction profile in the atmosphere?
8. What does Titan's surface look like? Are there lakes and oceans of methane? What are the major geologic features? Are they land features or ice masses of some sort (perhaps frozen methane)?
9. What was the condition of the protoplanetary solar nebula in the region in which Titan formed?

The instruments proposed for the Titan planetary science study are:

1. Ion mass spectrometer to determine the composition of the ionosphere.
2. Neutral mass isotopic spectrometer for a number density identification and ratios of neutral upper atmosphere constituents and electron temperatures.

Instruments proposed for the descent module are:

1. Neutral mass spectrometer for determining the number density, vertical profile, identification, and isotopic ratios of the atmospheric constituents.
2. Nepelometer to study the physical structure and location of the cloud layers.
3. Gas chromatograph for profiles of trace constituents including noble gases (Ne, Ar, and Kr); HCN, propane, acetylene, carbon monoxide, and so forth.
4. Descent imager/radiometer to determine the vertical distribution of atmospheric constituents such as methane, ammonia, and aerosols by measuring relative light levels at the near IR and visible wavelengths.

The mission profile involves a Galileo-like probe (see Jupiter section) carried to Titan aboard a flyby spacecraft or on a Saturn orbiter. Trip times are expected to be of the order of 3.5 y for a Probe/Flyby or about 6.4 y for a Probe/Orbiter mission.

THE OUTER PLANETS FLYBY/PROBES (SATURN, URANUS, NEPTUNE)

The giant outer planets differ from the terrestrial planets in many ways, and they offer an opportunity to investigate their internal structures and bulk compositions through a detailed study of the compositions of their atmospheres. In situ determinations of isotopic and molecular compositions in the outer planet atmospheres can also provide diagnostic information on the protoplanetary conditions and radial properties of the solar nebula as they existed in that particular region of the Solar System. In addition, regions of the atmospheres of the Outer Planets have clouds of aerosols of an interesting chemical nature; further, the transport of energy within the atmosphere is important for understanding the internal structure and evolution of the planet. A list of the major driving questions follows:

1. What is the reason for the significant variation of the chemical and thermal properties of Jupiter and Saturn as a pair versus Uranus and Neptune as a pair, and why are Uranus and Neptune so different from each other?

2. What are the abundances of helium and hydrogen and trace constituents? What are the isotopic ratios of the major elements?
3. What is the structure and composition of the cloud and aerosol layers?
4. What is the transport and energy deposition mechanism in the atmosphere?
5. What are the relative isotopic abundances in the atmosphere?
6. What was the condition of the protoplanetary nebula in these areas of the Solar System?
7. How did the planet evolve?
8. What is the nature of the moons, rings, magnetic fields, and atmospheric dynamics?

The instruments to achieve the above objectives are a neutral mass spectrometer, nephelometer, He abundance detector and, if possible, a gas chromatograph.

The base-line mission would include a Galileolike probe and a Probe Carrier spacecraft. The mission to Saturn would require about 3.5 y. To arrive at the very outer planets would take about 5.5 y to Uranus and about 9.5 y to Neptune.

SATURN ORBITER

The spacecraft for this mission would be launched using the Shuttle Centaur and would arrive after a 3.5-y flight. On arrival at Saturn, the spacecraft would be propulsively decelerated, taking advantage of Titan's gravity field. During an orbital phase of about 2 y, many flybys of the Saturnian moons would occur. The rings of Saturn would be studied from a complete range of observational phase angles, including numerous stellar and radio occultations. Imaging of the Saturnian clouds would take place on each orbit from high altitudes, while IR soundings of the atmosphere would be done from periapsis.

The objectives are many. They involve the developing of an understanding of the behavior of the assembly of moons, field phenomena, the rings, and the planet itself. Included in the plans are orbital changes over the duration of the mission in order to do detailed mapping of Titan. I have already mentioned that it is an aerosol-shrouded body, perhaps very like our own prebiotic Earth.

Some of the specific scientific objectives are:

1. Determination of the composition of the satellite surfaces (minerals and ices) and geological history of each object.
2. Determination of the nature and origin of the dark material on the moon Iapetus' leading hemisphere.
3. Determination of the time variability of Titan's clouds/hazes?
4. Characterization of Titan's surface on a regional scale.

Among the compelling scientific questions are:

1. What is the three-dimensional fine structure of Saturn's rings?
2. What causes the transient spokes observed in the rings?
3. What are the size distribution, chemical composition, and physical state of the ring particles?
4. What does the surface of Titan look like? Are there actually lakes or oceans of methane?
5. What are the compositions of the moons of Saturn? What minerals and ices are present on their surfaces?
6. What is the composition of Saturn's atmosphere?

The proposed instruments for such a mission would be an imager, IR radiometer, UV radiometer, dust particle detector, and radar.

SUMMARY

I have briefly described a core program for planetary exploration through the year 2000. This program represents the best judgment of the Solar System Exploration Committee, which is made up of 25 members of the space science community with an impressive record of accomplishment in the area of manned and unmanned exploration of the planets. Much of what I have written here would have appeared, not many years ago, to be grist for the mill of writers of science fiction, were it not for the actual accomplishments discussed in this volume. It has been a period of extraordinary achievement and an exciting scientific adventure.

REFERENCE

1. *Planetary Exploration through the Year 2000,* S.S.E.C., NASA (1983).

GLOSSARY

In view of the fact that this is a volume in a monograph series devoted to analytical chemistry and will most likely be read by chemists, there are a number of terms with which the reader may be unfamiliar. Accordingly, this glossary is appended to make some of these terms clear. A number of the terms come from geology and mineralogy, and some are characteristic of the space program. Many of the definitions are brief and, necessarily, not complete. Additional information can be found in standard geological glossaries, space dictionaries, and other reference materials.

ACCESSORY MINERALS: Minerals occurring in small amounts in a rock.

ACCRETION: A process by which small particles and gases come together to form larger bodies, eventually of planetary size.

ACHONDRITES: Stony meteorites lacking chondrules.

AEON: One billion years.

AGGLUTINATE: A common particle type in lunar soils. Agglutinates are made up of comminuted rock, mineral, and glass fragments cemented by glass.

ALBEDO: Refers to the percentage of the incoming radiation reflected by a natural surface.

ANORTHITE (AN): $CaAl_2 Si_2 O_8$, the most Ca-rich member of the plagioclase (feldspar) series of minerals (see plagioclases).

ANORTHOSITE: An igneous rock made up almost entirely of plagioclase feldspar.

APHELION: The point in the orbit of a planet, asteroid, comet, or satellite farthest from the sun.

BAR: The international unit of pressure (= 0.987 atm)

BASALT: A fine-grained, dark-colored igneous rock, composed mainly of plagioclase (feldspar) and pyroxene; usually contains other minerals such as olivine or ilmenite.

BASALTIC ACHONDRITE: Calcium-rich stony meteorites, lacking chondrules and nickel–iron. They show some similarity to terrestrial and lunar basalts; examples are euchrites and howardites.

BRECCIA: A rock consisting of angular, coarse fragments embedded in a fine-grained matrix.

CARBONACEOUS CHONDRITES: The most primitive of the stony meteorites. The abundances of the nonvolatile elements are thought to most closely approximate those in the primordial solar nebula.

CHALCOPHILE ELEMENT: An element that preferentially enters sulfide minerals.

CHONDRULE: Small, rounded bodies in meteorites (generally less than 1 mm in diameter) commonly composed of olivine and/or orthopyroxene.

CLAST: A discrete particle or fragment of rock or mineral usually found included in a larger rock.

CLINOPYROXENE: Minerals of the pyroxene group (e.g., augite, pigeonite), which crystallize in the monoclinic system.

COSMIC ELEMENT ABUNDANCES: The abundances of the chemical elements in the solar nebula before the formation of the Sun, planets, and meteorites.

CUMULATE: A plutonic igneous rock composed mainly of crystals accumulated by sinking or floating from a magma.

CURIE TEMPERATURE: The temperature above which a ferromagnetic material becomes nonmagnetic.

DUNITE: A mineral (peridotite) that consists almost entirely of olivine and contains accessory chromite and pyroxene.

ECLOGITE: A dense rock consisting of garnet and pyroxene, similar chemically to basalt.

EJECTA: Materials ejected from the crater by meteorite impact or volcanic action.

EUCRITE: A meteorite composed essentially of feldspars and augites (pyroxene group).

EXPOSURE AGE: Period of time during which a sample has been at or near the lunar surface, determined on the basis of cosmogenic rare gas contents, particle track densities, short-lived radioisotopes, or agglutinate content in the case of soil samples.

EXTRUSIVE: An igneous rock that solidified at the surface.

FINES: Lunar material arbitrarily defined as less than 1 cm in diameter, in the lunar case synonymous with the soils.

FRACTIONATION: The separation of chemical elements from an initially homogeneous state into different phases or systems.

GARDENING: The process of turning over of the lunar soil or regolith by meteorite bombardment.

GRANITE: An igneous rock made up mostly of quartz and alkali feldspar.

GROUNDMASS: The fine-grained rock material between the larger mineral grains; synonymous with matrix.

HALF-LIFE: The time interval in which a number of radioactive atoms decay to half their number.

IGNEOUS: Applied to rocks or processes involving the solidification of hot molten material.

ILMENITE: A mineral $FeTiO_3$, slightly magnetic, abundant on the Moon; common but less abundant on the Earth's surface.

INTERSERTAL: A term used to describe the texture of igneous rocks in which a base or mesostasis of glass and small crystals fills the interstices between unoriented feldspar lathes.

ISOCHRON: A line on a diagram passing through plots of samples with the same age but different isotope ratios. The line connects points of equal time or age.

IRON METEORITES: A class of meteorites composed chiefly of iron or iron–nickel.

KREEP: Potassium, rare-earth elements, and phosphorus.

LAVA: Fluid rock coming from the Earth's interior (or Moon) from volcanoes and the like; on solidification it forms extrusive igneous rocks.

LIL: Large-ion lithophile elements. Those lithophile elements (e.g., K, Rb, REE, U, and Th) having ionic radii larger than common lunar rock-forming elements and which usually behave as trace elements in meteorites.

LITHOPHILE ELEMENT: An element tending to concentrate in oxygen-containing compounds, particularly silicates.

LM: Lunar Module.

LRL: Lunar Receiving Laboratory, NASA L. B. Johnson Space Center.

KAMACITE: Naturally occurring body-centered-cubic α iron, occurs commonly in meteorites.

MAGMA: Term applied to molten rock in the interior of a planet.

MAGMATIC DIFFERENTIATION: production of rocks of differing chemical composition during cooling and crystallization of silica melt or magma by processes such as removal of early formed mineral phases.

MANTLE: The solid region of the Earth extending from the base of the crust (10–60 km depth) to top of the core (2900-km depth); principal constituents are iron, magnesium, silicon, and oxygen, which occur as mineral phases of increasing density with depth.

MARE (PL. MARIA): The dark, generally flat areas of the Moon formerly thought to be seas.

MASCON: Mass concentration.

MASS SPECTROMETER: Analytical instrument used to determine the amounts of various isotopes and molecular species in a sample by separating them according to their mass to charge ratio.

MATRIX: The fine-grained material in which larger mineral or rock fragments are embedded (used interchangeably with groundmass).

METAMORPHIC: A term used to describe rocks that have recrystallized in solid state as a result of drastic changes in temperature, pressure, and chemical environment.

MESOSTASIS: The interstitial, generally fine-grained material, between larger mineral grains in a rock; may be used synonymously with matrix or groundmass.

MICROCRATER (ZAP PIT): Crater produced by impact of interplanetary particles generally having masses of less than 10^{-3} g.

NORITE: A type of gabbro in which orthopyroxene is dominant over clinopyroxene.

NUCLEON: Subatomic nuclear particles (mainly protons and neutrons).

NUCLIDES: Atoms characterized by the number of protons (Z) and neutrons (N).

OCEANIC THOLEITE: Basaltic rock characterized by low concentrations of potassium and related elements; probably the most common lava erupted at the midoceanic ridges.

OPHITIC: A rock texture composed of elongated feldspar crystals embedded in pyroxene or olivine.

ORTHOPYROXENE: An orthorhobic member of the pyroxene mineral group.

PERIDOTITE: An igneous rock characterized pyroxene and olivine (but no feldspar).

PHENOCRYST: A large, early-formed crystal in igneous rock, surrounded by a fine-grained groundmass.

PHOTOSPHERE: An outer layer of the sun, about 350 km thick—the source of most solar radiation.

PLAGIOCLASE: A subgroup (or series) of the feldspar group minerals.

PLUTONIC: A term applied to igneous rocks that have crystallized at depth usually with coarsely crystalline texture.

PORPHYRITIC: Having larger crystals set in a finer groundmass.

ppb: Parts per billion.

ppm: Parts per million.

PYROXENE: A closely related group of minerals that includes augite, pigeonite, and the like.

RADIOGENIC LEAD: Lead sisotopes formed by the radioactive decay of uranium and thorium (^{206}Pb from ^{238}U; ^{207}Pb from ^{235}U; ^{208}Pb from ^{232}Th).

REGOLITH: Loose surface material, composed of rock fragments and soil that overlies consolidated bedrock.

REMANENT MAGNETIZATION: The component of rock magnetization "frozen in" to an igneous or impact melt as its ferromagnetic minerals were cooled through their Curie points.

RESIDUAL LIQUID: The material remaining after most of the magma has crystallized; it is sometimes characterized by an abundance of volatile constituents.

r-PROCESS NUCLIDES: Those nuclides produced under conditions of intense neutron flux associated with a supernovae explosion.

SOIL (LUNAR): Fine-grained lunar material arbitrarily defined as less than 1 cm in diameter.

SOLAR NEBULA: The primitive disk-shaped cloud of dust and from which all bodies of the solar system are thought to have originated.

SOLAR WIND: The stream of charged particles (mainly ionized hydrogen) moving outward from the sun with velocities in the range of 300–500 km/sec.

TAENITE: Naturally occurring face centered cubic gamma iron; commonly occurs in meteorites.

TEKTITES: Small, glassy objects of wide geographic distribution. Their origin is controversial. One viewpoint holds them to be due to the splashing of melted terrestrial country rock as a consequence of meteoroid, asteroid, or cometary impact. Another viewpoint is that they are extraterrestrial in origin, perhaps from the Moon.

TERMINATOR: The line separating illuminated and dark portions of a celestial body.

TRACE ELEMENT: An element found in very low (trace) amounts, generally less than 0.1%.

TROILITE: A mineral composed of FeS. A common accessory mineral in iron and most stony meteorites.

VOLATILE ELEMENT: An element volatile at temperatures below 1300°C.

BIBLIOGRAPHY AND SUPPLEMENTAL READING

The following is a list of references and sources that can be used to obtain additional information as well as greater detail on some of the material in this text. The list is made up of NASA publications as well as texts and papers. In a number of instances they have already been cited but are being repeated here in the interest of completeness.

Adler, I. and J. I. Trombka, *Geochemical Exploration of the Moon and Planets,* Springer Verlag, New York, 1970.

Adler, I. et al., Apollo 15 and 16: Results of the Integrated Geochemical Experiment, *The Moon,* **7,** 487 (1973).

Alfven, H. and G. Arrhenius, Origin and Evolution of the Earth–Moon System, *The Moon,* **5,** 210 (1972).

Anders, E. *Meteorites and the Early History of the Solar System,* in Jastrow & Cameron (Eds.), *Origin of the Solar System,* Academic, New York, 1963.

Baldwin, R. B. *The Measure of the Moon,* University of Chicago Press, Chicago, 1963.

Baldwin, R. B. *A Fundamental Survey of the Moon,* McGraw-Hill, New York, 1965.

Baldwin, R. B. Summary of Arguments for a Hot Moon, *Science,* **170,** 1297, (1970).

Barnes, V. E. (Ed.), Tektites, *Scientific American,* **205,** 58 (1961).

Best, J. B. Apollo 15 Lunar Samples, *Third Lunar Science Conference,* Lunar Science Institute, Houston, 1972.

Binder, A. B. Internal Structure of Mars, *J. Geophys. Res.,* **74,** 3110 (1969).

Binder, A. B. et al., Observations of the Moon and of Terrestrial Rocks in the Infrared. *Icarus,* **4,** 415 (1965).

Burnett, D. S., W. A. Fowler, and F. Hoyle, Nucleosynthesis in the Early History of the Solar System. *Geochim. Cosmochim. Acta.,* **29,** 1209 (1965).

Cavaretta, G. et al., Geochemistry of the Green Glass Spheres from Apollo 15 Samples, *Third Lunar Science Conference,* Lunar Science Institute, Houston, 1972, p. 202.

Chamberlain, J. W. (Ed.), *Post Apollo Lunar Science,* Lunar Science Institute, Houston, 1972.

Chao, E. C. T., J. A. Boreman, J. A. Minkin, O. B. James, and G. A. Desborough, Lunar Glasses of Impact Origin: Physical and Chemical Characteristics and Geologic Interpretations, *J. Geophys. Res.,* **75,** 7445 (1970).

Christian, R. P. et al., Chemical Composition of Some Apollo 15 Igneous Rocks, *Third Lunar Science Conference,* Lunar Science Institute, Houston, 1972.

Compston, W. B. et al., The Chemistry and Age of Apollo 11 Lunar Material, Proceedings Apollo 11 Lunar Science Conference, Suppl. 1, *Geochim. Cosmochim. Acta* (1970).

Crozaz, G. R. et al., Solar Flare and Cosmic Ray Studies of Apollo 14 and 15 Samples, Proceedings Third Lunar Science Conference, Suppl. 3, *Geochim. Cosmochim. Acta* (1972).

Engle, A. E. J. and C. G. Engle, Lunar Rock Compositions and Some Interpretations, *Science,* **167,** 527 (1970).

Engle, A. E. J., C. G. Engle, A. L. Sutton and A. T. Myers, Composition of Five Apollo 11 and Apollo 12 Rocks and One Apollo 11 Soil and Petrogenic Considerations, Proceedings Apollo 11 Lunar Science Conference, Suppl. 1, *Geochim. Cosmochim. Acta* (1970).

Eschelman, V. R. The Atmospheres of Mars and Venus, *Scientific American,* **220,** 78 (1969).

Faul, H. *Ages of Rocks, Planets, and Stars,* McGraw-Hill, New York, 1966.

Frondel, C., C. Klein, Jr., J. Ito, and J. C. Drake, Mineralogical and Chemical Studies of Apollo 11 Lunar Fines and Selected Rocks, Proceedings Apollo 11 Lunar Science Conference, Suppl. 1, *Geochim. Cosmochim. Acta* (1970).

Gast, P. W. The Chemical Composition and Structure of the Moon. *The Moon,* **4,** 121 (1972).

Gast, P. W., N. J. Hubbard, and H. Weismann, Chemical Composition and Petrogenesis of Basalts from Tranquillity Base, Proceedings Apollo 11 Lunar Science Conference, Suppl. 1, *Geochim. Cosmochim. Acta* (1970).

Goldstein, J. W. and A. S. Doan, The Effects of Phosphorus on the Formation of the Widmanstatten Pattern in Iron Meteorites, *Geochim. Cosmochim. Acta,* **36,** 51 (1972).

Hanel, R. B. et al., Investigation of the Martian Environment by Infrared Spectroscopy on Mariner 9, *Icarus,* **17,** 423 (1972).

Hapke, B. W. Lunar Surface: Composition Inferred From Optical Properties, *Science,* **159,** 76, (1968).

Haskins, L. A. et al., Rare Earths and Other Trace Elements In Apollo 11 Lunar Samples, Proceedings Apollo 11 Lunar Science Conference, Suppl. 1, *Geochim. Cosmochim. Acta* (1970).

Hess, W. N., D. H. Menzel, and J. A. O'Keefe (Eds.), *The Nature of the Lunar Surface,* Johns Hopkins Press, Baltimore, 1966.

Hohenberg, C. M. Radioisotopes and the History of Nucleosynthesis in the Galaxy, *Science,* **166,** 211, (1969).

Keil, K., M. Prinz, and T. E. Bunch, Mineralogy, Petrology and Chemistry of Some Apollo 12 Samples, Levinson A. A. and S. R. Taylor *Moon Rocks and Minerals,* Pergamon, New York, 1971.

King, E., D. Heymann, and D. R. Criswell (Eds.), Proceedings Third Lunar Science Conference, Suppl. 3, *Geochim. Cosmochim. Acta* (1972).

Klein, C. Jr., Lunar Materials: Their Mineralogy, Petrology and Chemistry, *Earth Sci. Rev.,* **8,** 169, (1972).

Levinson, A. A. (Ed.), Proceedings of the Apollo 11 Lunar Science Conference, Suppl. 1, *Geochim. Cosmochim. Acta* (1970).

Lipschutz, M. E. Cosmochemical Thermal Fractionation Information by Neutron Activation Analysis, *Proc. Int. Symposium on Artificial Radioactivity,* Poona, India, 1984.

Logan, L. M. et al., Midinfrared Emission Spectra of Apollo 14 and 15 Soils and Remote Compositional Mapping of the Moon, Proceedings of the Third Lunar Science Conference, Suppl. 3, *Geochim. Cosmochim. Acta* (1972).

Lowman, P. D., Jr., *The Geology of the Moon,* Weldflug, Zurich, 1969.

Lowman, P. D., Jr., Composition of the Lunar Highlands, Possible Implications for the Evolution of the Earth's Crust, *J. Geophys. Res.,* **74,** 495 (1969).

Lowman, P. D., Jr., The Geological Evolution of the Moon, *J. Geol.,* **80,** 125 (1972).

Luna 16, Special Issue. *Earth Planet. Sci. Lett.,* **13,** (1972).

.ıariner to Mars, Collection of Papers, *Science,* **175,** 293 (1971).

Mason, B. *Meteorites,* John Wiley & Sons, New York, 1962

Mason, B. The Lunar Rocks, *Scientific American,* **225,** 49, (1971).

Mason, B. and W. G. Melson, *The Lunar Rocks,* Wiley-Interscience, New York, 1970.

Melson, W. G. and B. Mason, Lunar "Basalts," Some Comparisons With Terrestrial and Meteoritic Analogs, and a Proposed Classification and Nomenclature, *Moon Rock and Minerals,* Pergamon, New York, 1971.

Moon Issue, *Science,* **167,** (1970).

Morgan, J. W. et al., Trace Elements in Apollo 15 Samples: Implications for Meteorite Influx and Volatile Depletion in the Moon, Proceedings Third Lunar Science Conference, Suppl. 3, *Geochim. Cosmochim. Acta* (1972).

Mueller, R. F. Planetary Probe: Origin of the Atmosphere of Venus, *Science,* **163,** 1321 (1969).

Muller, P. M. and W. L. Sjogren, Mascons: Lunar Mass Concentrations. *Science,* **161,** 680 (1968).

National Aeronautics and Space Administration, *Surveyor Program Results (Final Report),* NASA SP-184, 1969.

National Aeronautics and Space Administration, *Apollo 11: Preliminary Science Report,* NASA SP-214, 1969.

National Aeronautics and Space Administration, *Apollo 12: Preliminary Science Report,* NASA SP-235, 1970.

National Aeronautics and Space Administration, *Apollo 14: Preliminary Science Report,* NASA SP-272, 1971.

National Aeronautics and Space Administration, *Apollo 15: Preliminary Science Report,* NASA SP-289, 1972.

National Aeronautics and Space Administration, *Apollo 16: Preliminary Science Report,* NASA SP-315, 1972.

National Aeronautics and Space Administration, *Apollo 17: Preliminary Science Report,* NASA SP-330, 1973.

O'Keefe, J. A. (Ed.), *Tektites,* University of Chicago Press, Chicago, 1973.

O'Keefe, J. A. Origin of the Moon, *J. Geophys. Res.,* **75,** 6565 (1970b).

O'Keefe, J. A. Apollo 11: Implications for the Early History of the Solar System, *Trans. Amer. Geophys. Union,* **51,** 633 (1970a).

O'Kelley, G. D., J. S. Eldrige, E. Schonfeld, and K. J. Northeutt, Primordial Radioelements and Cosmogenic Radionuclides in Lunar Samples from Apollo 15, *Science,* **175,** 440 (1972).

Papanastassiou, D. A. and G. J. Wasserburg, Rb–Sr Ages from the Ocean of Storms, *Earth Planetary Sci. Lett.,* **8,** (1970).

Philpotts, J. A. and C. C. Schnetzler, Apollo 11 Lunar Samples: K, Rb, Sr, Ba and Rare Earth Concentrations in Some Rock and Separated Phases, Proceedings Apollo 11 Lunar Science Conference, Suppl. 1, *Geochim. Cosmochim. Acta* (1970).

Pugh, M. J. and J. A. Bastin, Infrared Observations of the Moon and Their Interpretation, *The Moon,* **4,** (1972).

Reid, A. M. et al., The Major Element Compositions of Lunar Rocks as Inferred from Glass Compositions in the Lunar Soils, Proceedings Third Lunar Science Conference, Suppl. 3, *Geochim. Cosmochin. Acta* (1972).

Reynolds, J. H. The Age of the Elements in the Solar System, *Scientific American,* 202, 1960.

Schnetzler, C. C., and J. A. Philpotts, Alkali, Alkaline Earth, and Rare-Earth Element Concentrations in Some Apollo 12 Soils, Rocks and Separated Phases, Proceedings Apollo 11 Lunar Science Conference, Suppl. 1, *Geochim. Cosmochim. Acta* (1970).

Schonfeld, E. and C. Meyer Jr., The Abundances of Components of the Lunar Soils by a Least Squares Mixing Model and the Formation Age of KREEP, Proceedings Third Lunar Science Conference, Suppl. 3, *Geochim. Cosmochim. Acta* (1972).

Silver, L. T. Lead Volatilization and Volatile Transfer Processes on the Moon, *Third Lunar Science Conference,* Lunar Science Institute, Houston, 1972.

Steinbacher, R. H. and S. Gunther, The Mariner Mars 1971 Experiments, Introduction, *Icarus,* **12,** 3 (1971).

Tatsumoto, M. Age of the Moon: An Isotope Study of U–Th–Pb Systematics of Apollo 11 Lunar Samples, Proceedings Apollo 11 Lunar Science Conference, Suppl. 1, *Geochim. Cosmochim. Acta* (1970).

Tatsumoto, M. et al. U–Th–Pb and Rb–Sr Measurements on Some Apollo 14

Lunar Samples, Proceedings Third Lunar Science Conference, Suppl. 3, *Geochim. Cosmochim. Acta* (1972).

Tera, F., D. A. Papanastassiou and G. J. Wasserburg, A Lunar Cataclysm at about 3.95 AE and the Structure of the Lunar Crust, *Fourth Lunar Science Conference,* Lunar Science Institute, Houston, 1973.

Toksoz, M. N. et al., Lunar Crust: Structure and Composition, *Science,* **176,** 1012 (1972).

Toksoz, M. N. and S. C. Solomon, Thermal History and Evolution of the Moon, *The Moon,* **7,** 251 (1973).

Watkins, C. *Third Lunar Science Conference* (revised abstracts), Lunar Science Institute, Houston, 1972.

Weston, C. R. A Strategy for Mars (Search for Life), *Amer. Scientist,* **53,** 495 (1965).

Wood, J. A. The Lunar Soil, *Scientific American,* **223,** 14 (1970).

INDEX

(*continued from front*)